ALLE ZEIT WACH
SJ
1842

Progress in Sensory Physiology

Editors: H. Autrum, D. Ottoson,
E.R. Perl, R.F. Schmidt, H. Shimazu, W.D. Willis
Editor-in-Chief: D. Ottoson

Volume 5
1985. 83 figures. V, 174 pages
ISBN 3-540-15339-X

Contents: *R.C. Hardie:* Functional Organization of the Fly Retina. – *H. Machemer, J.W. Deitmer:* Mechanoreception in Ciliates. – *M. Lindauer, H. Martin:* The Biological Significance of the Earth's Magnetic Field. – *H. Bleckmann:* Perception of Water Surface Waves: How Surface Waves are Used for Prey Identification Communications.

Volume 6
1986. 86 figures. V, 227 pages
ISBN 3-540-15340-3

Contents: *T. Sato:* Receptor Potential in Rat Taste Cells. – *K.B. Døving:* Functional Properties of the Fish Olfactory System. – *J.A. Coles:* Homeostasis of Extracellular Fluid in Retinas of Invertebrates and Vertebrates. – *S. Mense:* Slowly Conducting Afferent Fibers from Deep Tissues: Neurobiological Properties and Central Nervous Actions.

Volume 7
D.R. Irvine
The Auditory Brainstem
1986. 60 figures. XI, 143 pages
ISBN 3-540-16299-2

Contents: Introduction and Plan of the Review. – The Performance of the Auditory Central Processor. – Auditory Nerve Input to the Central Processor. – Cochlear Nucleus: Anatomy and Physiology. – Superior Olivary Complex: Anatomy and Physiology. – Nuclei of Lateral Lemniscus: Anatomy and Physiology. – The Auditory Midbrain: Anatomy and Physiology. – Auditory Brainstem Processing: Integration and Conclusions.

Volume 8
1987. 56 figures. V, 207 pages
ISBN 3-540-16300-X

Contents: *W. Skrandies:* The Upper and Lower Visual Field of Man: Electrophysiological and Functional Differences. – *J. Le Magnen:* Central Processing of Sensory Information in the Control of Feeding. – *A. Faurion:* Physiology of the Sweet Taste.

Progress in

Sensory Physiology 9

Editors: H. Autrum D. Ottoson
E. R. Perl R. F. Schmidt H. Shimazu W. D. Willis

Editor-in-Chief: D. Ottoson

With Contributions by
G. N. Akoev G. N. Andrianov R. D. Foreman
A. Gallego O. A. Mora Y. Oomura
J. E. Sánchez-Criado J. Syka

Springer-Verlag
Berlin Heidelberg New York
London Paris Tokyo

Editor-in-Chief

Professor Dr. David Ottoson
Karolinska Institutet, Fysiologiska Institutionen II
Solnavägen 1, 10401 Stockholm 60, Sweden

Editors

Professor Dr. Hansjochem Autrum
Zoologisches Institut der Universität München
Luisenstrasse 14, 8000 München 2, Germany

Professor Dr. Edward R. Perl
Department of Physiology
University of North Carolina at Chapel Hill
Chapel Hill, NC 27514, USA

Professor Dr. Robert F. Schmidt
Physiologisches Institut der Universität
Röntgenring 9, 8700 Würzburg, Germany

Professor Dr. Hiroshi Shimazu
Department of Neurophysiology
University of Tokyo, Institute of Brain Research
7.3.1. Hongo, Bunkyo Ku, Tokyo, Japan

Professor Dr. William D. Willis
The Marine Biomedical Institute
University of Texas Medical Branch
Galveston, TX 77550, USA

With 71 Figures

ISBN-13: 978-3-642-74060-2 e-ISBN-13: 978-3-642-74058-9
DOI: 10.1007/978-3-642-74058-9

The Library of Congress has cataloged this serial publication as follows:
Progress in sensory physiology. – 1- – Berlin; New York: Springer-Verlag, 1981 –
v.: ill.; 25 cm. Irregular. Editor: D. Ottoson. ISSN 0721-9156 = Progress in sensory physiology. 1. Senses and sensation-Collected works. 2. Sense-organs-Collected works. 3. Neurophysiology-Collected works. I. Ottoson, David, 1918 –.
[DNLM: 1. Neurophysiology-periodicals. 2. Sensation-physiology-periodicals.
W1 PR681G] QP431.P73 599'.01'8-dc19 85-647743 AACR 2 MARC-S [8509]

Softcover reprint of the hardcover 1st edition 1989

Typesetting: K+V Fotosatz GmbH, Beerfelden

2121/3140-543210 – Printed on acid-free paper

Contents

Organization of the Spinothalamic Tract as a Relay for Cardiopulmonary Sympathetic Afferent Fiber Activity

R. D. Foreman

Department of Physiology and Biophysics, Oklahoma University Health Sciences Center, P.O. Box 26901, Oklahoma City, OK 73190, USA

Progress in Sensory Physiology. Vol 9

1 Introduction

Sympathetic afferent fibers originate from a visceral organ, course in the thoracolumbar rami communicantes, have cell bodies located in dorsal root ganglia, and terminate in the gray matter of the spinal cord. Sympathetic afferent fibers from the heart transmit information about noxious stimuli associated with myocardial ischemia, i.e. angina pectoris. Previous reviews have described the characteristics of cardiovascular sympathetic afferent fibers (Bishop et al. 1983; Malliani 1982). This review summarizes that work and focuses on the neural mechanisms underlying the complexities of angina pectoris.

In order to understand anginal pain, cells forming the classical pain pathway, the spinothalamic tract (STT), were chosen for study. These cells were chosen to address questions about anginal pain because they transmit nociceptive information to brain regions that are involved in the perception of pain. Antidromic activation of STT cells provided a means of identifying cells involved with transmission of nociceptive information in anesthetized animals. Other ascending pathways may also transmit nociceptive information, but many studies show that the STT plays an important role. Visceral pain is commonly referred to overlying somatic structures. The pain of angina pectoris can be sensed over a wide area of the thorax: in the retrosternal, precordial anterior thoracic, and anterior cervical regions of the chest; in the left or sometimes even the right shoulder, arm, wrist, or hand; or in the jaw and teeth (Harrison and Reeves 1968). This description of referred pain is typical of heart disease, but it can also typify other noncardiac diseases, such as gastrointestinal disorders, esophageal diseases, pulmonary diseases, and musculoskeletal disorders (Christie and Conti 1981). Two major theories have been proposed to explain the phenomenon of referred pain and, especially, angina pectoris. MacKenzie (1909) proposed the convergence-facilitation theory. This proposes that sympathetic afferent fibers have no direct connection with somatic nociceptive pathways, but rather that they set up an "irritable focus near the cells of this pathway." Ruch (1961) presented the convergence-projection theory to explain referred pain, suggesting that some visceral afferent fibers subserving transmission of nociceptive information and cutaneous afferent fibers converge onto the same neuron at some point in the sensory pathway. Impulses resulting from this convergence would reach the pain-perception areas of the brain and be interpreted as having originated in the somatic structures, an interpretation probably learned from previous experiences when the same fiber tract was stimulated by somatic nociceptive afferent fibers. This is the most likely theory to explain referred pain. This review will show that visceral and somatic afferent information excites the same cells of the STT.

Anginal-like pain is a common complaint of patients who experience gallbladder disease. STT cells may be used for transmission of anginal-like pain from the heart, as well as pain due to gallbladder distension. This review will demonstrate that afferent fibers from two different visceral organs can ultimately converge onto the same STT cells at a given segment; the complexity of the integration of sympathetic afferent information in the spinal cord will also be described.

Sympathetic afferent information is modulated in the STT. The activity of STT cells can be changed by descending pathways that originate from several different regions of the brain stem and from the vagus nerve. The role of vagal afferents in modulating cell activity in the spinal cord is of particular interest because it may provide a partial explanation of the neurophysiological mechanisms underlying silent myocardial ischemia, i.e., the presence of myocardial ischemia without angina pectoris (Cohn 1986; Rutishauser and Roskamm 1984). Vagal-sympathetic interactions will be described by examining the effects of vagal stimulation on the response of STT cells located in the upper thoracic spinal segments to activation of cardiopulmonary sympathetic afferent fibers. In addition, several sites in the brain stem have been studied to determine which nuclei or regions might modulate sympathetic afferent information coming onto STT cells.

In summary, in this review the effects of sympathetic afferent information arising from noxious stimulation of the heart on STT cell activity will be described, and the modulation of STT cells by inputs from other regions of the spinal cord, from the vagus, and from supraspinal structures will also be discussed.

2 Characteristics of Cardiac Receptors and Their Afferent Fibers

2.1 Morphology

Our understanding of the innervation characteristics and anatomical features of cardiac receptors remains limited in spite of many studies. Several reasons (Bishop et al. 1983) have been suggested for this lack of knowledge: (a) specific methods for staining afferents and their receptors are not available; (b) no precise studies are available to differentiate vagal from sympathetic receptor endings in the heart (c) the course and termination of individual afferent fibers are difficult to follow and (d) the sensory functions of structures seen by electron microscopy are uncertain.

Many early investigators suggested the existence of sensory nerve endings in the heart (Berkley 1894; Holmes 1957; Khabarova 1963; Nettleship 1936; Nonidez 1939; Woollard 1926). Sensory nerve endings found in the heart are classified as either diffuse or compact unencapsulated endings (Miller and Kasahara 1964; Nonidez 1939). Several investigators have described an end-net or terminal reticulum formed from apparently different myelinated fibers whose branches anastomose in the endocardium (Holmes 1957; Johnston 1968; Meyling 1953; Miller and Kasahara 1964; Mitchell 1953). Since the differentiation between unencapsulated endings and end-nets may be artificial, these findings should be viewed with some caution (Floyd 1979). Complex unencapsulated endings are found primarily in the epicardium and are branched endings of medium-sized to large fibers. No sensory endings are found in the myocardium but bundles of mixed myelinated and unmyelinated nerves can be traced through connective tissues of the septa between the muscles (Miller and Kasahara 1964).

Since the sensory terminals of the heart can utilize either sympathetic or vagal afferent fibers, differential ablation of these nerves is necessary to locate the paths along which various types of information are transmitted. Woollard (1926) and Nonidez (1939) suggested that a large proportion of the cardiac sensory endings are innervated by afferents found in the vagus. However, Nettleship (1936) showed that bilateral removal of the stellate ganglia causes degeneration of the endocardial plexus near the apex of the ventricles and in one-half to three-

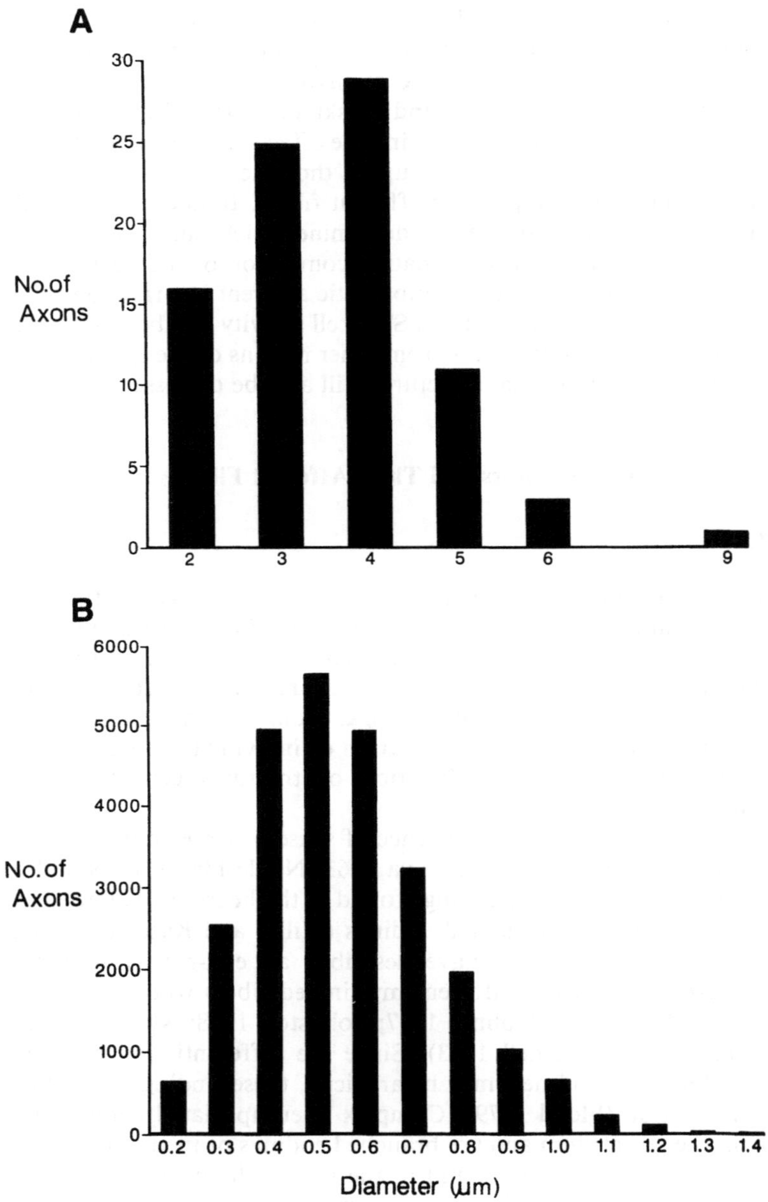

Fig. 1A, B. Histograms of diameters of myelinated (**A**) and unmyelinated (**B**) fibers found in a representative left inferior cardiac nerve of the cat. Note the differences in scales on the ordinate and abscissa

fourths of the coronary artery plexus. Vagal and sympathetic afferent fibers appear to intermingle in the heart, as described in anatomical (Khabarova 1963) and electrophysiological data (Bishop et al. 1983).
The sympathetic sensory system of the heart is composed of both myelinated and unmyelinated afferent nerve fibers. In the inferior cardiac nerve about 50 – 150 fibers are myelinated and 25000 – 40000 are unmyelinated (Fig. 1; Emery et al. 1976). Degeneration studies (Emery et al. 1978) show that 90% of the myelinated fibers are sensory, but the proportion of unmyelinated sensory fibers could not be determined for certain because of the degeneration of a large number of postganglionic efferent fibers. The largest myelinated sympathetic sensory fibers are A-delta fibers (Emery et al. 1976). Other anatomical studies have demonstrated unmyelinated afferent fibers in the cardiopulmonary sympathetic afferent innervation (Oldfield and McLachlan 1978; Seagard et al. 1978).

2.2 Electrophysiology of Receptors

Punctate mechanical stimulation of the epicardial surface of the heart was shown by Ueda et al. (1969) to activate sympathetic afferent fibers (Fig. 2). These investigators recorded afferent impulses from axons of the left upper thoracic white rami communicantes and found that receptors were distributed to the surface of the left ventricle, left atrium, left coronary artery, ascending and descending thoracic aorta, pulmonary artery, left pulmonary veins, pericardium, pleura of the left side, and in some cases on the surface of the right ventricle near the anterior descending artery.

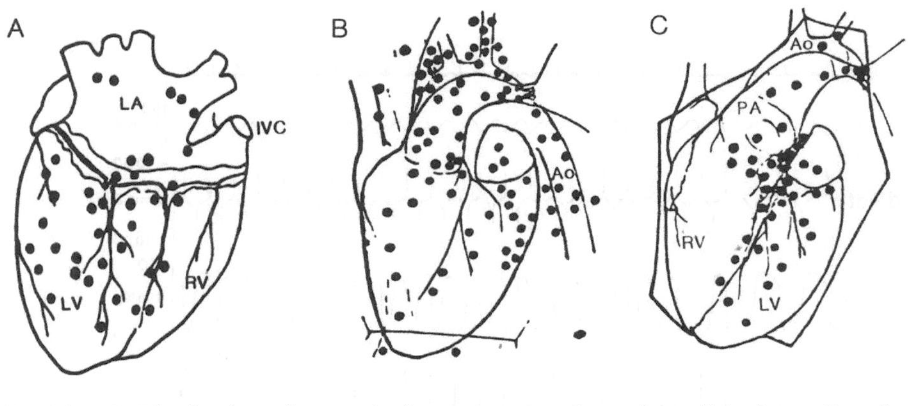

Fig. 2A – C. Distribution of sympathetic receptors on the surface of the heart. Impulses were recorded from the left rami communicantes while the heart was probed with a blunt probe. Receptors were distributed over the posterior (**A**) and anterior (**B, C**) surface of the heart. (Adapted from Ueda et al. 1969 with permission of Japanese Heart Journal). *Ao*, aorta; *IVC*, inferior vena cava; *LA*, left atrium; *LV*, left ventricle; *PA*, pulmonary artery; *RV*, right ventricle

2.2.1 *Atrial Receptors*

Sympathetic afferent nerve fibers innervating the atrial walls are both myelinated (Malliani et al. 1973b; Uchida and Murao 1974c; Uchida 1975) and unmyelinated (Uchida and Murao 1974c; Uchida 1975). The receptors can be excited both by muscular contraction and by stretch, so their mechanism of activation is still in question (Malliani 1982).

2.2.2 *Ventricular Receptors*

Ventricular sympathetic sensory receptors that have myelinated nerve fibers seem to be mechanosensitive because they fire spontaneous impulses with cardiac periodicity (Hess et al. 1974; Malliani et al. 1973b, 1975; Uchida et al. 1974; Uchida 1975). Background discharge usually consists of one action potential per cardiac cycle following the onset of the Q waves of the ECG (Fig. 3). For most fibers an action potential does not occur with all cardiac cycles, but cardiac periodicity is detectable. By contrast, in unmyelinated nerve fibers there does not appear to be a fixed temporal relationship between receptor impulses and ventricular dynamics (Fig. 4) (Casati et al. 1979; Lombardi et al. 1981; Uchida et al. 1974; Uchida 1975). Spontaneous impulses commonly occur with ventricular systole. When ventricular pressure is elevated by mechanical obstruction of the aorta, impulses occur during more cardiac cycles.

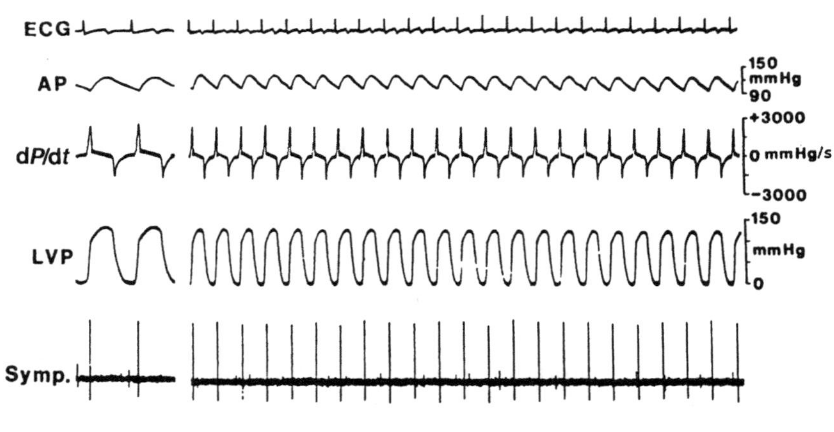

Fig. 3. Single-unit recording of spontaneous activity of a sympathetic afferent myelinated fiber originating from the left ventricle (*bottom trace*). Other tracings, from *top* to *bottom*, are ECG, arterial blood pressure, left ventricular pressure differential d*P*/d*t*, and left ventricular pressure. Tracings were recorded at different speeds; the *left bar* at the bottom represents 0.5 s, and the *right bar* 1.0 s. (From Malliani 1982, with permission of Springer-Verlag)

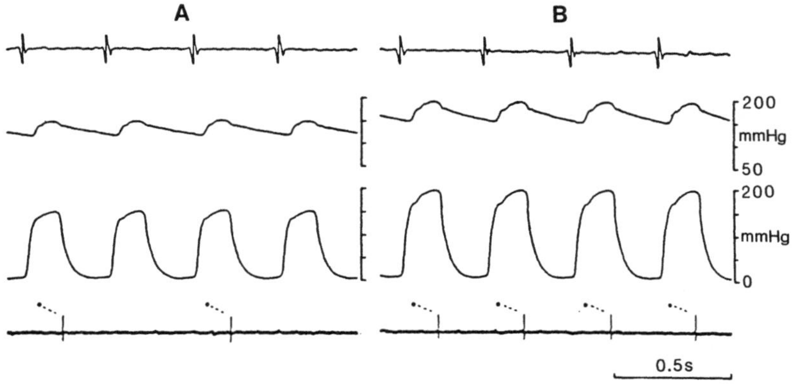

Fig. 4A, B. Action potentials recorded from a sympathetic afferent unmyelinated fiber with its receptor located in the left ventricle (*bottom trace*). Other tracings from *top* to *bottom* are ECG, systemic arterial pressure, and left ventricular pressure. Recordings were made during spontaneous activity (**A**) and occlusion of the aorta (**B**). *Dots* show the calculated conduction time between the initiation of action potentials and their recording. (From Casati et al. 1979, with permission of Springer-Verlag)

2.2.3 Coronary Artery Occlusion

Mechanical interventions, coronary artery occlusions, and chemical injections can stimulate receptors of the heart. Since this review focuses on nociception, the effects of coronary artery occlusion and injections of noxious chemicals will be considered. Myocardial ischemia can be produced by interrupting blood flow in the left coronary artery (Brown and Malliani 1971; Lombardi et al. 1981; Malliani et al. 1973b). The activity of the sympathetic afferent fibers increases approximately 10–15 s after onset of the occlusion (Fig. 5A). Both myelinated (Bosnjak et al. 1979; Brown 1967; Malliani et al. 1973b; Uchida and Murao 1974b) and unmyelinated (Casati et al. 1979; Lombardi et al. 1981; Uchida and Murao 1974b) ventricular sympathetic afferent fibers respond to ischemia. Myelinated ventricular afferent fibers are regarded as predominantly mechanosensitive because their impulse pattern is closely related to mechanical events of the ischemic left ventricular wall (Bosnjak et al. 1979; Malliani et al. 1973b; Uchida and Murao 1974b). In contrast, the unmyelinated fibers most commonly discharge with an irregular pattern during occlusion (Uchida and Murao 1974b; Lombardi et al. 1981).

2.2.4 Chemical Stimulation

Sympathetic ventricular receptors with unmyelinated fibers respond markedly to chemical substances such as bradykinin, potassium, acids, and veratridine. These substances have been applied topically to the epicardial surface (Baker et al. 1980; Coleridge and Coleridge 1980; Uchida and Murao 1974a, c, d, 1975), injected i.v. (Nishi et al. 1977), or injected directly into the coronary arteries

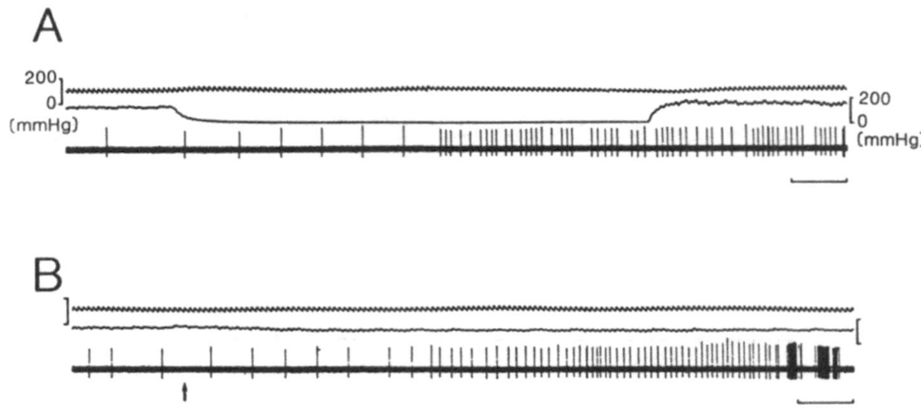

Fig. 5A, B. Increased sympathetic activity after applying noxious stimuli to the heart. Action potentials were recorded from a sympathetic afferent unmyelinated nerve fiber with a left ventricular receptive field. Tracings from *top* to *bottom* are systemic arterial pressure, coronary perfusion pressure, and neural activity. **A** Ischemia of the myocardium was generated by interrupting perfusion of the left main coronary artery. The decreased perfusion pressure represents the period of occlusion. **B** The algesic chemical bradykinin (30 ng/kg) was injected into the left main coronary artery at the time marked by the arrow. The *bars* represent 4 s. (Adapted, with permission of American Heart Journal, from Lombardi et al. 1981)

(Brown and Malliani 1971; Lombardi et al. 1981). Injecting bradykinin into the coronary circulation increases activity of the sympathetic afferent fibers approximately 10 – 15 s after the onset of the injection (Fig. 5B). The increased activity could be attributed to mechanosensitive and chemosensitive endings; however, the existence of both types of receptors in the heart is controversial. Malliani (1982) argues that receptors with unmyelinated sympathetic afferent fibers (Lombardi et al. 1981) and myelinated fibers (Nishi et al. 1977) possess characteristics similar to polymodal receptors (Burgess and Perl 1973). This argument is based on the observation that the receptors discharge spontaneously when hemodynamic conditions are normal and possess some degree of mechanosensitivity (Malliani 1982). Coronary artery occlusion or intracoronary injections of small amounts of bradykinin markedly increase their impulse activity because the same polymodal receptors are excited. Malliani (1986) has hypothesized that cardiac nociception occurs when a spatially restricted population of polymodal receptors in the heart is strongly excited.

The existence of both mechanosensitive and chemosensitive units has also been suggested (Baker et al. 1980; Nerdrum et al. 1986). In these studies mechanosensitive endings are identified by their cardiac discharge pattern and by their immediate and vigorous response when the receptor site is gently stroked with a fine probe or bristle. By contrast, chemosensitive endings have a sparse and irregular discharge without cardiac modulation. Activity of the chemosensitive units changes little when vascular pressure is increased. Although both mechanosensitive and chemosensitive sympathetic cardiac afferents are stimulated following bradykinin application, epicardial application of prostaglandin E_1 (PGE_1)

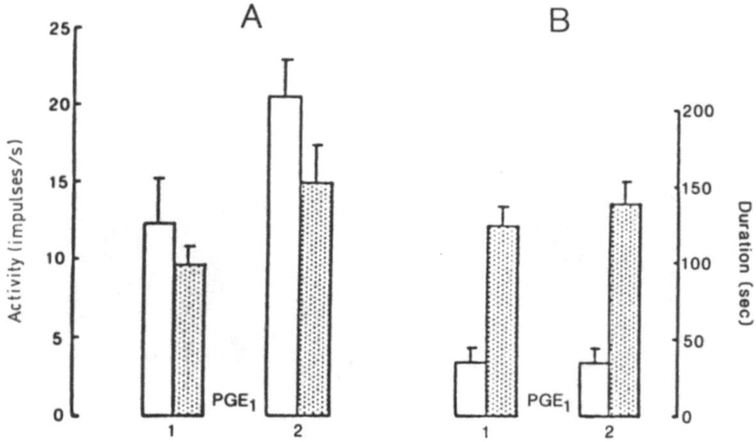

Fig. 6A, B. Responses of chemosensitive (**A**) and mechanosensitive (**B**) sympathetic afferent fibers to a local epicardial application of bradykinin (1.0 μg/ml) before and after PGE_1 was applied to the heart. *Bars* show mean ± SE. *Open bars*, increases in impulse activity; *stippled bars*, duration of the neural responses. Bradykinin (0.3 ml/min) was dripped slowly onto the receptive field of the heart for 15 min before (*1*) and after (*2*) PGE_1 (0.1 μg/ml). (From Nerdrum et al. 1986, with permission of the American Physiological Society)

significantly increases the magnitude and duration of the response of the chemosensitive endings to bradykinin but does not affect the response of mechanosensitive endings (Fig. 6). This suggests that there are both mechanosensitive and chemosensitive endings in the heart.

3 Dorsal Root Ganglion Cells

Cell bodies of the afferent fibers from the heart and coronary arteries are classically described as being in the dorsal root ganglia of the T_1-T_5 segments (Mitchell 1953; White 1957). Studies using horseradish peroxidase (HRP) have more clearly defined the location of the cell bodies of sympathetic afferents arising from the cardiopulmonary region (Kuo et al. 1984; Oldfield and McLachlan 1978; Vance and Bowker 1983). These studies show that afferent fibers have cell bodies occurring from C_8 to T_9, but the major concentration is found between segments T_2 and T_6. Vance and Bowker (1983) demonstrated that the distribution of ganglionic cells remains the same when HRP is injected onto the surface of heart near the left anterior descending branch of the coronary artery (Fig. 7). Thus, the distribution of the sympathetic afferent fibers extends over a wide range of segments but is primarily focused on segments T_2-T_6. Cell bodies for these fibers are much smaller than those found at random in the dorsal root ganglia (Fig. 7). Clinical studies (Lindgren and Olivecrona 1947; White 1957; White and Bland 1948) and experimental animal studies (White et al. 1933) have demonstrated that sympathetic afferent fibers found in the T_1-T_4 rami

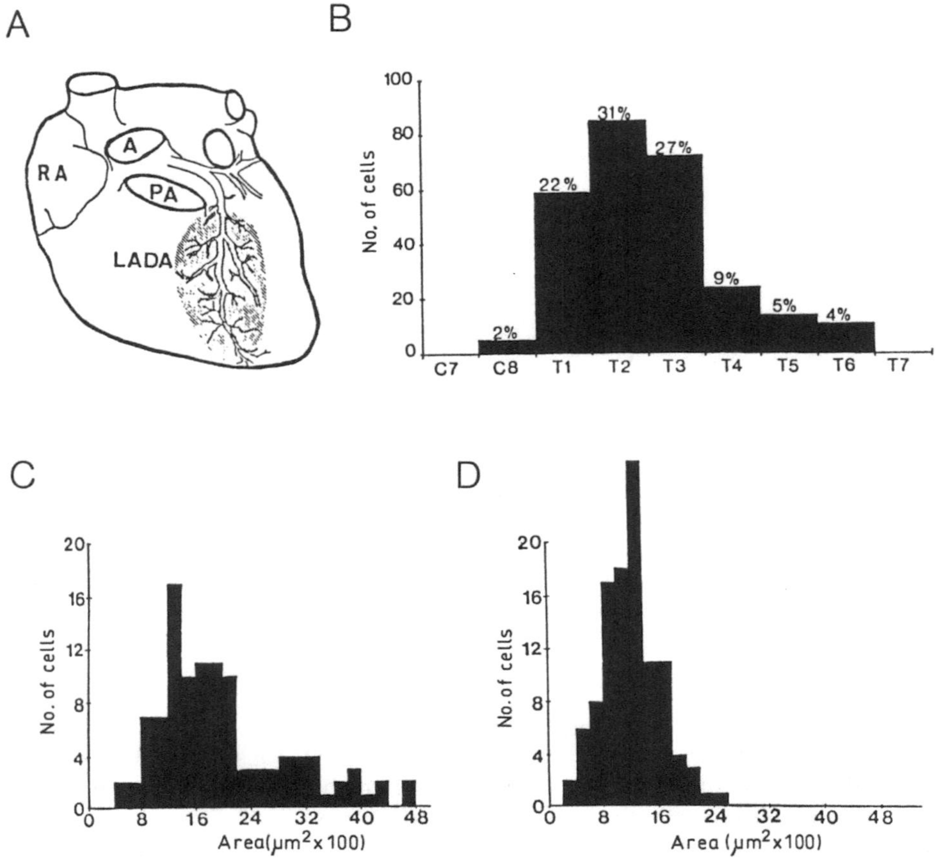

Fig. 7 A – D. Segmental distribution of dorsal root ganglion cells with sympathetic afferent fibers originating from the heart. **A** The shaded area outlines the size of the typical injection of HRP on the surface of the heart. *A*, aorta; *LADA*, left anterior descending artery; *PA*, pulmonary artery; *RA*, right atrium. **B** Histogram illustrating the distribution of HRP-filled dorsal root ganglion cells in each spinal segment. **C** and **D** Histograms showing the cross-sectional areas of randomly chosen cells (**C**) and retrogradely labeled cells (**D**). (From Vance and Bowker 1983, with permission of Elsevier Biomedical Press)

communicantes transmit information that produces the symptoms of angina pectoris.

4 Termination of Sympathetic Afferent Fibers in the Spinal Cord Gray Matter

The general distribution of terminals of somatic sensory afferent fibers has been studied for several years using degeneration methods and HRP techniques (Grant and Ygge 1981; Smith 1983; Ygge and Grant 1983). Only recently have HRP studies compared the termination of visceral and somatic inputs (Cervero and Connell 1984). Terminations of somatic afferent fibers are located primarily in

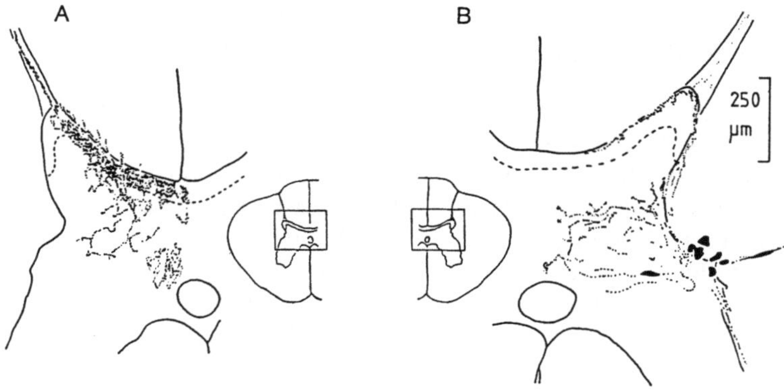

Fig. 8 A, B. Terminal distributions of somatic (**A**) and visceral (**B**) afferent fibers. **A** HRP was applied to the T_9 intercostal nerve, and the projections were reconstructed from three transverse serial sections of the dorsal horn of the T_9 spinal segment. **B** After HRP was applied to the central end of the splanchnic nerve, and the visceral terminations were reconstructed from seven serial sections of the T_9 dorsal horn segment. The spinal cord areas represented in the reconstruction are illustrated in the small insets. (From Cervero and Connell 1984, with permission of Alan R. Liss, Inc.)

laminae I – V of the ipsilateral dorsal horn and in the ipsilateral Clarke's column (Fig. 8 A). Bundles of somatic fibers join the dorsal horn from the dorsal roots via the dorsal columns and Lissauer's tract. These bundles are arranged somatotopically in a mediolateral direction (Cervero and Connell 1984). The medial dorsal horn receives afferent input from the ventral regions of the dermatome, and the lateral dorsal horn receives afferents from the dorsal regions of the dermatome.

The termination sites for sympathetic afferent fibers are quite different from those of somatic afferent fibers (Fig. 8 B). The density of visceral projections to the dorsal horn is lower than that of somatic projections. All visceral afferent axons in the thoracic (Cervero and Connell 1984; Kuo et al. 1984), lumbar (Morgan et al. 1986; Neuhuber 1982) and sacral (Nadelhaft et al. 1983) segments of the spinal cord project laterally in small bundles from Lissauer's tract through lamina I along the outer margin of the dorsal horn into laminae V and VII. A weak projection of visceral afferents is also found along the medial edge of the dorsal horn. The substantia gelatinosa does not appear to have any sympathetic afferent fibers. The innervation needed for nociception and autonomic reflexes fits with this pattern of terminations.

5 Spinal and Supraspinal Reflexes

Several reflexes can be activated by sympathetic afferent impulses entering the gray matter of the spinal cord. Responses may represent cardio-cardiac reflexes, or they may involve efferent outflow of different segments of the spinal cord, such as the outflow to the kidney (Purtock et al. 1977; Weaver 1977). These afferent impulses increase the discharge rate of sympathetic preganglionic efferent fibers either directly or indirectly (Brown and Malliani 1971; Kostreva et al. 1976; Malliani et al. 1972, 1973a, 1975; Pagani et al. 1974; Peterson and Brown 1971). Such increased discharge of sympathetic preganglionic fibers increases the heart rate and myocardial contractility. Preganglionic fibers are excited either during occlusion of the coronary artery or coronary sinuses, or during increased pressure of the coronary artery (Brown and Malliani 1971). Volume infusion activates sympathetic afferent fibers and causes reflex tachycardia and pressor responses (Bishop et al. 1976). Detailed descriptions of reflexes associated with sympathetic afferent stimulation can be found in other reviews (Abboud and Thames 1983; Bishop et al. 1983; Brown 1979; Coleridge and Coleridge 1980; Donald and Shepherd 1978; Longhurst 1984; Malliani 1982).

6 Ascending Spinal Pathways

Multiple ascending pathways project to supraspinal regions. This section will focus on those pathways that transmit nociceptive information from sympathetic afferent fibers to the brain stem and thalamus. Major ascending pathways found in the anterior lateral white matter include the spinothalamic tract (STT), the spinoreticular tract (SRT) and the spinomesencephalic tract. Ascending nociceptive tracts are reviewed by Willis (1985). In addition to the ascending pathways for nociception in the anterior lateral quadrant, recent studies have shown that neurons originating in lamina I of the gray matter project via the dorsolateral funiculus to the thalamus and midbrain (Jones et al. 1985; Swett et al. 1985).
The STT has been divided into two separate systems (Fig. 9). One system is the lateral spinothalamic tract (lSTT) system which ascends to the ventral and ventral posterior lateral regions of the thalamus and responds to a wide range of stimuli

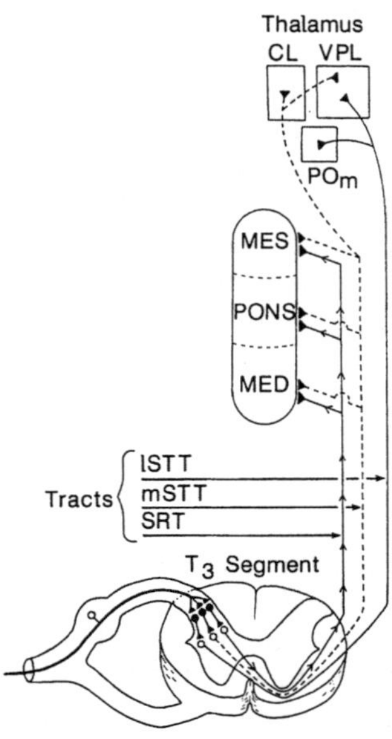

Fig. 9. Diagram showing the nociceptive pathways in the spinal cord projecting to the brain stem and thalamus. *lSTT*, lateral spinothalamic tract; *mSTT*, medial spinothalamic tract; *SRT*, spinoreticular tract; *MES*, mesencephalon; *MED*, medulla; *CL*, centralis lateralis nucleus of the medial thalamus; *VPL*, ventral posterior posterolateral nucleus of the thalamus; PO_m, medial portion of the posterior thalamus

(Dennis and Melzack 1977; Price and Mayer 1974). Neurons of the lSTT may mediate the sensory discriminative aspects of pain (Mehler et al. 1960; Melzack and Wall 1982; Price and Dubner 1977). The lSTT is particularly suited to transmitting phasic information rapidly to the brain and is important in signaling the onset of injury or sudden changes in the damaged area. This information helps the organism to locate precisely the area of pain and to respond and thereby prevent any further damage. The other system is the medial spinothalamic tract (mSTT). This relatively small pathway projects directly to the medial and intralaminar nuclei of the thalamus. The mSTT system is well suited to transmitting tonic information about the state of an organism, because it signals the actual presence of peripheral damage and has the capacity to transmit the message as long as the area of injury is vulnerable to reinjury (Melzack and Wall 1982). It has been proposed that this pathway may be partially responsible for the aversive drive or affective component of pain and may form a distinct motivational-affective pain system (Melzack and Casey 1968; Melzack and Wall 1982).

Another system that transmits sympathetic afferent information from the cardiopulmonary region is the SRT (Fig. 9) (Foreman et al. 1984; Haber et al. 1982; Kevetter et al. 1982). Axons in this system ascend in the ventrolateral quadrant of the spinal cord and course medially into the brain stem reticular formation in the medullary region (Melzack and Wall 1982). The reticular formation can be involved in aversive drive and similar pain-related behavior (Casey et al. 1974). This input also provides a mechanism for activating the autonomic nervous system.

Although these pathways have been discussed as separate entities, there is a marked overlap between the systems. Several studies, for instances have shown that ascending neurons can project both to the medial and the lateral thalamus (Ammons et al. 1985a, b; Giesler et al. 1981; Kevetter and Willis 1983) or that these ascending projections can go to the reticular formation as well as to the thalamus (Foreman et al. 1984; Hayes and Rustioni 1980; Kevetter and Willis 1983; Menétrey et al. 1980). In this review, the lSTT system is described more completely because it is better understood, but, to a lesser extent, information on the mSTT system has also been included. Other ascending pathways that may transmit information include the spinomesencephalic tract (Kerr 1975; Mehler 1969; Menétrey et al. 1982; Yezierski and Schwartz 1986), the dorsal column postsynaptic system (Bennett et al. 1983), the spinoparabrachial tract (Cechetto et al. 1985; Hylden et al. 1985), the spinohypothalamic tract (Burstein et al. 1987), and the spinosolitary tract (Menétrey and Basbaum 1987).

7 Thalamic Sites

The lSTT of primates projects primarily to the ventral posterolateral nucleus (Berkley 1980; Boivie 1979; Bowsher 1961; Chang and Ruch 1947; Clark 1936; Kerr 1975; Kerr and Lippman 1974; Mehler et al. 1960; Pearson and Haines 1980). These connections are similar to those found in humans (Bowsher 1957; Mehler 1962; Walker 1940). The spinothalamic projection to the ventral posterolateral nucleus is somatotopically organized. Fibers from the lumbar cord are distributed laterally in the thalamus while fibers originating in the cervical

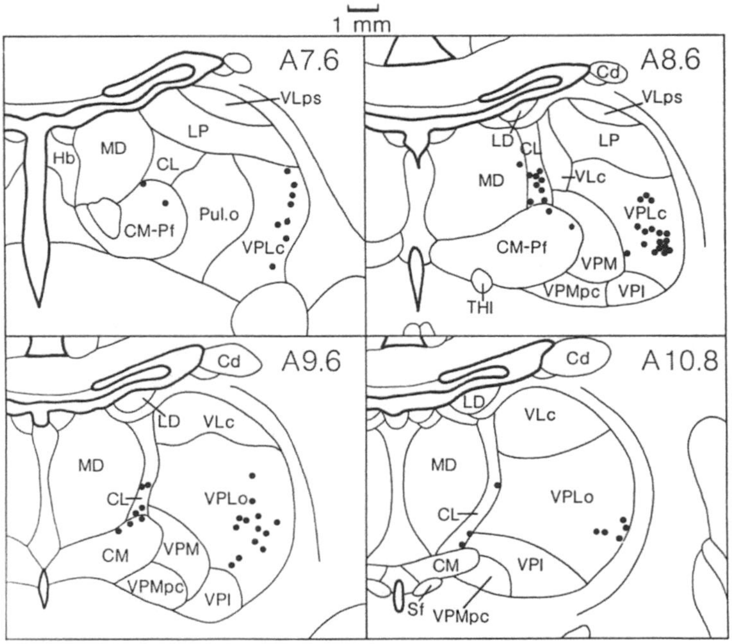

Fig. 10. Antidromic stimulation sites in the thalamus. All sites were determined from lesion marks produced at the end of each experiment. Sections were redrawn from the atlas of Szabo and Cowan (1984). *Cd*, caudate nucleus; *CL*, nucleus, centralis lateralis; *CM*, centrum medianum; *Hb*, habenula complex; *LD*, lateral dorsal nucleus; *LP*, lateral posterior nucleus; *MD*, dorsomedial nucleus; *Pf*, parafascicular nucleus; *Pul. o*, oral pulvinar nucleus; *VLc*, ventral lateral nucleus, caudal part; *VLps*, ventral lateral nucleus, pars postrema; *VPI*, ventral posteroinferior nucleus; *VPLc*, ventral posterolateral nucleus, caudal part; *VPLo*, ventral posterolateral nucleus, oral part; *VPM*, ventral posteromedial nucleus; *VPM pc*, ventral posteromedial nucleus, parvocellular part; *THI*, habenulo-interpeduncular tract; *Sf*, subfascicular nucleus. (From Ammons et al. 1985, with permission of the American Physiological Society)

cord are found medially (Boivie 1979; Chang and Rue 1947; Clark 1936; Mehler 1960; Walker 1940). The lSTT cells that receive sympathetic afferent input in the upper thoracic spinal cord project primarily to the middle regions of the ventral posterolateral nucleus (Fig. 10). These sites are more medial than the projections of those neurons that are antidromically activated in the lumbar or sacral regions of the spinal cord (Applebaum et al. 1975; Milne et al. 1981).

The intralaminar complex, including the centralis lateralis nucleus and adjacent related parts of the mediodorsal nucleus, also receives input from the STT (Berkley 1980; Boivie 1971, 1979; Jones and Burton 1974; Kerr 1975; Mehler et al. 1960; Mehler 1962; Pearson and Haines 1980). Some spinothalamic terminals are also found in parts of the centrum medianum, parafascicular, paraventricular, paracentral, central medial, reunions (Boivie 1979; Burton and Craig 1983), and submedial nuclei of the thalamus (Craig and Burton 1981). The mSTT cells that receive sympathetic afferent fiber input are activated from stimulation sites located primarily in the centralis lateralis, and the centrum medianum-parafascicular region (Fig. 10) (Ammons et al. 1985a). This coincides well with the results of other electrophysiological studies (Giesler et al. 1981). Stimulation sites were found either in the ventral posterolateral nucleus or in the intralaminar nuclei, although some axons project to both areas (Ammons et al. 1985a, b; Giesler et al. 1981).

8 The Spinothalamic Tract System (STT)

8.1 Anatomical Studies

The locations of the cells of origin of the STT have been shown by retrograde chromotolysis, antidromic mapping, and retrograde labeling with substances such as HRP. Figure 11A shows the distribution of the cells of origin of the STT in the thoracic spinal cord of the monkey as identified using HRP. STT cells are found in the marginal zone, the deeper layers of the dorsal horn, the intermediate region, and occasionally in the ventral horn. The lSTT cells are found more often in the marginal zone and the upper part of the dorsal horn, whereas a higher proportion of mSTT cells are found in the ventral horn (Carstens and Trevino 1978; Giesler et al. 1979; Willis et al. 1979).

8.2 Electrophysiological Studies

Antidromic activation of STT cells following stimulation within the thalamus or in the STT as it enters the thalamus has been used to identify STT cells (Albe-Fessard et al. 1974; Applebaum et al. 1975; Dilly et al. 1968; Fox et al. 1980; Price et al. 1978; Trevino et al. 1973). The criteria used for determining antidromic activation include a constant-latency discharge during repeated testing,

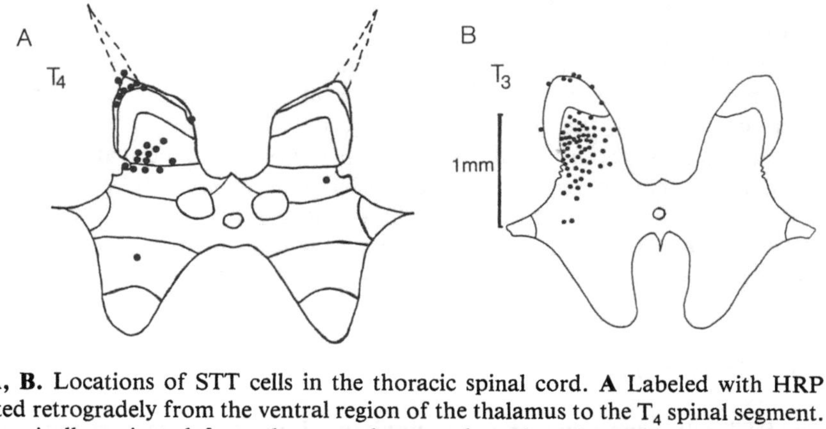

Fig. 11A, B. Locations of STT cells in the thoracic spinal cord. **A** Labeled with HRP transported retrogradely from the ventral region of the thalamus to the T_4 spinal segment. **B** Antidromically activated from the ventral posterolateral nucleus of the thalamus to segments T_3-T_4. (**A** redrawn, from Trevino et al. 1976, with permission of Pergamon Press)

the ability to follow a high-frequency train of discharges, and collision of antidromic with orthodromic spikes inside the critical interval (Trevino et al. 1972, 1973). Figure 11 B shows the locations of STT cells that are antidromically activated from the lateral thalamus or medial thalamus. The mSTT cells are usually localized in laminae I, IV, V, and VII, with their numbers increasing in the deeper laminae (Ammons et al. 1985a). The lSTT cells are most commonly found in lamina IV, whereas those neurons projecting to both the lateral and medial thalamus are most often found in lamina V (Ammons et al. 1985a).

8.3 Viscerosomatic Convergence (Angina Pectoris)

It is now appropriate to consider a likely neurophysiological basis of angina pectoris. Only rarely is this pain sensed as originating from a site in the diseased visceral structure; usually it is referred to overlying or nearby somatic structures. To explain this symptom Ruch (1961) proposed that "... some visceral afferents converge with cutaneous pain afferents to end upon the *same* neuron at some point in the sensory pathway – spinal, thalamic, or cortical – and that the system of fibers is sufficiently organized topographically to provide the dermatomal reference." The STT provides the ascending pathway for transmitting this information. There is convergence of somatic and visceral afferent inputs onto STT neurons (Ammons et al. 1985a, b; Blair et al. 1981).
The following discussion will focus on afferent systems that converge on the cells of origin of the STT. Visceral inputs arising from the cardiopulmonary region and somatic inputs arising from the upper arms and chest will be discussed separately, but an important characteristic with respect to viscerosomatic convergence must be kept in mind: all lSTT cells and 90% of the mSTT cells (Ammons et al. 1985a) studied to date receive both visceral and somatic input. These studies did not select only for cells that responded to sympathetic afferent input since input was always tested after a STT cell was found by antidromic stimulation. A very small population of STT cells receiving input from the splanchnic nerve was also tested, but 95% of these neurons received viscerosomatic input (Foreman et al. 1981). By contrast, only 84% of the lumbosacral lSTT cells responded to distension of the urinary bladder (Milne et al. 1981). This difference may be due to the fact that Milne et al. (1981) activated urinary bladder afferents by using the more natural bladder distension whereas sympathetic afferent nerves in the cardiopulmonary region were activated by electrical stimulation. This percentage of lumbosacral neurons activated by a natural stimulus agrees closely with the finding that approximately 75% – 80% of the lSTT cells in the thoracic cord are activated when a noxious chemical is injected into the heart (Blair et al. 1982).

8.3.1 Somatic Input

The STT cells generally respond to noxious stimuli and rarely respond only to innocuous stimuli. Many neurons respond to innocuous stimuli, but those same neurons respond much better to a noxious stimulus (Chung et al. 1979; Craig and

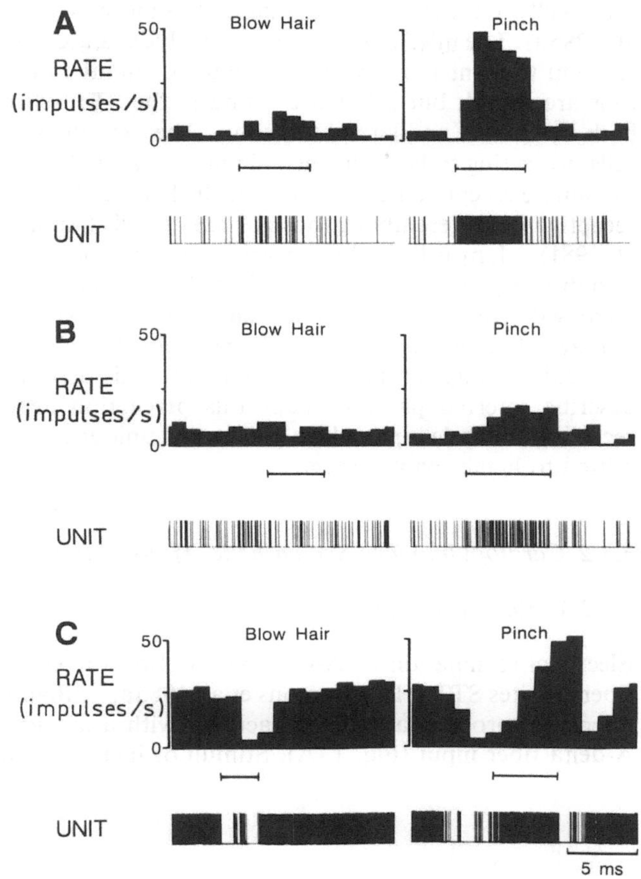

Fig. 12A–C. Somatic input onto STT cells. Responses of **A** a wide dynamic range cell, **B** a high-threshold cell, and **C** a high-threshold inhibitory cell. *Bars* show the time for which the stimulus was applied to the skin or skin and muscle

Kniffki 1985; Fox et al. 1980; Giesler et al. 1976; Price et al. 1978; Willis et al. 1974). STT cells are generally classified as low-threshold, wide dynamic range (or multireceptive), or high-threshold (or nociceptive specific) cells. Wide dynamic range and high-threshold cells are most likely to play an important role in signaling pain (Willis 1985). The most common STT cells are wide dynamic range cells; these neurons respond with a slow discharge rate to hair movement or touching of the skin but discharge vigorously when pinch is applied to the skin or to the skin and muscle (Fig. 12A). High-threshold cells are not excited by blowing hairs but increase their discharge rate during the pinch (Fig. 12B). Another category of cells respond vigorously to a noxious pinch applied to the skin or to skin and muscle, but spontaneous activity of the cells is suppressed completely when hairs are moved (Fig. 12C). These cells are classified as high-threshold inhibitory (Blair et al. 1981).

Somatic receptive fields of lSTT cells are usually restricted to the left chest and forelimb and do not cross the midline (Ammons et al. 1985a). By contrast, mSTT cells often have bilateral or large unilateral receptive fields (Ammons et al. 1985a). Both the mSTT and lSTT cell populations have many wide dynamic range cells, but the mSTT population has more high-threshold cells (Ammons et al. 1985a). The mSTT cell group also includes a greater number of deep cells that respond to joint movement or to muscle stimulation. No low-threshold mSTT cells are found, but a few are found in the STT cell populations projecting to both the medial and lateral thalamus. The excitatory receptive fields of the STT cells projecting to both the medial and lateral thalamus are most often similar to the simple receptive fields found for the lSTT cells, although some have complex receptive fields resembling those of mSTT cells (Ammons et al. 1985a; Giesler et al. 1981). Many cells in the thoracic segments not only respond to skin pinch but also discharge more vigorously when skin and muscle are pinched together. The increased effect of muscle input may contribute to understanding the sensations that are often associated with referred pain. Muscle pain is commonly described as a dull, aching sensation (Lewis 1942). This description is also often used to describe referred pain. Perhaps this perception arises from many STT cells receiving a muscle input which produces a unique pattern of activity that is transmitted to higher pain centers.

8.3.2 Cardiopulmonary Sympathetic Afferent Input

8.3.2.1 Electrical Stimulation

Electrical stimulation of A-delta and C fibers in the cardiopulmonary afferent fibers excites STT cells (Ammons et al. 1985a, b; Blair et al. 1981). Low stimulus intensities produce a volley of activity with a latency that correlates well with A-delta fiber input (Fig. 13A). Stimuli of increased intensity and duration gen-

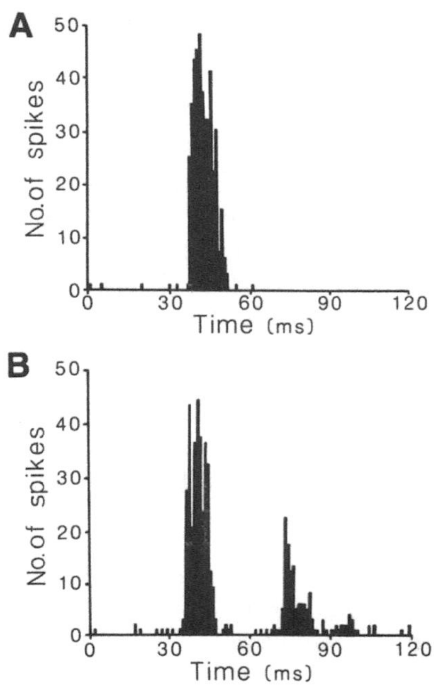

Fig. 13A – B. Histograms of responses of an STT cell to electrical stimulation of the cardiopulmonary sympathetic afferent fibers. Stimulus strengths and durations were sufficient to activate A-delta (**A**) and both A-delta and C fibers (**B**). The stimulus was applied at 30 ms

erate a later volley of activity with a latency coincidental with C fiber input (Fig. 13 B). The minimum afferent conduction velocity is approximately 9 m/s for the A-delta peak and approximately 1.0 m/s for the C fiber peak. Approximately 50% of the cells receive only A-delta fiber input, 40% receive both A-delta and C fiber input, and 10% receive only C fiber input (Blair et al. 1981). Two interesting features have developed from these studies. First, STT cells receiving both A-delta and C fiber input discharge at a higher spontaneous rate than cells receiving only A-delta fiber input (Blair et al. 1982). Secondly, cells with both A-delta and C fiber input are more likely to respond to a noxious stimulus such as coronary artery occlusion or application of an algesic chemical to the heart (Ammons et al. 1985 b; Blair et al. 1984 b). These differences in the responsiveness of neurons to various inputs are not understood at the present time, but they may be important in formulating a hypothesis about cardiac pain.

8.3.2.2 Natural Stimulation

Electrical stimulation demonstrates that afferent information from the cardiopulmonary region produces responses in tract neurons of the gray matter in the spinal cord. Stimulation of receptors under "natural" conditions is crucial for understanding how cells of the spinal cord respond to events occurring in the cardiopulmonary region. Since the STT and SRT are the best studied ascending systems for visceral input, potentially noxious stimuli are used to activate the cells of these pathways. Several cardiac stimuli can produce pain and reflex responses (Aviado and Schmidt 1955); examples include the following: (a) reduced coronary arterial pressure distal to an occlusion affects coronary artery mechanosensitive receptors; (b) ischemia stimulates myocardial mechanosensitive receptors; (c) liberation of chemical substances formed by tissue breakdown or platelet disintegration stimulates nociceptors; and (d) ischemia stimulates myocardial nociceptors. Since angina pectoris is generally associated with myocardial ischemia, coronary artery occlusion is a logical procedure for activating these receptors. Transient occlusions in anesthetized animals do not imitate classical clinical manifestations of coronary heart disease because diseased vessels are usually gradually occluded, and angina pectoris is experienced during increased exertion. However, certain patients experience angina pectoris from coronary artery spasm (Maseri and Chierchia 1982).

Coronary Artery Occlusion. Coronary artery occlusion causes myocardial ischemia, which activates STT and SRT cells and produces varied responses to the ischemic episode (Fig. 14) (Blair et al. 1984 b). Cells are categorized into four groups: those in which (a) activity increases during the ischemia produced by experimental occlusion; (b) activity increases at the onset of occlusion, adapts, and then increases again as ischemia develops; (c) activity increases rapidly at the onset or the release of occlusion and adapts or remains elevated throughout the occlusion (considered an innocuous response); and (d) activity does not change during occlusion. Cells responding immediately to occlusion may be activated by mechanosensitive receptors located in or near the coronary artery. However, cell activity usually increases approximately 13 s after the onset of the occlusion.

Fig. 14. Increased activity of an SRT neuron during occlusion of the left descending coronary artery, an example of a response to the onset of ischemia. There was an initial increase in cell activity followed by adaptation for several seconds and then increased activity again at about 15 s, when ischemia occurred. The tracings from *top* to *bottom* are rate of cell discharge (*Rate*), the time line (divided into 1-s intervals), action potentials from a window discriminator (*Cell*), ECG, systemic arterial pressure (*AP*), left ventricular pressure (*LVP*), and differential of left ventricular pressure (d*P*/d*t*). Occlusion for 29 s is denoted by the *horizontal bar*. (From Blair et al. 1984b, with permission of the American Physiological Society)

These changes are often associated with marked changes in the electrocardiogram and with falls in systemic blood pressure, left ventricular pressure, and contractility. Thus, ischemia produced during an experimental occlusion may activate mechanosensitive and chemosensitive cardiac receptors which, in turn, excite STT and SRT cells (Blair et al. 1984b).

STT and SRT neurons were analyzed according to their responses to an innocuous (i.e., mechanical distortion of the receptors) or noxious stimulus (i.e., during the ischemic episode) and to their responses to the type of sympathetic afferent input (Fig. 15). Cells receiving only A-delta input are unresponsive to the ischemic period of the coronary artery occlusion; however, cells receiving both A-delta and C fiber input are more likely to be excited by myocardial ischemia (Fig. 15A). Cells receiving A-delta and those receiving both A-delta and C fiber input respond to both innocuous and ischemic phases of the coronary occlusion, but cells receiving A-delta and C fiber input are more likely to respond to any type of stimulus (Fig. 15B). These results suggest that C fiber input is necessary for an ischemic response to occur, although not every cell responded to this stimulus.

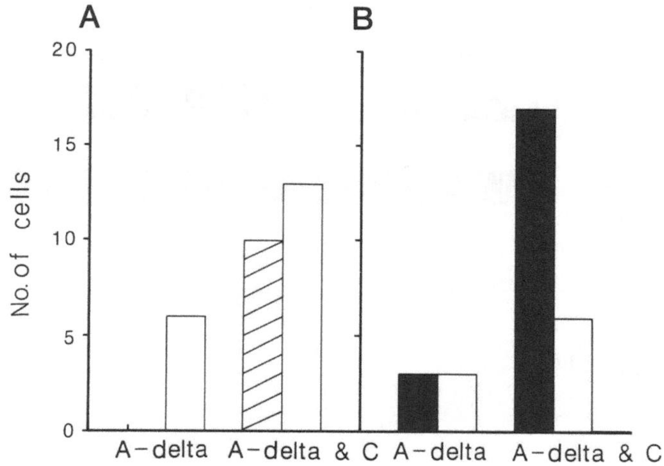

Fig. 15A, B. Relationship of cell response to coronary artery occlusion with the type of visceral afferent input to STT and SRT neurons. **A** Separated according to ischemic response; *open bar*, no response or firing at onset; *striped bar*, firing with ischemia, or at onset and with ischemia. The *A*-delta and C fiber input was related significantly ($P<0.05$) to cell's response to ischemia. **B** Responses divided according to whether a cell was non-responsive (*open bar*) or responsive (*filled bar*), with any type of response, to occlusion. (From Blair et al. 1984b, with permission of the American Physiological Society)

Chemical Activation. Several chemicals have been used to activate cardiac receptors (Baker et al. 1980; Coleridge and Coleridge 1980; Lombardi et al. 1981; Nishi et al. 1977; Uchida and Murao 1974a, d, 1975). Bradykinin was used to activate the receptors in studies on STT and SRT cells for several reasons: bradykinin is an endogenous algesic chemical which produces pain in animals (Besson et al. 1974; Guzman et al. 1962) and in humans (Armstrong et al. 1953, 1957; Bleehen and Keele 1977; Chahl 1979; Chahl and Kirk 1975; Elliot et al. 1960; Keele and Armstrong 1969; Lim and Guzman 1968); increased amounts of bradykinin are found in the effluent of the coronary sinus following coronary artery occlusion (Hashimoto et al. 1977; Kimura et al. 1973); and finally, sympathetic afferent fibers are excited when bradykinin is injected into the heart or dripped onto the epicardium (Baker et al. 1980; Lombardi et al. 1981; Nerdrum et al. 1986; Nishi et al. 1977).

The central pathways of nociceptive systems were studied by injecting bradykinin into the left atrial appendage to activate receptors either directly in the endocardium or throughout the coronary distribution of the heart (Fig. 16) (Blair et al. 1982, 1984a). Approximately 75% of the STT cells tested showed increased activity following injections of bradykinin (Blair et al. 1982). Average latency for the onset of increased cell activity was approximately 15 s. This coincides closely with the response time required for bradykinin when single-unit recordings were made from sympathetic afferent fibers (Malliani 1982). The response was sustained for varying lengths of time and then gradually returned to control levels. To demonstrate that STT cells were excited by cardiac afferents and not by receptors located in the general systemic circulation, bradykinin was

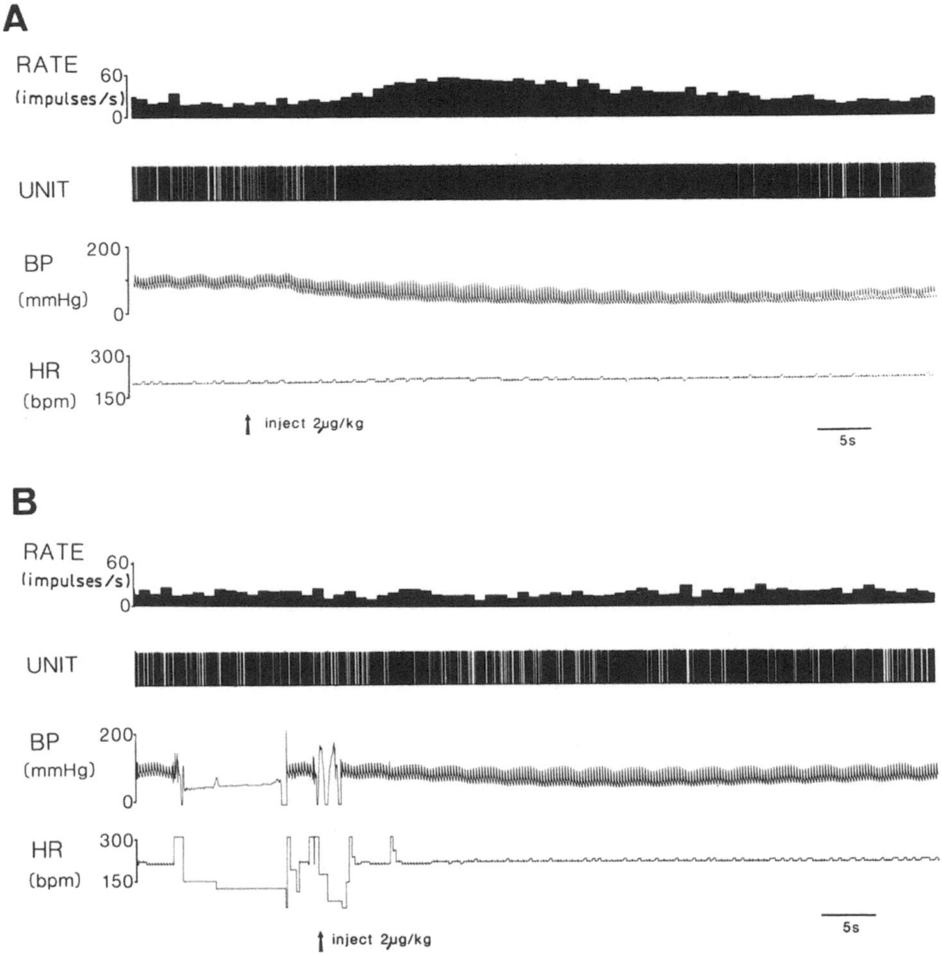

Fig. 16A, B. Increased discharge rate of an STT cell after injecting bradykinin (2 μg/kg) into the heart. Bradykinin was injected into the left atrium (**A**) and the aorta (**B**) just distal to the aortic arch. Tracings from *top* to *bottom* are rate of cell discharge, output from the window discriminator (*Unit*), representing one cellular action potential per deflection unit, systemic arterial pressure (*BP*), and heart rate. (From Blair et al. 1982, with permission of the American Heart Association)

injected into the aorta above the coronary ostia. Little if any response was observed following these injections (Fig. 16B). In addition, injection of 2% lidocaine into the pericardial sac surrounding the heart abolishes the STT cell response to bradykinin (Blair et al. 1982).

Sympathetic afferent input from C fibers increased the likelihood that a cell would respond to bradykinin (Fig. 17). Approximately 50% of cells receiving only A-delta fiber input respond to bradykinin, as opposed to about 80% when both A-delta and C fiber inputs are present. Baker et al. (1980) suggest that C fibers might transmit noxious information. It appears that C fiber input increases

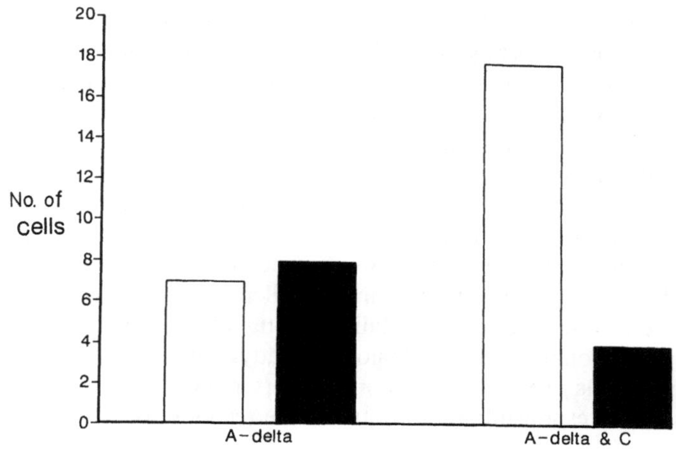

Fig. 17. STT cells responding (*open bars*) or not responding (*filled bars*) to bradykinin and classified according to the type of sympathetic afferent fiber input (A-delta or A-delta and C)

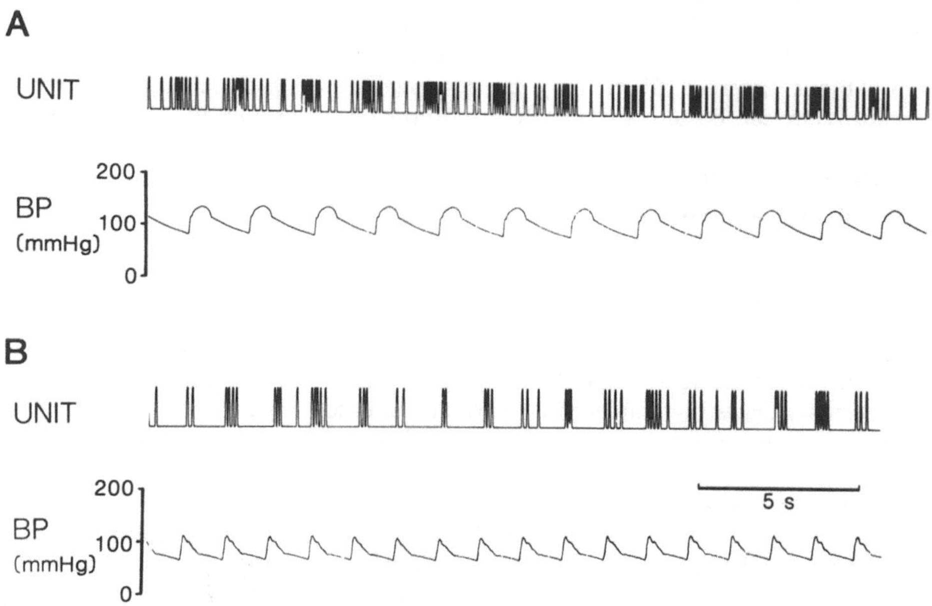

Fig. 18 A, B. STT cell activity entrained to the cardiac cycle following bradykinin injections. **A** An STT cell showing bursts of activity just before and during the upstroke of the arterial pressure pulse. **B** An STT cell discharging almost exclusively near the peak of the arterial pressure wave. *BP*, blood pressure. (From Ammons et al. 1985b, with permission of the American Physiological Society)

the likelihood that a cell will respond to bradykinin, although A-delta fiber input alone also excites some cells. However, as mentioned in the previous section, C fibers are required to excite STT cells during the ischemic phase of the coronary artery occlusion. This information coupled with the bradykinin responses emphasizes the important part that C fibers appear to play in the transmission of noxious information from the heart to the spinal cord.

The response of STT cells to bradykinin does not rule out other chemicals or other mechanisms which may be involved in activating these cells. Prostaglandins may constitute an intermediate step in increasing the impulses elicited in sympathetic afferent fibers by bradykinin (Needleman 1976; Nerdrum et al. 1986). Such an intermediate step may explain the long latency before the discharge rate increases with coronary artery occlusion or bradykinin injections. Further studies with other chemicals released during coronary artery occlusion may determine the stimulus or stimuli responsible for producing the response of STT cells to ischemia.

Cardiac Rhythm. When STT cells respond to bradykinin, there appears initially to be a random increase in neuronal activity; however, in approximately 40% of the cells, activity later becomes entrained with the cardiac cycle (Fig. 18) (Blair et al. 1982; Ammons 1985b). This later entrained activity may be due to the sensitization of mechanosensitive afferent fibers, since increased cell activity occurs at the peak of the systolic pressure when the receptors would be maximally excited if they were located in series with the cardiac muscle. Myelinated sympathetic afferent fibers may be responsible for producing this rhythm because they can be activated much later in the occlusion period (Malliani et al. 1973b). Thus, the entrainment of many STT cells to the cardiac cycle during the bradykinin response may reflect the ability of bradykinin to sensitize sympathetic afferents to mechanical events.

8.4 Viscero-Visceral Convergence (Anginal-Like Pain)

Many visceral disorders elicit chest pain which can be misinterpreted as angina pectoris. Some patients exhibit anginal-like pain although they do not have symptoms suggesting underlying cardiac disease; instead they may have esophageal or gastrointestinal disorders, including hiatal hernia, gastritis, gallbladder disease, splanchnic flexor syndrome, or peptic ulcer (Hampton et al. 1959; Henderson et al. 1978; Miller 1942; Palmer 1958; Ravdin et al. 1955; Wehrmacher 1964). Such pain referral with gastrointestinal disorders suggests that there may be convergence of input from the gastrointestinal tract and from the heart onto the same STT neurons in the upper thoracic segments. Not only would this provide an explanation for some of the anginal-like pain that occurs, but it would also help elucidate the organization of the spinal cord with respect to integration of afferent input from different visceral organs.

8.4.1 Splanchnic Nerve Stimulation

Dorsal horn cells of the lower thoracic spinal cord (segments $T_7 - T_{11}$) increase their activity in response to electrical stimulation of the splanchnic nerve

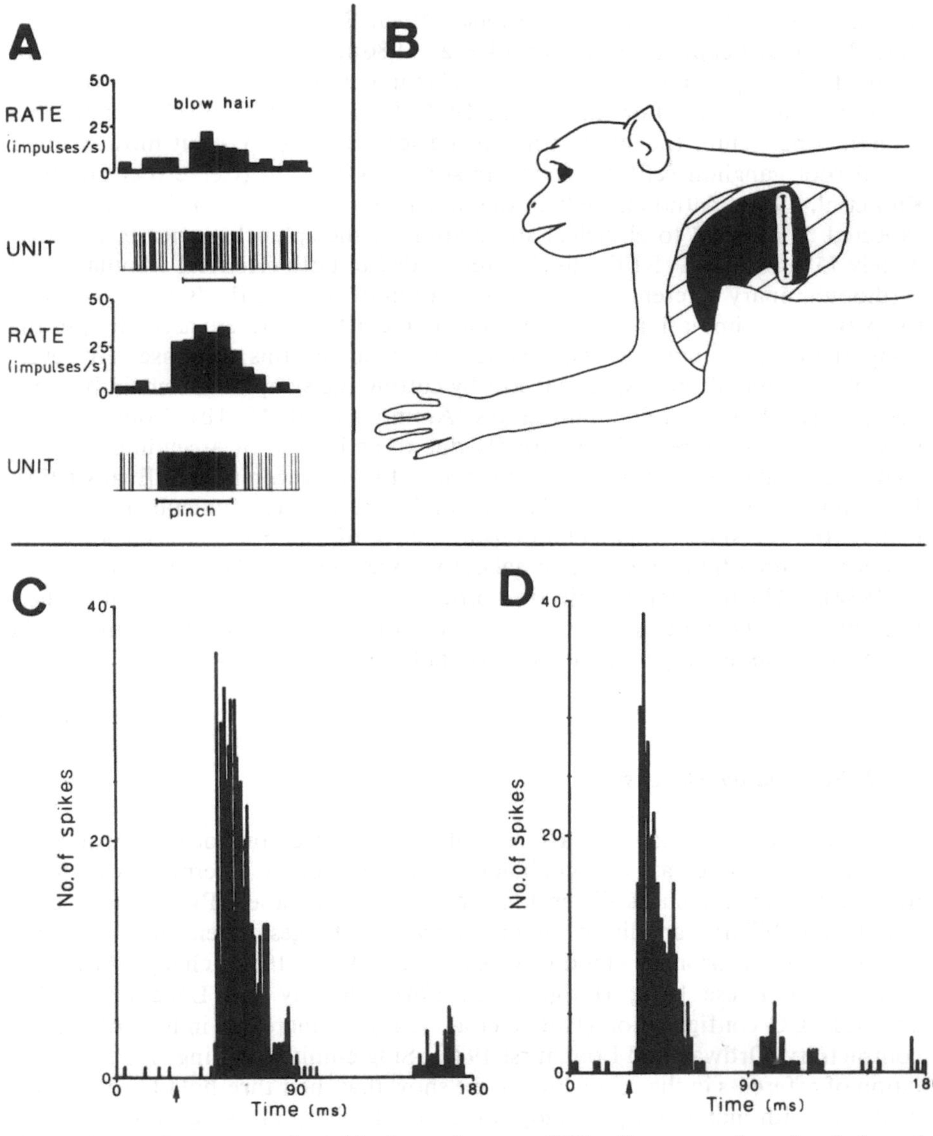

Fig. 19 A – D. Responses of a wide dynamic range STT cell to somatic, splanchnic and cardiopulmonary sympathetic afferent input. **A** Responses to blowing hair (*upper two traces*), and pinching skin and muscle of the chest region (*lower two traces*). The rate tracings are the number of cell discharges per second determined with a ratemeter. Unit tracings represent cell activity where one deflection represents one action potential. **B** Somatic receptive field. The *blackened area* is the region where blowing hair excited the cell. Responses to pinching skin and muscle were elicited from the region shown *in black* and *in stripes*. **C** Poststimulus histogram of cell response to SPL stimulation. The histogram is the cumulative response to 50 stimulus repetitions. The *arrow* shows the point at which the stimulus was applied. Bin width was 1 ms. **D** poststimulus histogram of cell response to cardiopulmonary sympathetic afferent stimulation. (From Ammons et al. 1984b, with permission of the American Physiological Society)

(Cervero 1983a, b; Cervero and Tattersall 1985; Fields et al. 1970; Gokin et al. 1977; Pomeranz et al. 1968; Tattersall et al. 1986a, b). In addition, STT cells of segments T_7-T_{11} also respond to electrical stimulation of A-delta and C fibers in the splanchnic nerve (Foreman et al. 1981; Hancock et al. 1975). Splanchnic afferent fibers enter the spinal cord in the segments T_3-T_{13} but most of their dorsal root ganglion cells are found in segments T_5-T_{11} (Cervero et al. 1984; Kuo et al. 1981). Although cells in the upper thoracic spinal cord would not be expected to respond to electrical stimulation of the splanchnic nerve, approximately 75% of T_2-T_5 STT cells that responded to both electrical stimulation of cardiopulmonary afferent fibers and activation of somatic fields also responded to splanchnic stimulation (Fig. 19). Splanchnic afferent fibers ascending in the sympathetic chain may be responsible for producing this response. However, only about 25% of the response is lost by cutting the sympathetic chain between the T_5-T_6 white rami communicantes (Ammons et al. 1984b). Consequently, most of this convergence is apparently due to activation of ascending propriospinal pathways (Bain et al. 1935; Downman 1955; Downman and Evans 1957). Interruption of the ipsilateral, lateral, and ventrolateral funiculi at approximately the T_6 spinal segment abolishes most of the STT cell responses to splanchnic input that enters the spinal cord in segments caudal to T_5 (Ammons et al. 1984b). The ipsilateral projection probably represents propriospinal connections between the upper and lower thoracic segments, although the pathway may be part of an ascending limb of a supraspinal loop (Cervero 1983c; Cervero et al. 1985).

8.4.2 Gallbladder Distension

The splanchnic nerve contains afferent fibers from receptors located throughout the abdominal viscera. The gallbladder was chosen to determine whether a natural stimulus, such as distension, could activate these STT neurons. Gallbladder and biliary duct distension or application of algesic chemicals to the gallbladder reduces coronary blood flow (Cullen and Reese 1952), changes heart rate (Cullen and Reese 1952; Hodge et al. 1947; Ordway and Longhurst 1983), changes ECG configuration (Hodge et al. 1947), or increases indices of cardiac contractility (Ordway and Longhurst 1983). Single-unit recordings from a population of afferents in the splanchnic nerve show that their threshold for activation coincides with maximum physiological levels of biliary pressure; more vigorous discharges are recorded when biliary pressure reaches noxious levels (Cervero 1982). Cervero (1982) suggested that the level of distension was noxious when blood pressure increased. Cervero (1983a) has also shown that gallbladder distension activates T_9 dorsal horn cells in lamina V, their expected segment for projection.

Gallbladder distension also activates T_2-T_5 STT cells (Ammons et al. 1984a), and produces phasic and tonic increases in the discharge rate of STT cells but no inhibition (Fig. 20). Most cells show a phasic response, with more activity during the change in gallbladder distension than during the period of maintained distension. Such responses are also found in lower thoracic spinal neurons which respond to gallbladder distension (Cervero 1983a); these responses are described

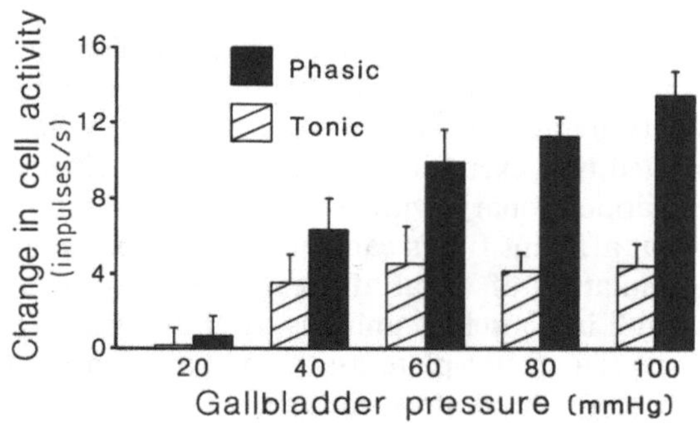

Fig. 20. Increasing STT cell activity with distension of the gallbladder. Cells were divided into those giving phasic (*filled bars*) or tonic responses (*striped bars*). (From Foreman 1986, with permission of Plenum Press)

as either rapidly or slowly adapting. The magnitude of the phasic responses are directly related to the degree of distension. The phasic response to gallbladder distension may provide the thalamus with information about the intensity of a sudden change in gallbladder pressure. In contrast, tonic responses reach a plateau and are maintained at this level. These responses may transmit information that indicates an abnormally high pressure in general. Phasic and tonic responses may be the neurophysiological correlates of sudden colic-type gallbladder pain and of chronic gallbladder pain, respectively.

An average gallbladder pressure of 50 mmHg is required to activate STT cells (Ammons et al. 1984a). Such a high activation threshold in STT cells fits their function of signaling pain; during acute colicystitis gallbladder pressures may rise to as high as 60 mmHg (Csendes and Sepulveda 1980). In addition, STT cells that receive gallbladder input are more likely to respond to intracardiac injections of bradykinin and to respond with a greater discharge frequency (Ammons et al. 1984a). The presence of this population of neurons may explain why patients who have both heart and gallbladder disease are likely to experience anginal-like pain during their gallbladder attacks (Ravdin et al. 1942). However, many patients who have gallbladder disease do not have pain referred to the same area as anginal pain. Therefore, nonconvergent projections from the spinal segments T_7-T_{11} are also important for signaling gallbladder pain.

9 Descending Modulation of Sympathetic Afferent Fibers

Sympathetic and parasympathetic efferent nerves exert a direct influence on the activity of the heart. Cardiac function is extrinsically controlled by a balance between the sympathetic and parasympathetic nervous systems, depending upon demands such as stress, exercise, and pain. Since efferent outflows from these two systems act in concert, the afferent fibers carried by sympathetic and vagal nerves perhaps also interact to modify the responses of sensory relay cells in the central nervous system.

9.1 Vagal Afferent Fibers

Vagal afferent fibers from the cardiopulmonary region tonically inhibit the vasomotor regions of the central nervous system that mediate circulatory changes including hypotension and decreased afferent sympathetic nerve activity (Kezdi et al. 1974; Thames and Abboud 1979; Thorén et al. 1976). Further evidence for such interactions comes from a study by Schwartz et al. (1973), who demonstrated that excitation of vagal afferent fibers inhibits the single-unit activity of cardiopulmonary sympathetic preganglionic fibers. These studies show that the vagal afferent fibers can suppress sympathetic afferent activity.

Stimulation of vagal afferent fibers can also inhibit the responses to noxious stimuli in conscious animals. Right cervical vagotomy reverses opioid-mediated analgesia in hot-plate assays of pain sensitivity in spontaneously hypertensive rats and reduces stress-induced analgesia resulting from exposure to intermittent foot shock (Maixner and Randich 1984; Maixner et al. 1982). Volume expansion produces long-lasting inhibition of the tail flick reflex to noxious radiant heat, and intravenous veratridine injections induce antinociception which is eliminated after bilateral cervical vagotomy (Maixner and Randich 1984; Randich et al. 1984). These studies strongly suggest that vagal afferent fibers can suppress responses to noxious stimuli. The ability of vagal afferents to modulate activity of STT and SRT neurons may be the cellular basis for the analgesia seen in conscious animals.

Both myelinated and unmyelinated vagal fibers transmit information from the heart to the brain stem. Myelinated fibers arise primarily from the atria and venoatrial junction (Coleridge et al. 1957; Paintel 1973; Thorén 1979). A few myelinated afferent fibers also arise from the ventricle and coronary arteries (Coleridge et al. 1964; Donald and Shepherd 1978). Myelinated fibers innervate a relatively localized area of the ventricle. On the other hand, vagal afferent C fibers have sensory endings distributed to all chambers of the heart (Baker et al. 1979; Coleridge et al. 1964; Muers and Sleight 1972; Sleight and Widdicombe 1965; Thorén 1977; Thorén 1979).

Unmyelinated afferent vagal fibers are classified as having mechanosensitive or chemosensitive endings. Mechanosensitive receptors are present in the atria (Thames et al. 1977; Thorén 1976) and ventricles (Oberg and Thorén 1973; Sleight and Widdicombe 1965; Thorén 1977). Chemosensitive endings are distributed similarly to mechanosensitive units. Various exogenous chemicals and naturally occurring substances, including bradykinin, stimulate these receptors (Kaufman et al. 1980). Both mechanosensitive and chemosensitive C fibers respond to veratrum alkaloids and nicotine (Coleridge and Coleridge 1979; Oberg and Thorén 1973). Thorough reviews of the receptor properties and reflexes associated with vagal afferent fibers have been published previously (Bishop et al. 1983; Coleridge and Coleridge 1979; Thorén 1979; Thorén et al. 1976).

9.2 Vagal Afferent Stimulation and the STT (Silent Myocardial Ischemia)

Activity of vagal afferent fibers can produce analgesia, but does it modify the activity of STT cells receiving input from the heart? Electrical stimulation of the vagal afferents can suppress the spontaneous activity of STT cells (Fig. 21; Ammons et al. 1983a, b). The activity of approximately 60% of STT cells is suppressed during left thoracic vagal stimulation, 13% are excited or excited and

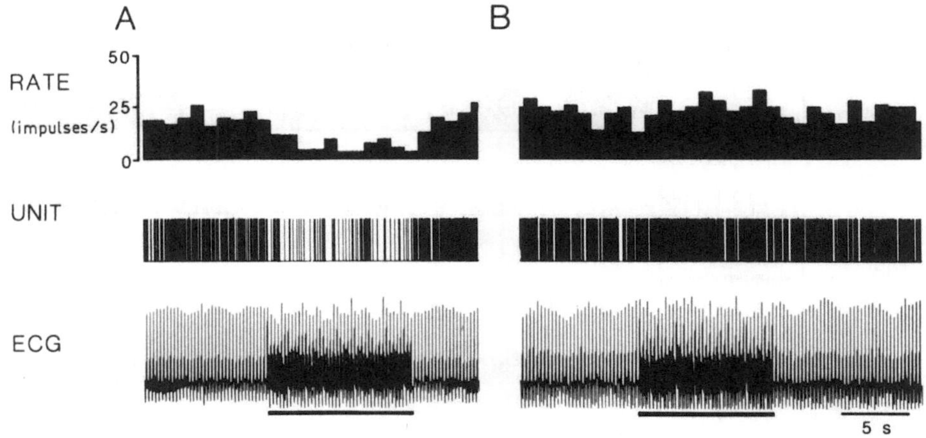

Fig. 21 A, B. Response of an STT neuron to vagal stimulation (horizontal bar) before (**A**) and after (**B**) bilateral cervical vagotomy. **A** Cell activity was inhibited when vagi were intact. **B** No inhibition of cell activity occurred after vagi were cut. (From Ammons et al. 1983a, with permission of the American Physiological Society)

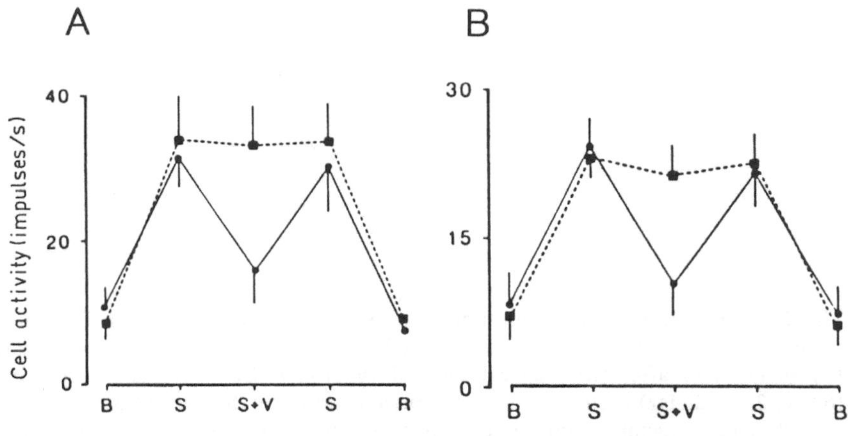

Fig. 22 A, B. Summary of responses of STT cells to determine source and origin of suppression of cell activity during stimulation of the left thoracic vagus. **A** Effect of vagal stimulation on cells responding to noxious somatic stimuli ($n = 7$). Significant inhibition ($P < 0.05$) occurred before (*continuous line*) but not after (*interrupted line*) bilateral cervical vagotomy. **B** Effect of left thoracic vagal stimulation above (*continuous line*) and below (*interrupted line*) the heart on cell responses to somatic stimuli ($n = 10$). Inhibition occurred when the stimulus was applied above the heart ($P < 0.01$) but not when stimulation occurred below the heart. Symbols along bottom of each graph: *B*, background activity; *S*, cell activity elicited by applying a noxious stimulus just before and after the vagal stimulation period; *S + V*, lowest cell activity during vagal stimulation with noxious stimulus maintained; *R* or *B*, cell activity after the stimulus period ended. (From Ammons et al. 1983a, adapted, with permission of the American Physiological Society)

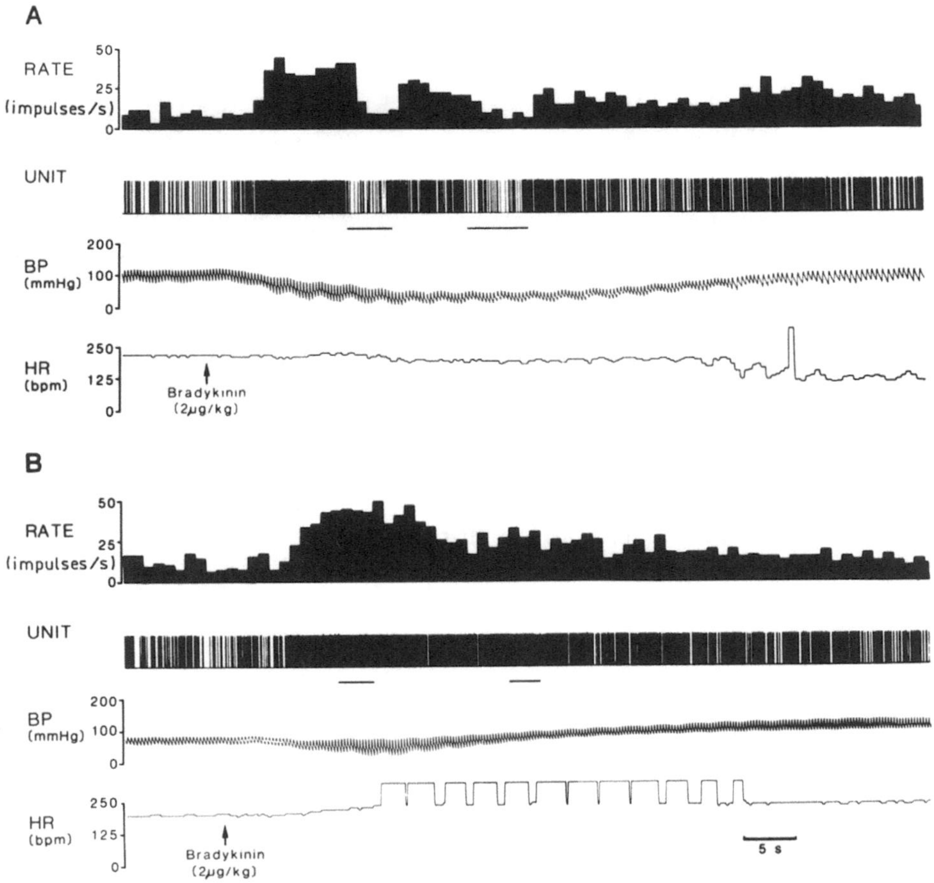

Fig. 23 A, B. Vagal stimulation suppresses STT cell activity resulting from a noxious visceral stimulus. Response of an STT cell to an intracardiac injection of bradykinin before (**A**) and after (**B**) bilateral cervical vagotomy. Left thoracic vagal stimulation (6.9 V, 2 ms, 20 Hz) occurred during the period shown by the *bars* below the second trace in each panel. Tracings from *top* to *bottom* are rate of cell discharge (*rate*), output of the window discriminator (*Unit*; each deflection represents one cell spike), blood pressure (*BP*), and heart rate (*HR*). (From Ammons et al. 1983b, with permission of the American Heart Association)

inhibited, and the remainder do not respond to stimulation (Ammons et al. 1983a, b). Left thoracic vagal stimulation produces less inhibition when cell activity is greater (Ammons et al. 1983a, b). Bilateral cervical vagotomy completely abolishes the response to left thoracic vagal stimulation. This demonstrates that the vagal afferent fibers are responsible for suppressing spontaneous cell activity in the STT.

In addition to suppression of spontaneous activity of STT cells, left vagal stimulation suppresses STT cell activity when a noxious stimulus is applied to the skin or skin and muscle (Fig. 22A). In this case left thoracic vagal stimulation inhibits all cells responding to noxious somatic stimulation but the inhibition is abolished after bilateral cervical vagotomy (Fig. 22A). In addition, vagal stimulation suppresses the increased activity of wide dynamic range cells when innocuous stimuli are used. Further experiments have distinguished between the effects of inhibition resulting from cardiopulmonary and abdominal vagal afferents. Electrical stimulation of the abdominal vagus does not suppress STT cell activity, whereas stimulation of the left thoracic vagus or the cardiac branch of the vagus markedly suppresses cell activity (Fig. 22B). The suppression of STT cell activity by vagal afferent stimulation appears to originate from the cardiopulmonary region and not from the abdominal region.

Activity of sympathetic afferent fibers arising from the heart can also be modulated by vagal afferent activity. Stimulation of the left thoracic vagus inhibits the marked increase in the activity of STT neurons following intracardiac injections of bradykinin (Fig. 23A). Interrupting the cervical vagus abolishes the effects on STT cell activity of stimulating the left thoracic vagus (Fig. 23B). In summary, the activity of vagal afferent fibers arising from the cardiopulmonary region modifies nociceptive responses to both somatic and visceral stimuli (Ammons et al. 1983a, b).

Preferential activation of vagal afferents, in conjunction with our results, may explain the numerous reports of cardiac ischemia with little or no pain (Cohn 1986; Margolis et al. 1973; Master and Geller 1969; Rutishauser and Roskamm 1984). Ischemia occurring in the restricted area of the myocardium may activate sympathetic or vagal afferents or both. Differences in the type of afferents activated by myocardial ischemia could markedly affect the responses of STT cells to noxious events. For example, preferential activation of vagal afferents could result in little or no increase in the activity of STT cells even when the sympathetic afferent fibers are excited (Ammons et al. 1983b). A series of studies by Thames and his coworkers (Felder and Thames 1979; Thames et al. 1978; Walker et al. 1978) show that preferential activation of vagal afferents can occur in the heart. Occlusion of the left circumflex artery is more effective in evoking vagal reflexes than is occlusion of the left anterior descending artery (Thames et al. 1978). Thus, the area of myocardium where ischemia occurs may determine the degree to which vagal afferents are activated. These findings may explain why myocardial infarction of the inferior wall results in bradycardia and hypotension, whereas anterior myocardial infarction results in tachycardia and hypertension (Randall et al. 1978; Webb et al. 1972). Vagal afferent activity perhaps predominates in patients with little or no pain during myocardial ischemia, and activation of the STT involving pain perception is prevented by increased vagal activity. This could explain the silent infarct in patients who experience severe myocardial ischemia.

9.3 Descending Inhibitory Pathways from Supraspinal Regions

Vagal afferent fibers inhibit the activity of STT cells and SRT cells. After vagal afferent information reaches the nucleus tractus solitarius (NTS) (Beckstead and Norgren 1979), the pathways and nuclei which mediate the inhibition are not known. Pain and noxious information can be modulated by activating descending spinal pathways that modulate the discharge rate of dorsal horn cells, including the STT cells. Descending modulation of STT and dorsal horn cell activity can be generated from the corticospinal pathway (Coulter et al. 1974; Yezierski et al. 1983), the periventricular gray (Ammons et al. 1986), medial and lateral hypothalamus, medial preoptic basal forebrain region (Carstens 1982; Carstens et al. 1982, 1983), somatosensory nuclei of the thalamus (Gerhart et al. 1981 b, 1983) periaqueductal gray, the dorsal raphe nucleus, the mesencephalic reticular formation (Bennett and Mayer 1979; Carstens et al. 1981; Duggan and Morton 1983; Hayes et al. 1979; Liebeskind et al. 1973; Oliveras et al. 1974), and the nucleus raphe magnus (Ammons et al. 1984c; Beall et al. 1976; Belcher et al. 1978; Duggan and Griersmith 1979; Edeson and Ryall 1983; Fields et al. 1977; Gerhart et al. 1981a; Guilbaud et al. 1977; McCreery et al. 1979; Rivot et al. 1980; Willis et al. 1977). In addition the retricular formation adjacent to the nucleus raphe magnus can also modulate spinal neurons (Belcher et al. 1978; Fields et al. 1977; Haber et al. 1978; McCreery et al. 1979; McCreery and Bloedel 1975; Rivot et al. 1980). It is likely that vagal inhibition of STT neurons may be mediated by activating connections between the NTS and descending systems that modulate nociception.

9.3.1 Subcoeruleus-Parabrachial Nuclei

Anatomical, electrophysiological, and behavioral evidence supports the hypothesis that the vagal afferent fibers can excite NTS neurons projecting to neurons of the subcoeruleus-parabrachial region that descend to produce their effect on STT cells. Anatomical studies have shown that vagal afferent fibers predominately terminate in the caudal region of the NTS in the primate (Beckstead and Norgren 1979; Fig. 24). The caudal NTS has its major projection ascending to the medial and lateral parabrachial nuclei, but smaller projections ascend to the nucleus ambiguus, the ventral lateral medulla, and the ventral basal thalamus (Beckstead et al. 1980; Fig. 24).

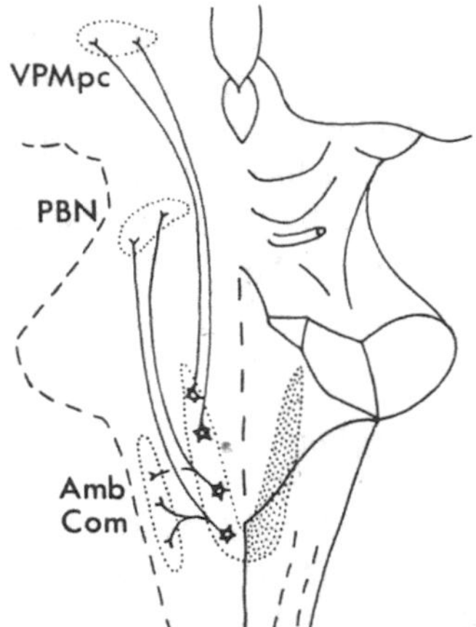

Fig. 24. Schematic diagram showing the dorsal view of the monkey's brain stem, with cell bodies shown in the NTS and projections to other regions. *Amb Com*, ambiguous complex; *PBN*, parabrachial nuclei; VPMpc, paracellular region of ventral posteromedial nucleus of the thalamus. (From Beckstead et al. 1980, with permission of Alan R. Liss, Inc.)

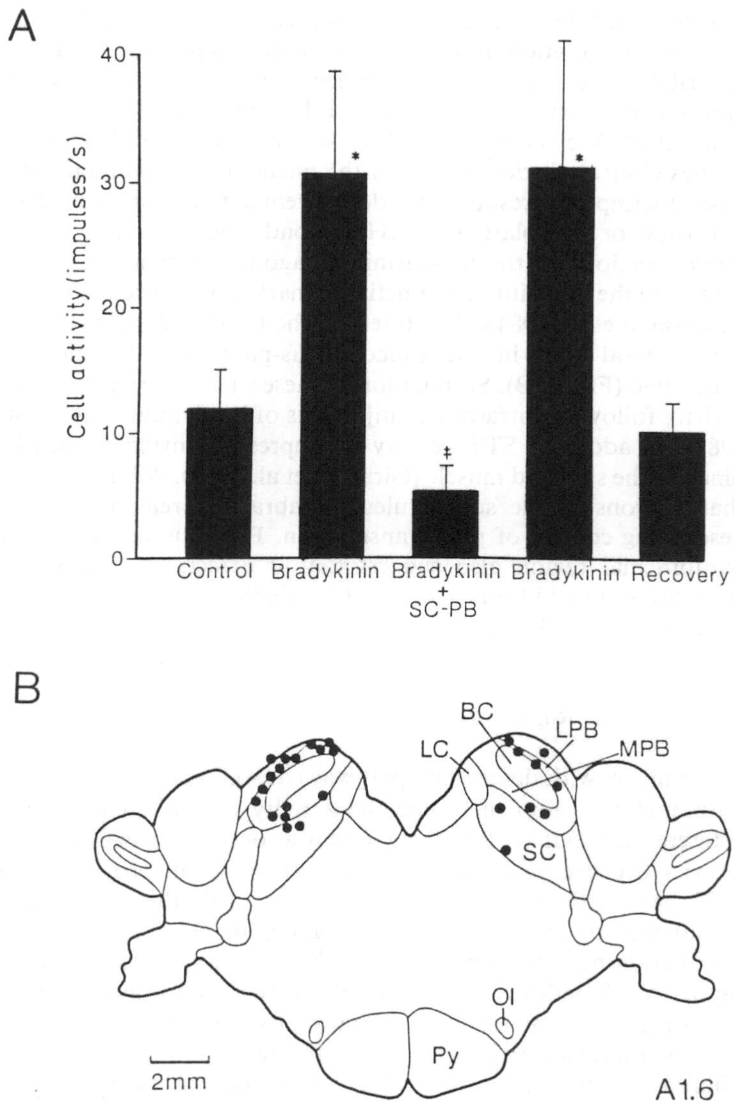

Fig. 25 A, B. Effects of electrically stimulating the subcoeruleus parabrachialis (SC-PB) on increased STT cell activity in response to intracardiac injections of bradykinin (4 μg/kg) (**A**) and location of stimulus sites producing the suppression of STT cell activity (**B**). **A** The increase in cell activity following bradykinin injections was significantly greater than control ($P<0.05$). SC-PB stimulation decreased cell activity below the bradykinin values ($P<0.05$). **B** SC-PB stimulation sites are marked by *black dots* on one brainstem section of the monkey drawn from the atlas of Szabo and Cowan (1984) and modified according to Westlund et al. (1984). The left side is ipsilateral to the spinothalamic tract cell. *BC*, brachium conjunctivium; *LPB*, lateral parabrachial nucleus; *MPB*, medial parabrachial nucleus; *SC*, nucleus subcoeruleus; *LC*, nucleus locus coeruleus; *Py*, pyramid; *OI*, inferior olivary nucleus. (From Brennen et al. 1987, with permission from Elsevier Biomedical Press)

Evidence that the subcoeruleus-parabrachial nucleus may be a possible relay for descending inhibition of noxious information is provided by retrograde transport of HRP, which shows direct projections from the subcoeruleus and medial parabrachial nuclei to the spinal cord (Hancock and Fougerousse 1976; Takeuchi et al. 1980; Westlund et al. 1984; Westlund and Coulter 1980). In behavioral studies electrolytic destruction of the medial parabrachial nucleus does not alter basal nociceptor thresholds but does attenuate morphine analgesia as assessed by tail flick or hot-plate tests (Hammond and Proudfit 1980). Furthermore, microinjections of the muscarinic antagonist, carbachol, in an area dorsal or ventral to the brachium conjunctivum markedly suppresses nociceptive responses (Katayama et al. 1984a, b). Based on the results of these studies, ipsilateral and contralateral sites in the subcoeruleus-parabrachial region were electrically stimulated (Fig. 25 B). Stimulation of these sites suppresses the increased STT cell activity following intracardiac injections of bradykinin (Fig. 25 A; Brennen et al. 1987). In addition, STT activity is suppressed during application of a noxious pinch to the skin and muscle (Girardot et al. 1987). All these experiments indicate that neurons in the subcoeruleus-parabrachial region may be important for descending control of pain transmission. Previous studies on the subcoeruleus-parabrachial region also suggest that it affects cardiovascular function and respiration, in addition to its possible role in pain transmission (Mraovitch et al. 1982; Von Euler 1986).

9.3.2 Nucleus Raphe Magnus

Other nuclei with descending pathways as well as the nuclei of the subcoeruleus-parabrachial region modulate the activity of STT cells receiving sympathetic afferent input but are less likely to play a role in the vagal suppression of STT cell activity because, at least in primates, the afferent projections from the caudal NTS have little or no direct connection with these systems. The effects of stimulation of the nucleus raphe magnus on STT cell activity during somatic stimulation have been well studied, but not the effects on visceral stimuli. The raphe-spinal pathway modulates the activity of upper thoracic STT neurons including the response to potentially noxious cardiac stimuli (Ammons et al. 1984c). Increased STT cell activity following intracardiac injections of bradykinin is strongly suppressed (Fig. 26A) during electrical stimulation of the nucleus raphe magnus (Fig. 26B). Increases in STT cell activity which arise from pinching the skin or skin and muscle are also suppressed (Ammons et al. 1984c).

9.3.3 Periventricular Gray

Electrical stimulation of the periventricular gray suppresses STT activity (Ammons et al. 1986). In humans the area surrounding the third ventricle is a particularly effective site for evoking stimulus-produced analgesia (Dieckmann and Witzmann 1982; Fairman 1979; Mayanagi et al. 1982; Mazars et al. 1979; Meyerson et al. 1979; Richardson and Akil 1977a, b). Electrical stimulation of the periventricular gray inhibits the responses of all STT cells in the thoracic

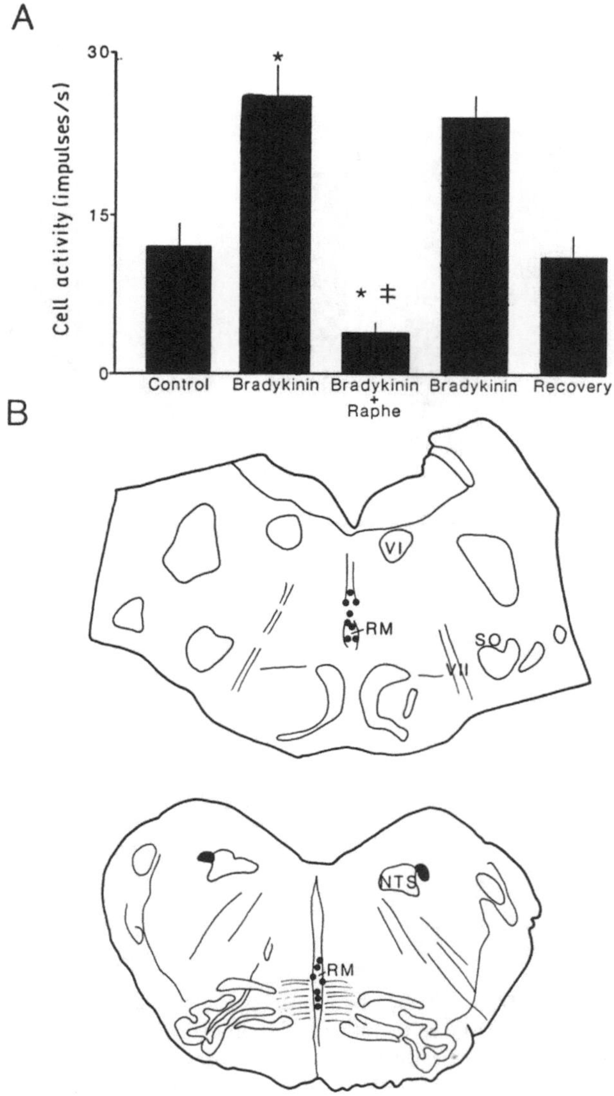

Fig. 26 A, B. Effects of stimulating the nucleus raphe magnus on the discharge rate of STT cells to intracardiac injections of bradykinin. **A** Mean cell response to bradykinin and effects of nucleus raphe magnus stimulation. The bradykinin values are the level of activity just before and after stimulation of the nucleus raphe magnus. *$P<0.05$ compared to control and recovery values, ‡$P<0.01$ compared to bradykinin values. **B** Black dots show the sites of nucleus raphe magnus where stimulation elicited suppression of STT cell activity to noxious visceral stimulation. The two sections represent the most caudal and rostral sites where stimulation electrodes were placed. *RM*, nucleus raphe magnus; *SO*, superior olive; *NTS*, solitary tract nucleus. (From Ammons et al. 1984, with permission of Elsevier Biomedical Press)

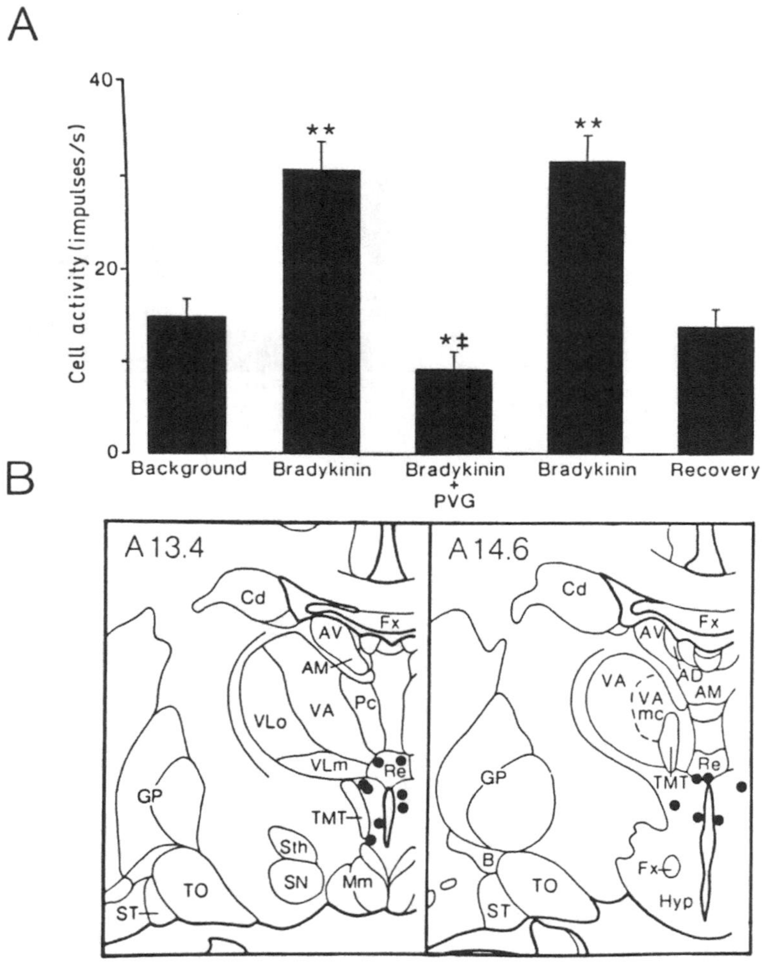

Fig. 27A, B. A Effects of stimulating the periventricular gray (PVG) on increased STT cell activity in response to intracardiac injections of bradykinin (2 µg/kg). *Background*, cell activity before bradykinin injection; *Bradykinin*, STT cell activity during response to bradykinin just before PVG stimulation; *Bradykinin + PVG*, lowest cell activity during PVG stimulation; *Second Bradykinin* value, peak of activity after PVG stimulation; *Recovery*, activity after the bradykinin response, ** $P<0.01$ compared with background, *$P<0.05$ compared with background, and ‡$P<0.001$ compared with bradykinin values. **B** PVG stimulation sites. *AD*, anterodorsal nucleus; *AM*, anteromedial nucleus; *AV*, anteroventral nucleus; *B*, basal nucleus; *Cd*, caudate nucleus; *Fx*, fornix; *GP*, globus pallidus; *Hyp*, hypothalamus; *Mm*, mamillary nuclei; *Pc*, paracentral nucleus; *Re*, ventromedical (reuniens) nucleus; *Sth*, subthalamic nucleus; *SN*, substantia nigra; *ST*, stria terminalis; *TMT*, mamillothalamic tract; *TO*, optic tract; *VA*, ventral anterior nucleus; *VAmc*, ventral anterior nucleus, magnocellular part; *VLm*, ventral lateral nucleus, medial; *VLo*, ventral lateral nucleus, oral. (From Ammons et al. 1986, with permission of the American Physiological Society)

spinal cord that respond to intracardiac injection of bradykinin (Fig. 27). In addition, periventricular stimulation also suppresses the responses of all neurons that respond to noxious pinch of the skin or to innocuous hair movement (Ammons et al. 1986). The degree of inhibition of cell responses to noxious pinch and to bradykinin are similar. The most effective stimulus sites are located within 0.5 mm of the third ventricle (Fig. 27) so the periventricular gray may participate in the descending inhibition of STT cell activity. This inhibitory system modulates STT cells projecting to both the medial and lateral thalamus. Direct projections from the periventricular gray to the spinal cord exist, but the exact descending projections that produce the descending inhibition are not known (Kuypers and Maisky 1975; Schwanzel-Fukuda et al. 1984; Swanson and Kuypers 1980).

9.4 Descending Excitatory Pathways

The emphasis of this review is on suppression of STT cell activity. Very few experiments have described excitation of cells by brain stem stimulation. The infrequent excitation of STT cells by vagal stimulation could result from disinhibition of supraspinal neurons. Electrical stimulation of the area surrounding the nucleus raphe magnus may excite some cells (Haber et al. 1978; Tattersall et al. 1986a, b). Stimulation of the motor region of the sensorimotor cortex may also excite STT cells (Yezierski et al. 1983). The role of this excitatory input to STT neurons is poorly understood.

10 Conclusions

The conclusions drawn here are illustrated in Fig. 28. The evidence presented in this review demonstrates that noxious information transmitted in sympathetic afferent fibers excites the STT cells of the spinal cord. Receptors for sympathetic afferent fibers are found in the heart, but their exact anatomical configuration is still uncertain. These receptors are distributed over all areas of the heart, including the atria, ventricles, and coronary arteries. Noxious stimuli, such as coronary artery occlusion and the application of bradykinin, can activate these receptors. It appears that both chemosensitive and mechanosensitive receptors are found in the heart. Noxious information arising from the cardiac receptors is transmitted in both A-delta and C fibers of the white rami, and their cell bodies are generally found in the dorsal root ganglia of segments T_1-T_5. Sympathetic afferent fibers enter the spinal cord and project primarily down the lateral aspect of the gray matter, but there is also a small group of fibers that travel medially. The afferent fibers terminate in the dorsal horn and intermediate gray matter where STT cells are located. Electrical stimulation of cardiopulmonary sympathetic afferent fibers and natural stimulation of cardiac receptors excite STT cells in segments T_1-T_5 of the spinal cord. The STT cells transmit information to both the medial and lateral thalamus. The same cells that receive input from sympathetic afferent fibers are also excited by mechanical manipulation of the somatic fields in structures overlying the heart. In addition to the convergence

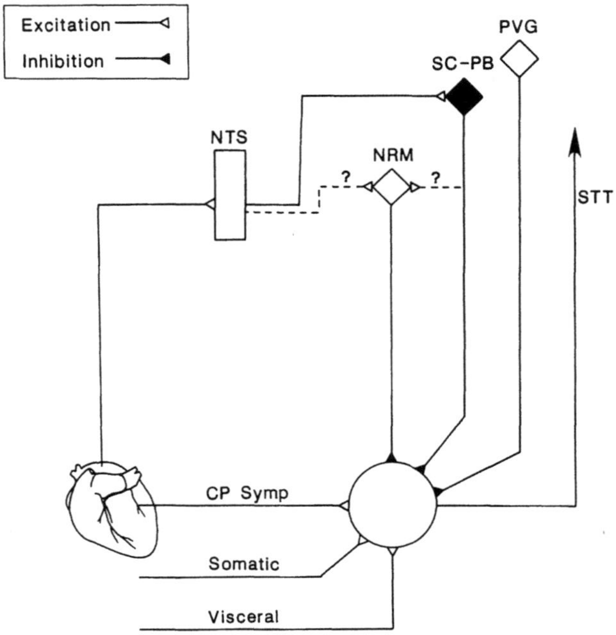

Fig. 28. A summary diagram showing possible connections of afferent input and ascending and descending pathways. The *filled diamond* emphasizes a likely site for effects of vagal inhibition. The *continuous lines* represent known pathways and the *interrupted lines* represent hypothetical or indirect pathways. The connections between the input and the representative cell nucleus or region may be monosynaptic or polysynaptic. Inhibition from descending pathways may be presynaptic or postsynaptic. Abbreviations: *CP Symp*, cardiopulmonary sympathetic afferent fibers; *NTS*, nucleus tractus solitarius; *NRM*, nucleus raphe magnus; *PVG*, periventricular Gray; *Somatic*, somatic afferent fibers; *STT*, spinothalamic tract system; *SC-PB*, subcoeruleus-parabrachial region; *Visceral*, visceral afferent fibers

of the visceral and somatic inputs, viscero-visceral interactions are found. Sympathetic afferent fibers arising from the gallbladder can also converge on cells receiving sympathetic afferent input from the cardiopulmonary region. This type of interactions may explain the anginal-like pain that occurs when certain patients have gallbladder disease.

Sympathetic afferent information received by STT cells is modulated by vagal mechanisms and by supraspinal descending pathways. Cardiopulmonary vagal afferent fibers and, in particular, afferents traveling in the cardiac vagal nerve modulate the activity of STT cells when noxious stimuli are applied to the heart. The neural mechanisms and pathways involved in the suppression of STT cell activity are unknown, but anatomical, electrophysiological, and behavioral results suggest potential pathways for producing this suppression. Vagal afferent information projects to the caudal portion of the NTS. The efferent projections from the NTS project to several areas, but especially to the subcoeruleus-parabrachial region. Axons of neurons in this area project down the spinal cord. This may be a pathway for the suppression of cell activity during vagal stimulation; however, one cannot exclude the roles that the nucleus raphe magnus, periventricular gray, and other nuclei may play in inhibition of STT activity. The interactions of vagal and sympathetic inputs may provide at least one possible mechanism for explaining silent myocardial ischemia, that is, the absence of pain during severe myocardial ischemia.

Improved understanding of the complex interactions occurring between visceral organs may eventually lead to new concepts for relating neuronal mechanisms to underlying visceral pain. These interactions show that modulation of information in the spinal cord depends upon the status of other visceral organs. The interactions and modulation of information in STT cells of the spinal cord demonstrate that the spinal cord is capable of processing and integrating afferent information before supraspinal regions are activated.

This review describes some aspects of the functional organization of the spinal cord. However, the processing and integration resulting from the convergence of visceral, viscero-visceral, and viscerosomatic inputs on neurons in the spinal cord are still poorly understood. Future studies using pharmacological, anatomical, physiological, and behavioral techniques must be employed in order to understand the complexities of the spinal cord and its projecting neurons. What receptors in the heart activate STT cells? What is the natural stimulus for activating the vagal afferent fibers that initiate the suppression of STT cell activity? What is the role of the various brain stem nuclei that suppress the activity of STT cells during visceral stimulation? Do visceral organs other than the gallbladder modulate activity of STT in the upper thoracic spinal cord? What neurotransmitters are involved? These and numerous other questions must be addressed if visceral pain and spinal cord function are to be understood.

11 References

Abboud FM, Thames MD (1983) Interaction of cardiovascular reflexes in circulatory control. In: Shepherd JT, Abboud FM (eds) Handbook of physiology, sect 2, The cardiovascular system, vol III, part 2. American Physiological Society, Bethesda, Maryland, pp 675–753

Albe-Fessard D, Levante D, Lamour Y (1974) Origin of spino-thalamic tract in monkeys. Brain Res 65:503–509

Ammons WS, Blair RW, Foreman RD (1983a) Vagal afferent inhibition of primate T_1-T_5 spinothalamic neurons. J Neurophysiol 50:926–940

Ammons WS, Blair RW, Foreman RD (1983b) Vagal afferent inhibition of spinothalamic cell responses to sympathetic afferents and bradykinin in the monkey. Circ Res 53:603–612

Ammons WS, Blair RW, Foreman RD (1984a) Responses of primate T_1-T_5 spinothalamic neurons to gallbladder distension. Am J Physiol 247:R995–R1002

Ammons WS, Blair RW, Foreman RD (1984b) Greater splanchnic excitation of primate T_1-T_5 spinothalamic neurons. J Neurophysiol 51:592–603

Ammons WS, Blair RW, Foreman RD (1984c) Raphe magnus inhibition of primate T_1-T_4 spinothalamic cells with cardiopulmonary visceral input. Pain 20:247–260

Ammons WS, Girardot M-N, Foreman RD (1985a) T_2-T_5 spinothalamic neurons projecting to medial thalamus with viscerosomatic input. J Neurophysiol 54:73–89

Ammons WS, Girardot M-N, Foreman RD (1985b) Effects of intracardiac bradykinin on T_2-T_5 spinothalamic cells. Am J Physiol 249:R147–R152
Ammons WS, Girardot M-N, Foreman RD (1986) Periventricular gray inhibition of thoracic spinothalamic cells projecting to medial and lateral thalamus. J Neurophysiol 55:1091–1103
Applebaum AE, Beall JE, Foreman RD, Willis WD (1975) Organization and receptive fields of primate spinothalamic tract neurons. J Neurophysiol 38:572–586
Armstrong D, Dry RML, Keele CA, Markham JW (1953) Observations on chemical excitation of cutaneous pain in man. J Physiol (Lond) 120:326–351
Armstrong D, Jepson JB, Keele CA, Stewart JW (1957) Pain producing substance in human inflammatory exudates and plasma. J Physiol (Lond) 135:350–370
Aviado DM, Schmidt CF (1955) Reflexes from stretch receptors in blood vessels, heart and lungs. Physiol Rev 35:247–300
Bain WA, Irving JT, McSwiney BA (1935) The afferent fibres from the abdomen in the splanchnic nerves. J Physiol (Lond) 84:323–333
Baker DG, Coleridge HM, Coleridge JCG (1979) Vagal afferent C fibres from the ventricle. In: Hainsworth R, Kidd C, Linden RJ (eds) Cardiac receptors. Cambridge University Press, Cambridge
Baker DG, Coleridge HM, Coleridge JCG, Nerdrum T (1980) Search for a cardiac nociceptor: stimulation by bradykinin of sympathetic afferent nerve endings in the heart of cat. J Physiol (Lond) 306:519–536
Beall JE, Martin RF, Applebaum AE, Willis WD (1976) Inhibition of primate spinothalamic tract neurons by stimulation in the region of the nucleus raphe magnus. Brain Res 114:328–333
Beckstead RM, Norgren R (1979) An autoradiographic examination of the central distribution of the trigeminal, facial, glossopharyngeal, and vagal nerves in the monkey. J Comp Neurol 184:455–472
Beckstead RM, Morse JR, Norgren R (1980) The nucleus of the solitary tract in the monkey: projections to the thalamus and brainstem nuclei. J Comp Neurol 190:259–282
Belcher G, Ryall RW, Schaffner R (1978) The differential effects of 5-hydroxytryptamine, noradrenaline and raphe stimulation on nociceptive and non-nociceptive dorsal horn interneurones in the cat. Brain Res 151:307–321
Bennett GJ, Mayer DJ (1979) Inhibition of spinal cord interneurons by narcotic microinjection and focal electrical stimulation in the periaqueductal central gray matter. Brain Res 172:243–257
Bennett GJ, Seltzer Z, Lu GW, Nishikawa N, Dubner R (1983) The cells of origin of the dorsal column postsynaptic projection in the lumbosacral enlargements of cats and monkeys. Somatosens Res 1:131–149
Berkley HJ (1894) The intrinsic nerve supply of cardiac ventricles in certain vertebrates. Johns Hopkins Hosp Ref 4:248–255
Berkley KJ (1980) Spatial relationships between the terminations of somatic sensory and motor pathways in the rostral brainstem of cats and monkeys. I. Ascending somatic sensory inputs to lateral diencephalon. J Comp Neurol 193:283–317
Besson JM, Guilbaud G, Lombard MC (1974) Effects of bradykinin intra-arterial injection into the limbs upon bulbar and mesencephalic reticular unit activity. Adv Neurol 4:207–215
Bishop VS, Lombardi F, Malliani A, Pagani M, Recordati G (1976) Reflex sympathetic tachycardia during intravenous infusions in chronic spinal cats. Am J Physiol 230:25–29
Bishop VS, Malliani A, Thorén P (1983) Cardiac mechanoreceptors. In: Shepherd JT, Abboud FM (eds) Handbook of physiology, sect 2. The cardiovascular system, vol III,

peripheral circulation and organ blood flow, part 2. American Physiological Society, Bethesda MD, pp 497–556

Blair RW, Weber RN, Foreman RD (1981) Characteristics of primate spinothalamic tract neurons receiving viscerosomatic convergent inputs in T_3-T_5 segments. J Neurophysiol 46:797–811

Blair RW, Weber RN, Foreman RD (1982) Responses of thoracic spinothalamic neurons to intracardiac injection of bradykinin in the monkey. Circ Res 51:83–94

Blair RW, Weber RN, Foreman RD (1984a) Responses of spinoreticular and spinothalamic cells to intracardiac bradykinin. Am J Physiol 246:H500–H507

Blair RW, Ammons WS, Foreman RD (1984b) Responses of thoracic spinothalamic and spinoretricular cells to coronary artery occlusion. J Neurophysiol 51:636–648

Bleehen T, Keele CA (1977) Observations on the algogenic actions of adenosine compounds on the human blister base preparation. Pain 3:367–377

Boivie J (1971) The termination of the spinothalamic tract in the cat. An experimental study with silver impregnation methods. Exp Brain Res 12:331–353

Boivie J (1979) An anatomical reinvestigation of the termination of the spinothalamic tract in the monkey. J Comp Neurol 186:343–370

Bosnjak ZK, Zuperku EJ, Coon RL, Kampine JP (1979) Acute coronary artery occlusion and cardiac sympathetic afferent nerve activity. Proc Soc Exp Biol Med 161: 142–148

Bowsher D (1957) Termination of the central pain pathway in man. The conscious appreciation of pain. Brain 80:606–622

Bowsher D (1961) The termination of secondary somatosensory neurons within the thalamus of *Macaca mulatta*:an experimental degeneration study. J Comp Neurol 117:213–227

Brennen TJ, Oh U-T, Girardot M-N, Ammons WS, Foreman RD (1987) Inhibition of cardiopulmonary input to thoracic spinothalamic tract cells by stimulation of the subcoeruleus-parabrachial region in the primate. J Autonom Nerv Sys 18:61–72

Brown AM (1967) Excitation of afferent cardiac sympathetic nerve fibres during myocardial ischaemia. J Physiol (Lond) 190:35–53

Brown AM (1979) Cardiac reflexes. In: Berne RM, Sperelakis N (eds) Handbook of physiology. The cardiovascular system. The heart, sect 2, vol I. American Physiological Society, Bethesda MD, pp 677–689

Brown AM, Malliani A (1971) Spinal sympathetic reflexes initiated by coronary receptors. J Physiol (Lond) 212:685–705

Burgess PR, Perl ER (1973) Cutaneous mechanoreceptors and nociceptors. In: Iggo A (ed) Somatosensory system. Springer, Berlin Heidelberg New York, pp 29–78 (Handbook of sensory physiology, vol 2)

Burstein R, Cliffer KD, Giesler GJ Jr (1987) Direct somatosensory projections from the spinal cord to the hypothalamus and telencephalon. J Neurosci 7(12):4159–4164

Burton H, Craig AD (1983) Spinothalamic projections in cat, raccoon and monkey: a study based on anterograde transport of horseradish peroxidase. In: Macchi G, Rustioni A, Spreafico R (eds) Somatosensory integration in the thalamus. Elsevier, Amsterdam, pp 17–41

Carstens E (1982) Inhibition of spinal dorsal horn neuronal responses to noxious skin heating by medial hypothalamic stimulation in the cat. J Neurophysiol 48:808–822

Carstens E, Trevino DL (1978) Laminar origins of spinothalamic projections in the cat as determined by the retrograde transport of horseradish peroxidase. J Comp Neurol 182:151–166

Carstens E, Fraunhoffer M, Zimmerman M (1981) Serotonergic mediation of descending inhibition from midbrain periaqueductal gray, but not reticular formation, of spinal nociceptive transmission in the cat. Pain 10:149–167

Carstens E, MacKinnon JD, Guinan MJ (1982) Inhibition of spinal dorsal horn neuronal responses to noxious skin heating by medial preoptic and septal stimulation in the cat. J Neurophysiol 48:981 – 991
Carstens E, Fraunhoffer M, Suberg SN (1983) Inhibition of spinal dorsal horn neuronal responses to noxious skin heating by lateral hypothalamus stimulation. J Neurophysiol 50:192 – 204
Casati R, Lombardi F, Malliani A (1979) Afferent sympathetic unmyelinated fibres with left ventricular endings in cats. J Physiol (Lond) 292:135 – 148
Casey KL, Keene JJ, Morrow T (1974) Bulboreticular and medial thalamic unit activity in relation to aversive behavior and pain. In: Bonica JJ (ed) Pain. Raven, New York, pp 197 – 205 (Advances in neurology, vol 4)
Cechetto DF, Standaert DG, Saper CB (1985) Spinal and trigeminal dorsal horn projections to the parabrachial nucleus in the rat. J Comp Neurol 240:153 – 160
Cervero F (1982) Afferent activity evoked by natural stimulation of the biliary system in the ferret. Pain 13:137 – 151
Cervero F (1983a) Somatic and visceral inputs into the thoracic spinal cord of the cat: effects of noxious stimulation of the biliary system. J Physiol (Lond) 337:51 – 67
Cervero F (1983b) Mechanisms of visceral pain. In: Lipton S, Miles J (eds) Persistent pain: modern methods of treatment, vol 4. Grune and Stratton, London, pp 1 – 19
Cervero F (1983c) Supraspinal connections of neurones in the thoracic spinal cord of the cat: ascending projections and effects of descending impulses. Brain Res 275:251 – 261
Cervero F, Connell LA (1984) Distribution of somatic and visceral primary afferent fibres within the thoracic spinal cord of the cat. J Comp Neurol 230:88 – 98
Cervero F, Tattersall JEH (1985) Cutaneous receptive fields of somatic and viscerosomatic neurones in the thoracic spinal cord of the cat. J Comp Neurol 237:325 – 332
Cervero F, Connell LA, Lawson SN (1984) Somatic and visceral primary afferents in the lower thoracic dorsal root ganglia of the cat. J Comp Neurol 228:422 – 431
Cervero F, Lumb BM, Tattersall JEH (1985) Supraspinal loops that mediate visceral inputs to thoracic spinal cord neurones in the cat: involvement of descending pathways from raphe and reticular formation. Neurosci Lett 56:189 – 194
Chahl LA (1979) Pain induced by inflammatory mediators. In: Beers RF, Bennett EG (eds) Mechanisms of pain and analgesic compounds. Raven, New York, pp 273 – 284
Chahl LA, Kirk EJ (1975) Toxins which produce pain. Pain 1:3 – 49
Chang HT, Ruch TC (1947) Topographical distribution of spinothalamic fibres in the thalamus of the spider monkey. J Anat 81:150 – 164
Christie LG, Conti CR (1981) Systematic approach to evaluation of angina-like chest pain: pathophysiology and clinical testing with emphasis on objective documentation of myocardial ischaemia. Am Heart J 102:897 – 912
Chung JM, Kenshalo DR Jr, Gerhart KD, Willis WD (1979) Excitation of primate spinothalamic neurons by cutaneous C-fiber volleys. J Neurophysiol 42:1354 – 1369
Clark WEL (1936) The termination of ascending tracts in the thalamus of the macaque monkey. J Anat 71:7 – 40
Coggeshall RE, Galbraith SL (1978) Categories of axons in mammalian rami communicantes. Part II. J Comp Neurol 181:349 – 360
Cohn PF (1986) Silent myocardial ischemia and infarction. In: Denolin H, Swan HJC (eds) Basic and clinical cardiology. Dekker, New York, pp 17 – 28
Coleridge HM, Coleridge JCG (1980) Cardiovascular afferents involved in the regulation of peripheral vessels. Annu Rev Physiol 42:413 – 427
Coleridge HM, Coleridge JCG, Kidd C (1964) Cardiac receptors in the dog with particular reference to two types of endings in the ventricular wall. J Physiol (Lond) 174: 323 – 339

Coleridge JCG, Coleridge HM (1979) Chemoreflex regulation of the heart. In: Berne RM, Sperelakis N (eds) Handbook of physiology, sect 2, The cardiovascular system, vol 1, The heart. American Physiological Society, Bethesda MD, pp 653 – 676

Coleridge JCG, Hemingway A, Holmes RL, Linden RJ (1957) The location of atrial receptors in the dog: a physiological and histological study. J Physiol (Lond) 136:174 – 197

Coulter JD, Maunz RA, Willis WD (1974) Effects of stimulation of sensorimotor cortex on primate spinothalamic neurons. Brain Res 65:351 – 356

Craig AD, Burton H (1981) Spinal and medullary lamina I projection to nucleus submedius in medial thalamus: a possible pain center. J Neurophysiol 45:443 – 466

Craig AD, Kniffki K-D (1985) Spinothalamic lumbosacral lamina I cells responsive to skin and muscle stimulation in the cat. J Physiol (Lond) 365:197 – 221

Csendes A, Sepulveda A (1980) Intraluminal gallbladder pressure measurements in patients with chronic or acute cholecystites. Am J Surg 13:383 – 385

Cullen ML, Reese HL (1952) Myocardial circulatory changes measured by clearance of Na24-effect of common duct distension on myocardial circulation. J Appl Physiol 5:281 – 284

Dennis SG, Melzack R (1977) Pain-signalling systems in the dorsal and ventral spinal cord. Pain 4:97 – 132

Dieckmann G, Witzmann A (1982) Initial and long term results of deep brain stimulation for chronic intractable pain. Appl Neurophysiol 45:167 – 172

Dilly PN, Wall PD, Webster KE (1968) Cells of origin of the spinothalamic tract in the cat and rat. Exp Neurol 21:550 – 562

Donald DE, Shepherd JT (1978) Reflexes from the heart and lungs; physiological curiosities or important regulatory mechanisms. Cardiovasc Res 12:449 – 469

Downman CBB (1955) Skeletal muscle reflexes of splanchnic and intercostal nerve origin in acute spinal and decerebrate cats. J Neurophysiol 18:217 – 235

Downman CBB, Evans MH (1957) The distribution of splanchnic afferents in the spinal cord of cat. J Physiol (Lond) 137:66 – 79

Duggan AW, Griersmith BT (1979) Inhibition of the spinal transmission of nociceptive information by supraspinal stimulation in the cat. Pain 6:149 – 161

Duggan AW, Morton CR (1983) Periaqueductal grey stimulation: an association between selective inhibition of dorsal horn neurones and changes in peripheral circulation. Pain 15:237 – 248

Edeson RO, Ryall RW (1983) Systematic mapping of descending inhibitory control by the medulla of nociceptive spinal neurones in cats. Brain Res 271:251 – 262

Elliott DF, Horton EW, Lewis GP (1960) Actions of pure bradykinin. J Physiol (Lond) 153:473 – 480

Emery DG, Foreman RD, Coggeshall RE (1976) Fiber analysis of the feline inferior cardiac sympathetic nerve. J Comp Neurol 166:457 – 468

Emery DG, Foreman RD, Coggeshall RE (1978) Categories of axons in the inferior cardiac nerve of the cat. J Comp Neurol 177:301 – 310

Fairman D (1979) Thalamic and hypothalamic simulation. In: Bonica JJ, Ventafridda (eds) Advances in pain research and therapy, vol 2. Raven, New York, pp 493 – 498

Felder RB, Thames MD (1979) Interaction between cardiac receptors and sinoaortic baroreceptors in the control of efferent cardiac sympathetic nerve activity during myocardial ischemia in dogs. Cir Res 45:728 – 736

Fields HL, Partridge LD, Winter DL (1970) Somatic and visceral receptive field properties of fibers in ventral quadrant white matter of the cat spinal cord. J Neurophysiol 33:827 – 837

Fields HL, Basbaum AI, Clanton CH, Anderson SD (1977) Nucleus raphe magnus inhibition of spinal cord dorsal horn neurons. Brain Res 126:441 – 453

Floyd K (1979) Light microscopy of nerve endings in the atrial endocardium. In: Hainsworth R, Kidd C, Linden RJ (eds) Cardiac receptors. Cambridge University Press, Cambridge, UK, pp 3–26

Foreman RD, Hancock MB, Willis WD (1981) Responses of spinothalamic tract cells in the thoracic spinal cord of the monkey to cutaneous and visceral inputs. Pain 11: 149–162

Foreman RD, Blair RW, Weber RN (1984) Viscerosomatic convergence onto T_2-T_4 spinoreticular, spinoreticular-spinothalamic and spinothalamic tract neurons in the cat. Exp Neurol 85:597–619

Foreman RD (1986) Spinal substrates for visceral pain. In: Yaksh TL (ed) Spinal afferent processing. Plenum, New York, pp 217–242

Fox RE, Holloway JA, Iggo A, Mokha SS (1980) Spinothalamic neurones in the cat: some electrophysiological observations. Brain Res 182:186–190

Gerhart KD, Wilcox TK, Chung JM, Willis WD (1981a) Inhibition of nociceptive and non-nociceptive responses of primate spinothalamic cells by stimulation in medial brainstem. J Neurophysiol 45:121–136

Gerhart KD, Yezierski RP, Wilcox TK, Grossman AE, Willis WD (1981b) Inhibition of primate spinothalamic tract neurons by stimulation in ipsilateral or contralateral ventral posterior lateral (VPL_c) thalamic nucleus. Brain Res 229:514–519

Gerhart KD, Yezierski RP, Fang ZR, Willis WD (1983) Inhibition of primate spinothalamic tract neurons by stimulation in the ventral posterior lateral (VPL_c) thalamic nucleus: possible mechanisms. J Neurophysiol 49:406–423

Giesler GJ, Menétrey D, Guilbaud G, Besson JM (1976) Lumbar cord neurons at the origin of the spinothalamic tract in the rat. Brain Res 118:320–324

Giesler GJ, Menétrey D, Basbaum AI (1979) Differential origins of spinothalamic tract projections to medial and lateral thalamus in the rat. J Comp Neurol 184:107–126

Giesler GJ, Yezierski RP, Gerhart KD, Willis WD (1981) Spinothalamic tract neurons that project to medial and/or lateral thalamic nuclei: evidence for a physiologically novel population of spinal cord neurons. J Neurophysiol 46:1285–1308

Girardot M-N, Brennan TJ, Ammons WS, Foreman RD (1987) Effects of stimulating the subcoeruleus-parabrachial region on the non-noxious and noxious responses of T_2-T_4 spinothalamic tract neurons in the primate. Brain Res 409:19–30

Gokin AP, Kostyuk PG, Preobrazhensky NN (1977) Neuronal mechanisms of interactions of high-threshold visceral and somatic afferent influences in spinal cord and medulla. J Physiol (Paris) 73:319–333

Grant G, Ygge J (1981) Somatotopic organisation of the thoracic spinal nerve in the dorsal horn demonstrated with transganglionic degeneration. J Comp Neurol 202:357–364

Guilbaud G, Oliveras JL, Giesler G Jr, Besson JM (1977) Effects induced by stimulation of the centralis inferior nucleus of the raphe on dorsal horn interneurons in cat's spinal cord. Brain Res 126:355–360

Guzman F, Braun C, Lim RKS (1962) Visceral pain and the pseudaffective response to intra-arterial injection of bradykinin and other algesic agents. Arch Int Pharmacodyn Ther 136:353–383

Haber LH, Martin RF, Chatt AB, Willis WD (1978) Effects of stimulation in nucleus reticularis gigantocellularis on the activity of spinothalamic tract neurons in the monkey. Brain Res 153:163–168

Haber LH, Moore BD, Willis WD (1982) Electrophysiological response properties of spinoreticular neurons in the monkey. J Comp Neurol 207:75–84

Hammond DL, Proudfit HK (1980) Effects of locus coeruleus lesions on morphine-induced antinociception. Brain Res 188:79–91

Hampton AG, Beckwith JR, Wood JE Jr (1959) The relationship between heart disease and gallbladder disease. Ann Intern Med 50:1135–1148

Hancock MB, Fougerousse CL (1976) Spinal projections from the nucleus locus coeruleus and nucleus subcoeruleus in the cat and monkey as demonstrated by the retrograde transport of horseradish peroxidase. Brain Res Bull 1:229 – 234
Hancock MB, Foreman RD, Willis WD (1975) Convergence of visceral and cutaneous input onto spinothalamic tract cells in the thoracic spinal cord of the cat. Exp Neurol 47:240 – 248
Harrison TR, Reeves TJ (1968) Patterns and causes of chest pain. In: Principles and problems of ischemic heart disease. Year Book Medical Publishers, Chicago, pp 197 – 204
Hashimoto K, Hirose M, Furukawa S, Haykawa H, Kimura E (1977) Changes in hemodynamics and bradykinin concentration in coronary sinus blood in experimental coronary artery occlusion. Jpn Heart J 18:679 – 689
Hayes NL, Rustioni A (1980) Spinothalamic and spinomedullary neurons in macaques: a single and double retrograde tracer study. Neuroscience 5:861 – 874
Hayes RL, Price DD, Ruda M, Dubner R (1979) Suppression of nociceptive responses in the primate by electrical stimulation of the brain or morphine administration: behavioral and electrophysiological comparisons. Brain Res 167:417 – 421
Henderson RD, Wigle HD, Sample K, Manyat G (1978) Atypical chest pain of cardiac and esophageal origin. Chest 73:24 – 41
Hess GL, Zuperku EJ, Coon RL, Kampine JP (1974) Sympathetic afferent nerve activity of left ventricular origin. Am J Physiol 227:543 – 546
Hodge GB, Messer AL, Hill H (1947) Effect of distension of the biliary tract on the electrocardiogram. Arch Surg 55:710 – 722
Holmes RL (1957) Structures in the arterial endocardium of the dog which stain with methylene blue, and the effects of unilateral vagotomy. J Anat 91:259 – 266
Hylden JLK, Hayashi H, Bennett GJ, Dubner R (1985) Spinal lamina I neurons projecting to the parabrachial area of the cat midbrain. Brain Res 336:195 – 198
Johnston BD (1968) Nerve endings in the human endocardium. Am J Anat 122:621 – 629
Jones EG, Burton H (1974) Cytoarchitecture and somatic sensory connectivity of thalamic nuclei other than the ventrobasal complex in the cat. J Comp Neurol 154:395 – 432
Jones MW, Hodge DJ Jr, Apkarian AV, Stevens RT (1985) A dorsolateral spinothalamic pathway in cat. Brain Res 335:188 – 193
Katayama Y, Watkins LR, Becker DP, Hayes RL (1984a) Evidence for involvement of cholinoceptive cells of the parabrachial region in environmentally induced nociceptive suppression in the cat. Brain Res 229:348 – 353
Katayama Y, Watkins LR, Becker DP, Hayes RL (1984b) Non-opiate analgesia induced by carbachol microinjection into the pontine parabrachial region of the cat. Brain Res 296:262 – 283
Kaufman MP, Baker DG, Coleridge HM, Coleridge JC (1980) Stimulation by bradykinin of afferent vagal C-fibers with chemosensitive endings in the heart and aorta of the dog. Circ Res 46:476 – 484
Keele CA, Armstrong D (1969) Substances producing pain and itch. Arnold, London
Kerr FWL (1975) The ventral spinothalamic tract and other ascending systems of the ventral funiculus of the spinal cord. J Comp Neurol 159:335 – 356
Kerr FWL, Lippman HH (1974) The primate spinothalamic tract as demonstrated by anterolateral cordotomy and comissural myelotomy. Adv Neurol 4:147 – 156
Kevetter GA, Willis WD (1983) Collaterals of spinothalamic cells in the rat. J Comp Neurol 215:453 – 464
Kevetter GA, Haber LH, Yezierski RP, Chung JM, Martin RF, Willis WD (1982) Cells of origin of the spinoreticular tract in the monkey. J Comp Neurol 207:61 – 74
Kezdi P, Kordenat RK, Misra SN (1974) Reflex inhibitory effects of vagal afferents in experimental myocardial infarction. Am J Cardiol 33:853 – 860

Khabarova AY (1963) The afferent innervation of the heart. Consultants Bureau, New York
Kimura E, Hashimoto K, Furukawa S, Hayakawa H (1973) Changes in bradykinin level in coronary sinus blood after the experimental occlusion of coronary artery. Am Heart J 85:635–647
Kostreva DR, Zuperku EJ, Purtock RV, Coon RL, Kampine JP (1975) Sympathetic afferent nerve activity of right heart origin. Am J Physiol 229:911–915
Kostreva DR, Hess GL, Zuperku EJ, Neumark J, Coon RL, Kampine JP (1976) Cardiac responses to stimulation of thoracic afferents in the primate and canine. Am J Physiol 231:1279–1284
Kuo DC, Krauthamer GM, Yamasaki DS (1981) The organization of visceral sensory neurones in the thoracic dorsal root ganglia (DRG) of the cat studied by horseradish peroxidase reaction using the cryostat. Brain Res 208:187–191
Kuo DC, Oravitz JJ, de Groat WC (1984) Tracing of afferent and efferent pathways in the left inferior cardiac nerve of the cat using retrograde and transport of horseradish peroxidase. Brain Res 321:111–118
Kuypers HGJM, Maiski VA (1975) Retrograde axonal transport of horseradish peroxidase from spinal cord to brain stem cell groups in the cat. Neurosci Lett 1:9–14
Lewis T (1942) Pain. Macmillan, New York
Liebeskind JC, Guibaud G, Besson JM, Oliveras J-L (1973) Analgesia from electrical stimulation of the periaqueductal gray matter in the cat: behavioral observations and inhibitory effects on spinal cord interneurons. Brain Res 50:441–446
Lim KS, Guzman F (1968) Manifestations of pain in analgesic evaluation in animals and man. In: Soulairac A, Cohn J, Charpentier J (eds) Pain. Academic, London, pp 119–152
Lindgren I, Olivecrona H (1947) Surgical treatment of angina pectoris. J Neurosurg 4:19–39
Lombardi F, Della Bella P, Casati R, Malliani A (1981) Effects of intracoronary administration of bradykinin on the impulse activity of afferent sympathetic unmyelinated fibers with left ventricular endings in the cat. Circ Res 48:69–75
Longhurst JC (1984) Cardiac receptors: their function in health and disease. Prog in Cardiovasc Dis 27:201–222
MacKenzie J (1909) Symptoms and their interpretation. Shaw, London
Maixner W, Randich A (1984) Role of the right vagal nerve trunk in antinociception. Brain Res 298:374–377
Maixner W, Touw KB, Brody MJ, Gebhart GF, Long JP (1982) Factors regulating the altered pain perception in the spontaneously hypertensive rat. Brain Res 237:137–145
Malliani A (1982) Cardiovascular and sympathetic afferent fibers. Rev Physiol Biochem Pharmacol 94:11–74
Malliani A (1986) The elusive link between transient myocardial ischemia and pain. Circulation 73:201–204
Malliani A, Peterson DF, Bishop VS, Brown AM (1972) Spinal sympathetic cardiocardiac reflexes. Circ Res 30:158–166
Malliani A, Parks M, Tuckett RP, Brown AM (1973a) Reflex increases in heart rate elicited by stimulation of afferent cardiac sympathetic nerve fibers in the cat. Circ Res 32:9–14
Malliani A, Recordati G, Schwartz PJ (1973b) Nervous activity of afferent cardiac sympathetic fibres with atrial and ventricular endings. J Physiol (Lond) 229:457–469
Malliani A, Lombardi F, Pagani M, Recordati G, Schwartz PJ (1975) Spinal cardiovascular reflexes. Brain Res 87:239–246
Margolis JR, Kannel WS, Feinleib M, Dawber TR, McNamara PW (1973) Clinical features of unrecognized myocardial infarction silent and symptomatic. Eighteen year followup: the Framingham study. Am J Cardiol 32:1–7

Maseri A, Chierchia S (1982) Coronary artery spasm: demonstration, definition, diagnosis and consequences. Prog Cardiovasc Dis 25:169–192
Master AM, Geller AJ (1969) The extent of completely asymptomatic coronary artery disease. Am J Cardiol 23:173–179
Mayanagi Y, Sano K, Suzuki I, Kanazawa I, Aoyagi I, Miyachi Y (1982) Stimulation and coagulation of the posteriomedial hypothalamus for intractable pain, with reference to β-endorphins. Appl Neurophysiol 45:136–142
Mazars GJ, Merienne L, Cioloca C (1979) Comparative study of electrical stimulation of posterior thalamic nuclei, periaqueductal gray and other midline mesencephalic structures in man. In: Bonica JJ, Liebeskind JC, Albe-Fessard D (eds) Advances in pain research and therapy, vol 3. Raven, New York, pp 541–546
McCreery DB, Bloedel JR (1975) Reduction of the response of cat spinothalamic neurons to graded mechanical stimuli by electrical stimulation of the lower brain stem. Brain Res 97:151–156
McCreery DB, Bloedel JR, Hames EG (1979) Effects of stimulating in raphe nuclei and in reticular formation on response of spinothalamic neurons to mechanical stimuli. J Neurophysiol 42:166–182
Mehler WR (1962) The anatomy of the so-called 'pain tract' in man: an analysis of the course and distribution of the ascending fibers of the fasciculus anterolateralis. In: French JD, Porter RW (eds) Basic research in paraplegia. Thomas, Springfield, pp 26–55
Mehler WR (1969) Some neurological species differences – a posteriori. Ann NY Acad Sci 167:424–468
Mehler WR, Feferman ME, Nauta WJH (1960) Ascending axon degenerating following anterolateral cordotomy. An experimental study in the monkey. Brain 83:718–751
Melzack R, Casey KL (1968) Sensory, motivational and central control determinants of pain. In: Kenshalo D (ed) The skin senses. Thomas, Springfield, pp 423–443
Melzack R, Wall PD (1982) The challenge of pain. Basic Books, New York
Menétrey D, Babaum AI (1987) Spinal and trigeminal projections to the nucleus of the solitary tract: A possible substrate for somatovisceral and viscerovisceral reflex activation. J Comp Neurol 255:439–450
Menétery D, Chaouch A, Besson JM (1980) Location and properties of dorsal horn neurons at origin of spinoreticular tract in lumbar enlargement of the rat. J Neurophysiol 44:862–877
Menétrey D, Chaouch A, Binder D, Besson JM (1982) The origin of the spinomesencephalic tract in the rat: an anatomical study using the retrograde transport of horseradish peroxidase. J Comp Neurol 206:193–207
Meyerson BA, Boëthius J, Carlsson AM (1979) Alleviation of malignant pain by electrical stimulation in the periventricular-periaqueductal region: pain relief as related to stimulation sites. In: Bonica JJ, Liebeskind JC, Albe-Fessard D (eds) Advances in pain research and therapy, vol 3. Raven, New York, pp 525–533
Meyling HA (1953) Structure and significance of the peripheral extension of the autonomic nervous system. J Comp Neurol 99:495–543
Miller HR (1942) The interrelationship of disease of coronary arteries and gallbladder. Am Heart J 24:579–587
Miller MR, Kasahara M (1964) Studies on the nerve endings in the heart. Am J Anat 115:217–233
Milne RJ, Foreman RD, Giesler GJ, Willis WD (1981) Convergence of cutaneous and pelvic visceral nociceptive inputs onto primate spinothalamic neurons. Pain 11: 163–183
Mitchell GAG (1953) The innervation of the heart. Br Heart J 15:159–171

Morgan C, de Groat WC, Nadelhaft I (1986) The spinal distribution of sympathetic preganglionic and visceral primary afferent neurons that send axons into the hypogastric nerves of the cat. J Comp Neurol 243:23–40
Mraovitch S, Kamada M, Reis DJ (1982) Role of the nucleus parabrachialis in cardiovascular regulation in cat. Brain Res 232:57–75
Muers MF, Sleight P (1972) Action potentials from ventricular mechanoreceptors stimulated by occlusion of the coronary sinus in the dog. J Physiol (Lond) 221:283–309
Nadelhaft I, Roppolo J, Morgan C, de Groat WC (1983) Parasympathetic preganglionic neurones and visceral primary afferent in monkey sacral spinal cord revealed following application of horseradish peroxidase to pelvic nerve. J Comp Neurol 216:36–52
Needleman P (1976) The synthesis and function of prostaglandins in the heart. Fed Proc 35:2376–2381
Nerdrum T, Baker DG, Coleridge HM, Coleridge JCG (1986) Interaction of bradykinin and prostaglandin E_1 on cardiac pressor reflex and sympathetic afferents. Am J Physiol 250:R815–R822
Nettleship WA (1936) Experimental studies on the afferent innervation of the cat's heart. J Comp Neurol 64:115–131
Neuhuber W (1982) The central projections of visceral primary afferent neurones of the inferior mesenteric plexus and hypogastric nerve and the location of the related sensory and preganglionic sympathetic cell bodies in the rat. Anat Embryol (Berl) 164:413–425
Nishi K, Sakanashi M, Takenaka F (1977) Activation of afferent cardiac sympathetic nerve fibers of the cat by pain producing substances and by noxious heat. Pflugers Arch 372:53–61
Nonidez JF (1939) Studies on the innervation of the heart. I. Distribution of the cardiac nerves, with special reference to the identification of the sympathetic and parasympathetic postganglionics. Am J Anat 68:361–413
Oberg B, Thorén P (1973) Circulatory responses to stimulation of medullated and nonmedulated afferents in the cardiac nerve in the cat. Acta Physiol Scand 87:121–132
Oldfield BJ, McLachlan EM (1978) Localization of sensory neurons traversing the stellate ganglion of the cat. J Comp Neurol 182:915–922
Oliveras JL, Besson JM, Guilbaud G, Liebeskind JC (1974) Behavioral and electrophysiological evidence of pain inhibition from midbrain stimulation in the cat. Exp Brain Res 20:32–44
Ordway GA, Longhurst JC (1983) Cardiovascular reflexes arising from the gallbladder of the cat: effects of capsaicin, bradykinin, and distension. Circ Res 52:26–35
Pagani M, Schwartz PJ, Banks R, Lombardi F, Malliani A (1974) Reflex responses of sympathetic preganglionic neurones initiated by different cardiovascular receptors in spinal animals. Brain Res 68:215–225
Paintal AS (1973) Vagal sensory receptors and their reflex effects. Physiol Rev 53: 159–227
Palmer ED (1958) Chest pain of colon origin. Gastroenterologia 90:15–21
Pearson JC, Haines DE (1980) Somatosensory thalamus of a prosimian primate (Galago senegalensis) I. Configuration of nuclei and termination of spinothalamic fibers. J Comp Neurol 190:533–558
Peterson DF, Brown AM (1971) Pressor reflexes produced by stimulation of afferent fibers in cardiac sympathetic nerves of the cat. Circ Res 28:605–610
Pomeranz B, Wall PD, Weber WV (1968) Cord cells responding to fine myelinated afferents from viscera, muscle and skin. J Physiol (Lond) 199:511–532
Price DD, Dubner R (1977) Neurons that subserve the sensory-discriminative aspects of pain. Pain 3:307–338
Price DD, Mayer DJ (1974) Physiological laminar organization of the dorsal horn of M. mullata. Brain Res 79:321–325

Price DD, Hayes RL, Ruda MA, Dubner R (1978) Spatial and temporal transformations of input to spinothalamic tract neurons and their relation to somatic sensations. J Neurophysiol 41:933–947

Purtock RV, von Colditz JH, Seagard JL, Igler FO, Zuperku EJ, Kampine JP (1977) Reflex effects of thoracic sympathetic afferent nerve stimulation on the kidney. Am J Physiol 233:H580–H586

Randall DC, Hasson DM, Brady JV (1978) Acute cardiovascular consequences of anterior descending coronary artery occlusion in unanesthetized monkey. Proc Soc Exp Biol Med 58:135–140

Randich A, Simpson TA, Hanger PA, Fisher RL (1984) Activation of vagal afferents by veratrine induces antinociception. Physiol Psychol 12:293–301

Ravdin IS, Royster HP, Sanders GB (1942) Reflexes originating in the common duct giving rise to pain simulating angina pectoris. Ann Surg 115:1055–1062

Ravdin IS, Fitz-Hugh T Jr, Wolferth CC, Barbieri EA, Ravdin RG (1955) Relation of gallstone disease to angina pectoris. Arch Surg 70:333–342

Richardson DE, Akil H (1977a) Pain reduction by electrical brain stimulation in man. Part 1: acute administration in periaqueductal and periventricular sites. J Neurosurg 47:178–183

Richardson DE, Akil H (1977b) Pain reduction by electrical brain stimulation in man. Part 2: chronic self-administration in the periventricular gray matter. J Neurosurg 47:184–194

Rivot JP, Chaouch A, Besson JM (1980) Nucleus raphe magnus modulation of response of rat dorsal horn neurons to unmyelinated fiber inputs: partial involvement of serotonergic pathways. J Neurophysiol 44:1039–1057

Ruch TC (1961) Pathophysiology of pain. In: Ruch TC, Patton HD, Woodbury JW, Towe AL (eds) Neurophysiology. Saunders, Philadelphia

Rutishauser W, Roskamm H (1984) Silent myocardial ischemia. Springer, Berlin Heidelberg New York

Schwanzel-Fukuda M, Morrell JI, Pfaff DW (1984) Localization of forebrain neurons which project directly to the medulla and spinal cord of the rat by retrograde tracing with wheat germ agglutenin. J Comp Neurol 226:1–20

Schwartz PJ, Pagani M, Lombardi F, Malliani A, Brown AM (1973) A cardiocardiac sympathovagal reflex in the cat. Cir Res 32:215–220

Seagard JL, Pederson HJ, Kostreva DR, Van Horn DL, Cusik JF, Kampine JP (1978) Ultrastructural identification of afferent fibers of cardiac origin in thoracic sympathetic nerves in the dog. Am J Anat 153:217–232

Sleight P, Widdicombe JG (1965) Action potentials in fibres from receptors in the epicardium and myocardium of the dog's left ventricle. J Physiol (Lond) 181:235–258

Smith CL (1983) The development and post-natal organisation of primary afferent projections to the rat thoracic spinal cord. J Comp Neurol 220:29–43

Swanson LW, Kuypers HGJM (1980) The paraventricular nucleus of the hypothalamus: cytoarchitectonic subdivisions and organization of projections to the pituitary dorsal vagal complex, and spinal cord as demonstrated by retrograde fluorescence double-labeling methods. J Comp Neurol 194:555–570

Swett JE, McMahon SB, Wall PD (1985) Long ascending projections to the midbrain from cells of lamina I and nucleus of the dorsolateral funiculus of the rat spinal cord. J Comp Neurol 238:401–416

Szabo J, Cowan WMA (1984) A stereotaxic atlas of the brain of the cynomolgus monkey (*Macaca fascicularis*). J Comp Neurol 222:265–300

Takeuchi Y, Uemura M, Matsuda K, Matsushima R, Mizuno N (1980) Parabrachial nucleus neurons projecting to the lower brainstem and the spinal cord. A study in the cat by the Fink-Heimer and the horseradish peroxidase methods. Exp Neurol 70:403–413

Tattersall JEH, Cervero F, Lumb BM (1986a) Viscerosomatic neurons in the lower thoracic spinal cord of the cat: excitations and inhibitions evoked by splanchnic and somatic nerve volleys and by stimulation of brain stem nuclei. J Neurophysiol 56(5):1411 – 1423
Tattersall JEH, Cervero F, Lumb BM (1986b) Effects of reversible spinalization on the visceral input to viscerosomatic neurons in the lower thoracic spinal cord of the cat. J Neurophysiol 56:785 – 796
Thames MD, Abboud FM (1979) Inhibition of renal sympathetic nerve activity during myocardial ischemia mediated by left ventricular receptors with vagal afferents in dogs. J Clin Invest 63:395 – 402
Thames MD, Donald DE, Shepherd JT (1977) Behavior of cardiac receptors with nonmyelinated vagal afferents during spontaneous respiration in cats. Circ Res 41:694 – 701
Thames MD, Klopfenstein HS, Abboud FM, Mark AL, Walker JL (1978) Preferential distribution of inhibitory cardiac receptors with vagal afferents to the inferoposterior wall of the left ventricle activated during coronary occlusion in the dog. Circ Res 43:512 – 519
Thorén P (1976) Atrial receptors with non-medullated vagal afferents in the cat: discharge frequency and pattern in relation to atrial pressure. Circ Res 38:357 – 362
Thorén PN (1977) Characteristics of left ventricular receptors with nonmedullated vagal afferents in cats. Circ Res 40:415 – 421
Thorén P (1979) Role of cardiac vagal C-fibers in cardiovascular control. Rev Physiol Biochem Pharmacol 86:1 – 94
Thorén PN, Donald DE, Shepherd JT (1976) Role of heart and lung receptors with nonmedullated vagal afferents in circulatory control. Circ Res 38 (Suppl II):2 – 9
Trevino DL (1976) The origin and projections of a spinal nociceptive and thermoreceptive pathway. In Zotterman Y (ed) Sensory functions of the skin. Pergamon, pp 367 – 377
Trevino DL, Maunz RA, Bryan RN, Willis WD (1972) Location of cells of origin of the spinothalamic tract in the lumbar enlargement of cat. Exp Neurol 34:64 – 77
Trevino DL, Coulter JD, Willis WD (1973) Location of cells of origin of spinothalamic tract in lumbar enlargement of the monkey. J Neurophysiol 36:750 – 761
Uchida Y (1975) Afferent sympathetic nerve fibers with mechanoreceptors in the right heart. Am J Physiol 228:223 – 230
Uchida Y, Murao S (1974a) Potassium-induced excitation of afferent cardiac sympathetic nerve fibers. Am J Physiol 226:603 – 607
Uchida Y, Murao S (1974b) Excitation of afferent cardiac sympathetic nerve fibers during coronary occlusion. Am J Physiol 226:1094 – 1099
Uchida Y, Murao S (1974c) Afferent sympathetic nerve fibers originating in the left atrial wall. Am J Physiol 227:753 – 758
Uchida Y, Murao S (1974d) Bradykinin-induced excitation of afferent cardiac sympathetic nerve fibers. Jpn Heart J 15:84 – 91
Uchida Y, Murao S (1975) Acid-induced excitation of afferent cardiac sympathetic nerve fibers. Am J Physiol 228:27 – 33
Uchida Y, Kamisaka K, Murao S, Ueda H (1974) Mechanosensitivity of afferent cardiac sympathetic nerve fibers. Am J Physiol 226:1088 – 1093
Ueda H, Uchida Y, Kamisaka K (1969) Distribution and responses of the cardiac sympathetic receptors to mechanically induced circulatory changes. Jpn Heart J 10:70 – 81
Vance WH, Bowker RC (1983) Spinal origins of cardiac afferents from the region of the left anterior descending artery. Brain Res 258:96 – 100
Von Euler C (1986) Brain stem mechanisms for generation and control of breathing pattern. In: Fishman AP (ed) Handbook of physiology, sect 3, The respiratory system, vol II. Williams and Wilkins, Baltimore, Maryland, pp 1 – 67
Walker AE (1940) The spinothalamic tract in man. Archs Neurol Psychiat 43:284 – 298

Walker JL, Thames MD, Abboud FM, Mark AL, Klopfenstein HS (1978) Preferential distribution of inhibitory cardiac receptors in left ventricle of the dog. Am J Physiol 235:H188 – H192

Weaver LC (1977) Cardiopulmonary sympathetic afferent influences on renal nerve activity. Am J Physiol 233:H592 – H599

Webb SW, Adgey AA, Pantridge JF (1972) Autonomic disturbances at onset of acute myocardial infarction. Br Med J [Clin Res] 3:89 – 92

Wehrmacher EH (1964) Pains in the chest. Thomas, Springfield, II, pp 342 – 361

Westlund KN, Coulter JD (1980) Descending projections of the locus coeruleus and subcoeruleus/medial parabrachial nuclei in the monkey: axonal transport studies and dopamine-β-hydroxylase neurocytochemistry. Brain Res Rev 2:235 – 264

Westlund KN, Bowker RM, Ziegler MG, Coulter JD (1984) Origins and terminations of descending noradrenergic projections to the spinal cord of the monkey. Brain Res 292:1 – 16

White JC (1957) Cardiac pain. Anatomic pathways and physiologic mechanisms. Circulation 16:644 – 655

White JC, Bland EF (1948) The surgical relief of severe angina pectoris. Medicine (Baltimore) 27:1 – 42

White JC, Garrey WE, Atkins JA (1933) Cardiac innervation: experimental and clinical studies. Arch Surg 26:765 – 786

Willis WD (1985) The pain system – The neural basis of nociceptive transmission in the mammalian nervous system. Karger, Basel, pp 145 – 212

Willis WD, Trevino DL, Coulter JD, Maunz RA (1974) Responses of primate spinothalamic tract neurons to natural stimulation of hindlimb. J Neurophysiol 37:358 – 372

Willis WD, Haber LH, Martin RF (1977) Inhibition of spinothalamic tract cells and interneurons by brain stem stimulation in the monkey. J Neurophysiol 40:968 – 981

Willis WD, Kenshalo DR, Leonard RB (1979) The cells of origin of the primate spinothalamic tract. J Comp Neurol 188:543 – 574

Woollard HH (1926) The innervation of the heart. J Anat 60:345 – 373

Yezierski RP, Schwartz RH (1986) Response and receptive-field properties of spinomesencephalic tract cells in the cat. J Neurophys 55:76 – 96

Yezierski RP, Gerhart KD, Schrock BJ, Willis WD (1983) A further examination of effects of cortical stimulation on primate spinothalamic tract cells. J Neurophysiol 49:424 – 441

Ygge J, Grant G (1983) The organisation of the thoracic spinal nerve projection in the rat dorsal horn demonstrated with transganglionic transport of horseradish peroxidase. J Comp Neurol 216:1 – 9

Synaptic Transmission in the Mechano- and Electroreceptors of the Acousticolateral System

G. N. Akoev and G. N. Andrianov

Pavlov Institute of Physiology, Academy of Sciences, Nab. Makarova 6, Leningrad, 199034, USSR

1 Introduction

The acousticolateral system of vertebrates includes the lateral-line organs, vestibular apparatus, and cochlea. Integration of these organs into a single system is based on the similarity of their functional and morphological characteristics, innervation, and embryonal development. The specialized receptors for the perception of weak electric fields in the lateral-line organs of some lower aquatic vertebrates may also be considered part of the acousticolateral system.

It is well known that the receptors of the acousticolateral system are secondary sense cells, characterized by synaptic transmission from the receptor cell to the afferent nerve fiber. For many reasons it is important to understand the mechanisms of synaptic transmission. At the synaptic level the local electric potentials are transformed into impulses in nerve fibers. The presence of synaptic transmission makes it possible for a great number of receptor cells to converge on one afferent fiber, a factor important in increased sensory sensitivity.

Progress in Sensory Physiology. Vol 9

The neurotransmitters in the receptors of the acousticolateral system still remain to be identified despite the great interest in this problem (see reviews) by Guth et al. 1976, 1981; Klinke 1981; Guth and Melamed 1982; Bobbin et al. 1984; Foster and Fagg 1984). The reported data are often contradictory.
To a certain extent the lack of considerable progress in our understanding of the mechanisms of synaptic transmission arises from the complexity of the methods used for studying the receptors of the inner ear. The relatively large electroreceptors (ampullae of Lorenzini) in some fish are advantageous for this purpose; they can be isolated from the organism and studied under experimental conditions.

The present paper deals with the morphology, physiology, and biochemistry of synaptic processes in the receptors of auditory organs, vestibular apparatus, lateral-line organs, and electroreceptors. Emphasis is placed on results obtained in recent years. Earlier publications are only cited to present very fundamental data, without which the elucidation of the problems would be incomplete.

2 Anatomy of the Receptor Systems

Throughout the vertebrate acousticolateral system the sensory receptors are hair cells (Fig. 1). They are grouped in specialized sensory regions in the cochlea, the cristae ampullaris of the semicircular canals, the macula sacculi, macula utriculi and macula lagena in the vestibular apparatus, as well as in neuromasts and canals in the lateral-line organs. The hair cells in these receptor organs perform a distinct mechanoreceptor function and morphologically are fundamentally similar. Only the structure of the ciliary apparatus varies in different species (Flock 1965, 1971).
A common hair cell is pear shaped, 30 – 60 μm long, about 3 μm wide in the apical part of the cell, and 8 – 10 μm wide at the level of the cell nucleus located at the basal end of the cell. The apical surface of the receptor cell has a bundle of stereocilia and a single kinocilium always directed asymmetrically with respect to the stereocilia. Kinocilia are about 0.3 μm in diameter and, unlike stereocilia, have nine peripheral and two central fibrils in their core which arise from the basal body. They are 15 – 40 μm long. Stereocilia vary in length and are arranged in a staircase fashion, with the longest being located adjacent to the kinocilium. Stereocilia are thinner than kinocilia (0.1 – 0.2 μm) and composed of a fibrillar core bound by the continuation of the cell surface membrane.
In the basal part, stereocilia have roots which protrude into the granular cuticular plate of a cell and sometimes pass through the cuticle into the cytoplasm. The cytoplasm of the hair cells contains microtubules and microfilaments which, along with the mitochondria, are concentrated in the region of the cell nucleus.
Almost all hair cells are characterized by the presence of special crystalline inclusions in their cytoplasm and even in their nucleus (Mullinger 1964; Yamada and Hama 1972; Wickham 1973; Toyoshima and Shimamura 1982). They have no surrounding membrane and are seen as parallel electron-dense lamellar structures about 25 nm in diameter. The function of the crystalline inclusions still remains to be studied, but their presence in the cytoplasma of all mechanoreceptor elements and absence in the supporting cells and some other receptor structures

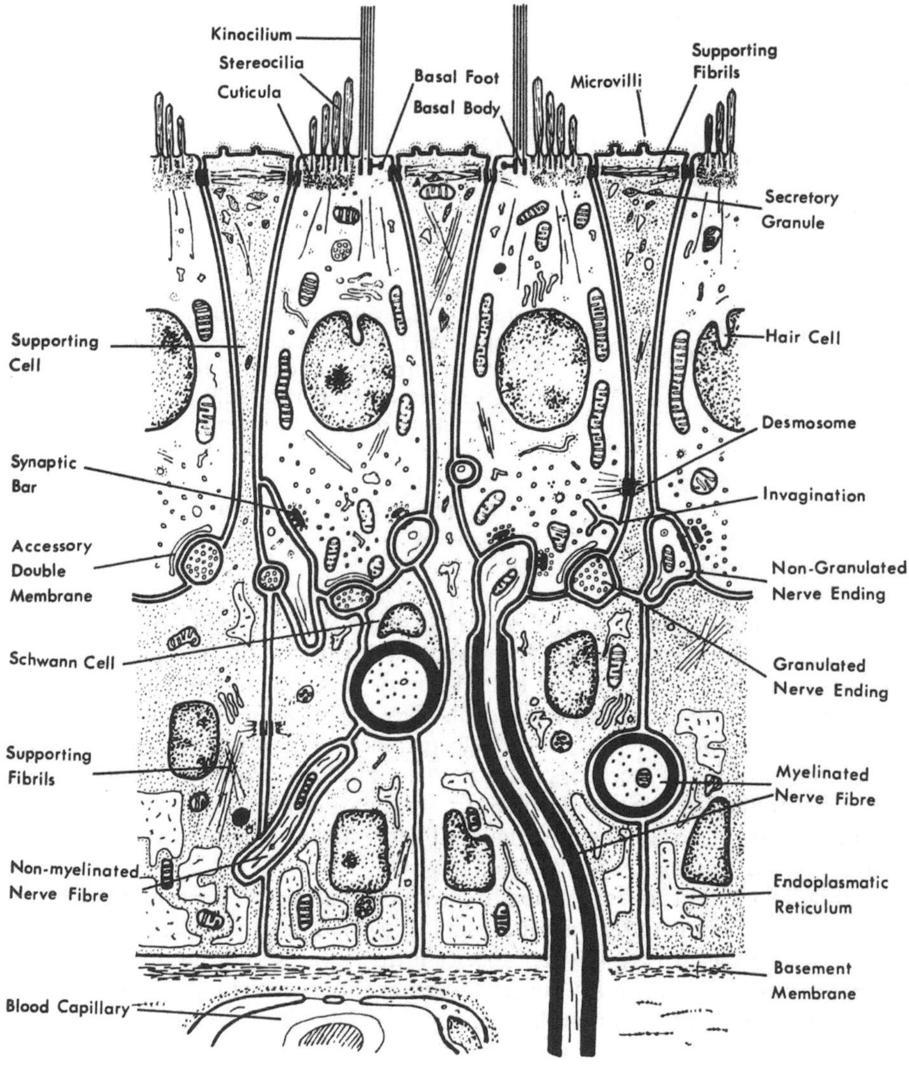

Fig. 1. Schematic drawing showing the ultrastructure of the sensory epithelium of the lateral-line canal organ of the *Lota vulgaris*. (Flock 1965)

not belonging to the acousticolateral system may indicate their involvement in the reception of mechanical stimuli by hair cells.

Very little is known about the structure and function of supporting cells. They are connected to the hair cells through a system of desmosomes, forming sensory epithelium. The supporting cells are usually longer than the receptors and also have cilia on their apical membrane. The cytoplasm of a cell, particularly in its apical part, contains granules 0.1 – 0.3 μm in diameter, microfilaments, and the usual set of cell organoids. There is evidence indicating that the supporting cells

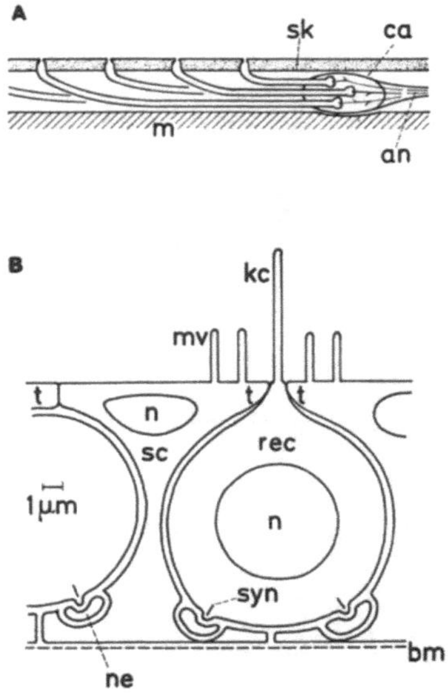

Fig. 2A, B. The anatomy of the ampullae of Lorenzini. **A** Section through skin showing canals ending in ampullae within a capsule (Murray 1974). **B** Section through the sensory epithelium, showing the receptor and supporting cells (Waltman 1966). *an*, ampullary nerve; *bm*, basement membrane; *ca*, capsule; *kc*, kinocilium; *m*, body muscles; *mv*, microvilli; *n*, nucleus; *ne*, nerve ending; *rec*, receptor cell; *sc*, supporting cell; *sk*, skin; *syn*, synapse; *t*, tight junction

in the lateral-line organs are involved in the secretory process in which a gelatinous cupula covers, the hair cells from above (Cornford and Barajas 1980).

The receptor cells of specialized lateral-line organs have a somewhat different structure. According to modern concepts they perform an electroreceptor function. There are two general types of electroreceptors, ampullary and tuberous. The ampullary type of receptors includes the ampullary organs of weakly electric fish, the ampullae of Lorenzinin of elasmobranchs, of chondrosteans, and of the marine catfish (Plotosus, the only marine teleost having electroreceptors), as well as the small pit organs of catfish.

The ampullary receptors are located in the epidermal cavity, communicating with the aquatic environment by way of a canal (Fig. 2). The length of the canal varies in different species. In fresh water species it does not exceed hundreds of micrometers; in sea fish it may reach several dozen centimeters. The sensory epithelium covers the bottom of the ampullary organ. It is 25 – 30 µm thick and composed of two types of cells, sensory and supporting cells. The receptor cells are usually pear shaped, 10 – 15 µm in diameter. On their apical surface they carry microvilli (in receptors of weakly electric fish) or a single kinocilium (in ampullae of Lorenzini), but no ciliary apparatus characteristic of the ordinary lateral-line organs is found here. To detect weak electric signals in the aquatic environment, the ampullary canal walls and the entire sensory epithelium must have a very high resistance. The electric isolation is provided by a system of tight junctions in the apical membrane area of receptor and supporting cells where current is channeled through the receptor cells (Waltman 1966; Sejnowski and Yodlowski 1982). Tight junctions were also found in the organ of Corti. Only a small part of the apical membrane is exposed. For instance, in the ampullae of Lorenzini the lumeral face of the apical membrane does not exceed 1% of the cell surface. The number of receptor cells in the receptor organs varies strongly, from 1 – 4 in weakly electric species to several thousands in marine fish. The structural features of ampullary receptors and their functional properties are treated in more detail in a number of basic studies and reviews (Szabo 1965, 1974; Waltman 1966; Andrianov et al. 1979; Andrianov and Ilyinsky 1983; Brown and Ilyinsky 1984; Zakon 1986).

Tuberous receptors are also located in epidermal capsules; however, the receptor cells have no direct contact with the outer medium since the external accessory cells extend between the sensory cells and the epidermal surface. There is great diversity in the structure of tuberous receptors in different species of electric fish (Szabo 1965, 1974; Zakon 1986).

In the acousticolateral system all the mechano- and electroreceptor cells make well-formed synapses with nerve terminals in which there are two types of endings: granular and nongranular (Flock 1965; Hama 1965). The basic difference between them is that the endings of type 1 – efferent – are characterized by the presence of a great number of vesicles (0.03 – 0.04 µm in diameter) in the synaptic area of the nerve and cell membranes, whereas in the endings of type 2 – afferent – the vesicles are grouped predominantly in the presynaptic part of the receptor cell. No efferent innervation is observed in separate hair cells of the lateral-line neuromasts and in all electroreceptors.

The number of nerve fibers innervating a single receptor cell varies greatly between organs and even within an organ. For example, in the organ of Corti in mammals, about 20 afferent fibers innervate one inner hair cell and a single afferent fiber innervates about 10 outer hair cells (Spoendlin 1972). However, the number of efferent terminals on the outer hair cells is greater than that on the inner cells (Wersäll et al. 1965). In the lateral-line organs of fish a single neuromast containing 10 – 50 cells is innervated by several afferent and efferent fibers (Flock and Wersäll 1962; Hama 1965). Amphibian neuromasts consist of approximately eight cells; several neuromasts (from two to six) are grouped into stitches, all innervated by one or two afferent and one or more efferent nerve fibers (Harris et al. 1970; Jørgensen and Flock 1973). In the ampullary organs of weakly electric fish a number of receptor cells in each electroreceptor is innervated by a single nerve fiber, whereas in the ampullae of Lorenzini a single nerve fiber can innervate many hundreds and sometimes thousands of receptor cells (Jørgensen et al. 1972; Szabo 1972).

The structure of synaptic contacts in various receptors of the acousticolateral system looks uniform. In the afferent synapse pre- and postsynaptic membranes are separated by a synaptic cleft 15 µm in diameter which appears to be more electron dense than the rest of the extracellular space between the hair cell and the nerve ending (Fig. 3). The synaptic cleft is often observed to have dark filaments bridging the pre- and postsynaptic membranes and apparently responsible for the constant distance between the membranes. (Benshalom and Flock 1980).

Due to postsynaptic density the postsynaptic membrane looks thicker than the presynaptic one. The thickness of the presynaptic membrane does not differ from that of the rest of the hair cell membrane.

An object of great interest is the structure of the synaptic bodies described in the lateral-line organs of fish (Hama 1965; Flock 1965; Benshalom and Flock 1980) and amphibians (Shelton 1971; Jørgensen and Flock 1973), in the electroreceptors of weakly electric fish (Mullinger 1964, 1969; Szabo and Wersall 1970; Szamier and Wachtel 1970), in the ampullae of Lorenzini (Barets and Szabo

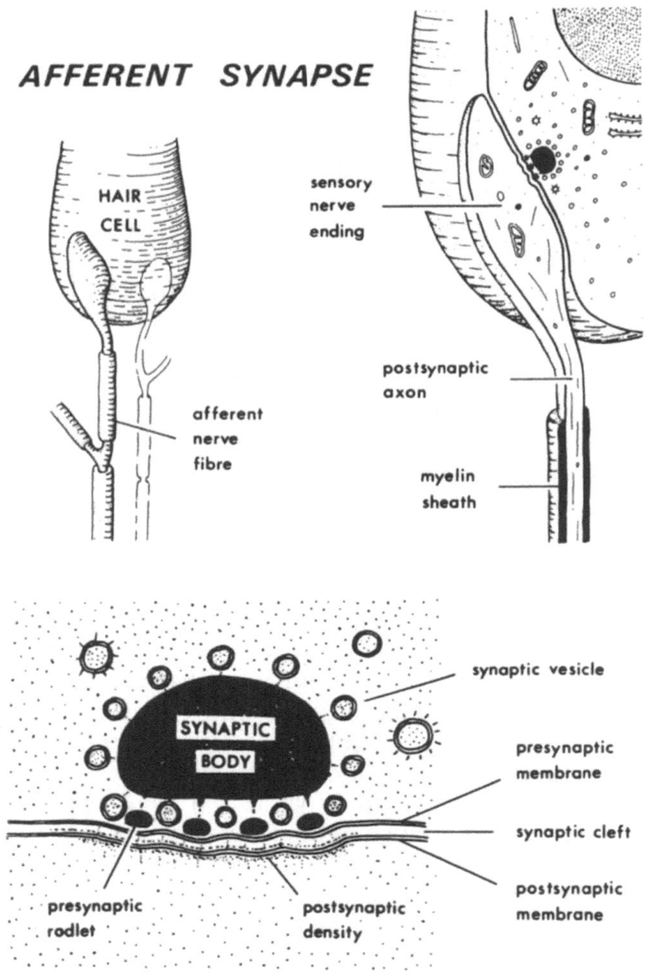

Fig. 3. Schematic drawing of the general structure of the afferent synapse on lateral-line hair cells. (Benshalom and Flock 1980)

1962; Waltman 1966; Derbin 1970; Jørgensen et al. 1972; Ilyinsky et al. 1974), in the organ of Corti (Smith and Sjöstrand 1961; Gleisner et al. 1973; Saito 1980), and in the vestibular apparatus (Hama and Saito 1977; Fayre and Sans 1979). The synaptic bodies are electron-dense formations of variable sizes and shapes (ribbons, bars, sheets, etc.) found in the receptor cell region directly adjacent to the presynaptic membrane. Synaptic body feet are elongated structures which fit into the grooves of the presynaptic membrane. A synaptic body has about 14 feet. In cross section they appear as round bodies and in longitudinal sections one can see that they consist of microtubules. The main mass of the synaptic body is connected to the feet by thin filaments. A considerable concentration of synaptic vesicles 30 nm in diameter without an electron-dense core is noted around the synaptic body. Vesicles with a dense core and coated vesicles can be seen in other parts of the cytoplasm. Since synaptic vesicles appear to assemble exclusively

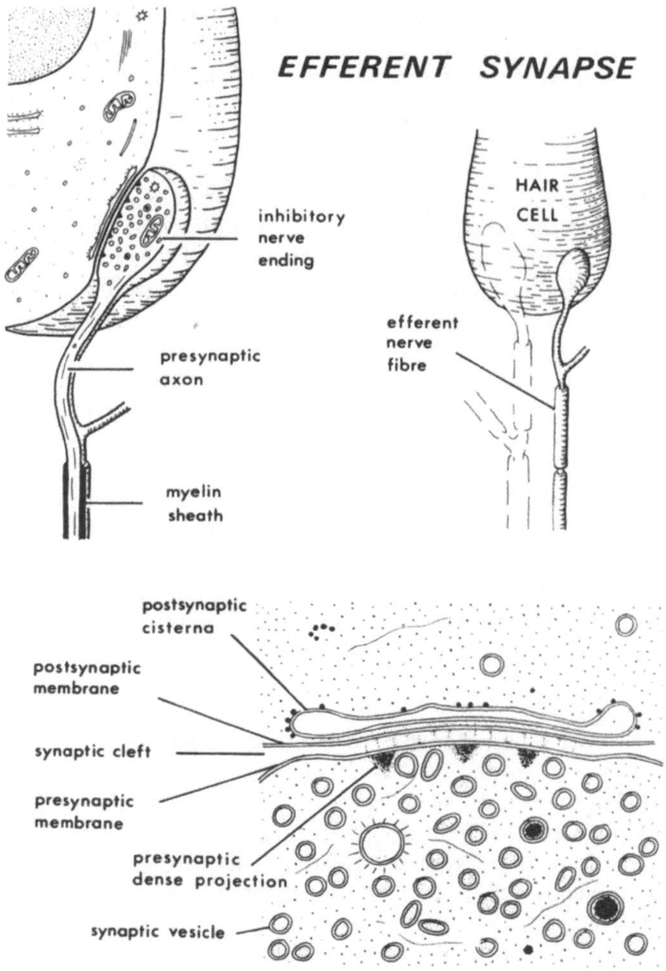

Fig. 4. Schematic representation of the efferent synapse on hair cells and its general structure. (Benshalom and Flock 1980)

around the synaptic body it seems likely that synaptic bodies indicate the loci in the hair cell where conventional chemical afferent synapses occur.

The functional role of synaptic bodies still remains obscure, but their presence at afferent synapses in sense organs suggests that in all cases they perform a similar function. One hypothesis purports that synaptic bodies may be a site for the storage of neurotransmitters or their precursors in the cell (Osborne and Thornhill 1972; Wagner 1973; Monagham 1975; Osborne 1977; Thornhill 1972).

Acoustic stimulation causes an increase in the size of synaptic bodies in the basilar papilla of the frog's ear (Frishkopf 1973). In lateral-line organs of salamander tadpoles, where synaptic bodies are seen in vivo in the transparent tail, the receptor stimulation can be observed to cause the bodies to move to the cell base (Flock et al. 1973b; Flock 1974). These data indicate that synaptic bodies play a distinct role in the plasticity function of sensory synapses. It has been also suggested that synaptic bodies control the guidance and access of synaptic vesicles to the presynaptic membrane (Benshalom and Flock 1980). Continuous stimulation, for instance, can cause the synaptic bodies to enhance the transmitting ability of an afferent synapse. Their size and number in this case increase, whereas under conditions of relative rest a reverse process takes place. The possibility that synaptic bodies are a site for calcium binding was rejected as a result of studies undertaken (Benshalom and Flock 1977).

The postsynaptic membrane of receptor cells consists of two layers. The inner one faces the cytoplasm of the nerve ending and contains a postsynaptic thickening. The outer layer facing the synaptic cleft is much thinner.

In the efferent synapse the vesicles are concentrated in the nerve ending. The presynaptic membrane of the nerve fiber does not differ from the rest of the membrane in thickness, but contains characteristic electron-dense projections 30 nm wide and 20 nm high (Fig. 4). The postsynaptic membrane is not distinguished from the rest of the membrane of a hair cell by special electron density. Pre- and postsynaptic membranes run parallel to each other, apparently due to intrasynaptic filaments. The efferent synapses in the area adjacent to the inner side of the postsynaptic membrane are characterized by the presence of special membrane structures called postsynaptic cisterns (Benshalom and Flock 1980). The distance of the cisterns from the postsynaptic membrane is considerably less than the width of the synaptic cleft. This distance is maintained constant by an electron-dense material. The cistern membrane facing the inner part of the hair cell carries ribosome-like granules on its surface, usually grouped in threes.

3 Electric Characteristics of Hair Cells: Receptor Potentials

During the last decade great achievements have been made in the investigation of the properties of hair cells in the acousticolateral system by means of a variety of physiological techniques, including intracellular recording with high-resistance microelectrodes, both membrane patch- and whole-cell recording with gigohm-seal electrodes.

In 1970 Harris et al. were the first to obtain intracellular recordings from the lateral-line hair cells of the mudpuppy *Necturus maculosus*. Subsequent intracellular recordings demonstrated the membrane potential of the hair cells to be -10 to -65 mV, the input resistance $10-100$ mΩ, and the membrane time constant 0.2 ms (Flock 1971; Flock et al. 1973a). Displacement of the sensory hairs toward the kinocilium evoked a depolarizing receptor potential up to 5 mV (Flock 1971). Displacement in the opposite direction produced hyperpolarizing receptor potential. A 15-mV shift in the potential caused no changes in the conductance of hair cells, i.e., the latter proved to be electrically inexcitable. The membrane potentials in the alligator auditory organs were also found to be fundamentally similar (on an average -74 mv) and these cells responded with a 1-mV potential change to the acoustic stimulation (Mulroy et al. 1974).

It is noteworthy that in the mammalian organ of Corti with two types of receptor cells, the average membrane and receptor potentials of the inner hair cells were -20 and -25 mV respectively (Russell and Sellick 1978) and in the outer hair cells -71 and -15 mV (Dallos et al. 1982). The input resistance of both cell types ranged from 10 to 34 mΩ. It should be noted that these cells did not differ in their sensitivity to acoustic stimuli. It has been recently demonstrated that the resting membrane potentials of mice' inner hair cells, outer hair cells, and supporting cells in vitro were −30 to −70 mV, −40 to −70 mV, and −60 to −75 mV respectively (Russell et al. 1986).

Additonally, the receptor potentials of inner hair cells differ in some respects from those of outer ones. Thus, the depolarizing receptor potentials of the outer hair cell saturate at a lower voltage (6.5 mV) than the inner hair cell. The most sensitive hair cells are depolarized by 0.8 mV when their stereocilia are displaced by 2 nm.

The studies on frog sacculus receptors have greatly contributed to current concepts concerning the primary processes in hair cells (Hudspeth and Corey 1977; Hudspeth 1982, 1983). The membrane potential of these cells was −50 to −60 mV, and in response to a mechanical stimulation a receptor potential of up to 15 mV was generated. The sign of the receptor potential was determined by the direction of the mechanical stimulus: a displacement of stereocilia toward a kinocilium gave rise to a depolarizing receptor potential and a decrease in input resistance; the stimulation in the opposite direction induced a hyperpolarizing receptor potential and an increase in input resistance. The maximum change in input resistance was 50 mΩ. The threshold sensitivity of the cells was equal to 5 Å (0.5 nm) and the latency of the response to rapid stimuli at 22 °C was 40 μs. In the intensity range up to 0.5 μm the amplitude of the receptor potential was linearly dependent on the intensities of mechanical stimuli.

It is important to note that even after the removal of kinocilium (Hudspeth and Jacobs 1979) stereocilia continued to generate receptor potentials with an amplitude up to 13 mV in response to mechanical stimuli. As in the case of intact cells, bending of stereocilia toward the kinocilium caused a depolarizing of receptor potentials. In normal cells mechanical stimulation of a stereocilium bundle alone generated receptor potentials; the stimulation of 10 stereocilia provided a receptor potential of 2 mV.

Thus, the results of the experiments indicate that an intact kinocilium is not required for the transduction process of vertebrate hair cells. One may assume that these structures are essential to provide polarization of hair cells in ontogenesis, and that their role in transduction may be limited only to a simple transmission of mechanical displacement from the cupula or otolithic membrane to the stereocilia. It is known, for instance, that the receptor cells of the organ of Corti in adult mammals contain no kinocilia at all. This probably accounts for the improved receptor responses to high frequency stimuli. It should be noted here that there is a fundamental difference between the hair cells of invertebrates and vertebrates. In the former the apical surface has only kinocilia which transform mechanical stimuli. Conversely, in vertebrates, these processes are apparently associated with stereocilia.

Interesting findings were obtained when recording extracellular potentials along the stereocilia of a hair cell. Hudspeth (1982) discovered that the transduction channels are located at or near the tips of the stereocilia (Fig. 5). The calculation shows that the receptor potential recorded from a hair cell accounts for 90% of the potential emerging in the stereocilium terminals. Other researchers also came to the conclusion that the transduction channels of the receptor cell are located in the area of the stereocilium apical membrane (Ohmori 1984a, b). The peak trans-

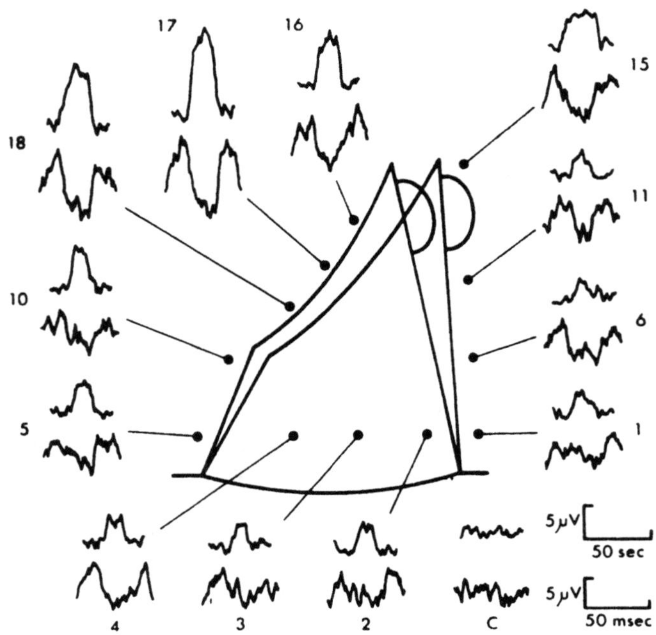

Fig. 5. Representative records for 12 recording sites around a hair bundle (Hudspeth 1982). The cell is stimulated by driving the hair bundle through 1.0-μm, 25-Hz sinusoidal deflections; the hair bundle is diagrammed at rest and at a deflection to +1 μm. The *upper record* at each recording site is the output of the lock-in analyzer. For comparison, the *lower records* are computer-averaged extracellular responses from 5-s samples of electrode output. The control records (*C*) show the noise level in each type of trace. Note that responses are consistently larger at the hair bundle's top than at its bottom and that the largest signals occur on the beveled top surface of the hair bundle

duction conductance of these channels in hair cells is, on the average, 2 – 5 nS, whereby the conductance of a single channel is 50 pS. Considering the fact that a single hair cell of the vestibular apparatus and the organ of Corti has 40 – 150 stereocilia on its apical surface, the peak transduction conduction of 2 – 5 nS could be due to the simultaneous opening of as few as 40 – 100 channels or about one channel per stereocilium (Hudspeth 1982; Russell 1983).

These transduction channels are nonspecific in nature and a wide variety of monovalent and divalent cations were permeable throught these hair cell channels in the lateral-line organs of the mudpuppy, of the bullfrog sacculus, and the vestibular organ of chicks (Sand 1975; Corey and Hudspeth 1979a, b; Ohmori 1985). The receptor current is carried in vivo by whatever ions are available. This would normally be K^+, whereas Ca^{2+} is a necessary cofactor in transduction since reduction of its concentration below 10 μ*M* abolished the receptor current. In contrast, Valli et al. (1977) showed that in the absence of K^+ in the endolymph the discharges from the ampullary nerve were completely suppressed, irrespective of its ionic content. In the hydrated state the transduction channel of hair cells has a diameter of about 0.6 nm. Voltage-clamp experiments on isolated hair cells

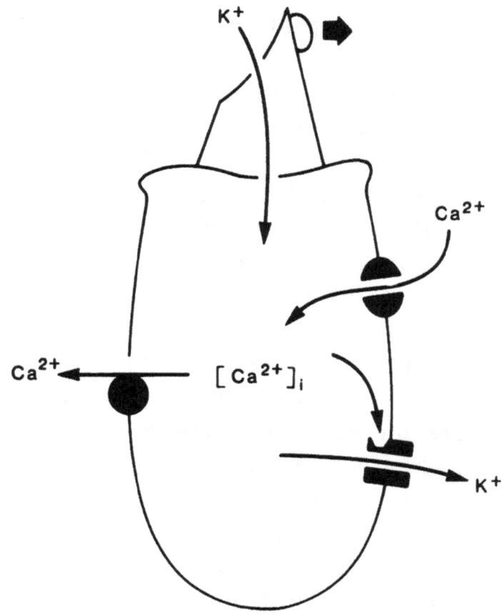

Fig. 6. Schematic diagram of the proposed mechanism for resonant frequency tuning in the bullfrog's hair cell (Hudspeth 1986). Depolarization of the cell by a stimulus opens voltage-sensitive Ca^{2+} channels. The influx of Ca^{2+} regeneratively depolarizes the cell; $[Ca^{2+}]$ then rises, activating Ca^{2+}-sensitive K^+ channels. The resultant efflux of K^+ repolarizes the cell to and beyond the steady-state potential. Ca^{2+} within the cytoplasm is buffered and rapidly extruded

from the bullfrog sacculus and vestibular hair cells of the chick revealed the existence of inward and outward currents (Lewis and Hudspeth 1983; Hudspeth 1986; Ohmori 1984a) (Fig. 6). The inward Ca^{2+} current is activated at a potential more positive than −60 mV. The main characteristics of this noninactivated Ca^{2+} conductance are that it is activated at the resting potential and involved in the release of the afferent synaptic transmitter. There are between 200 and 500 Ca^{2+} channels in the hair cells (Ohmori 1984b); these channels and anomalous rectifying ones exist mainly on the basal part of the hair cell membrane. Depolarization of hair cells also evokes two outward K^+ currents, namely, transitory current (Lewis and Hudspeth 1983) and Ca^{2+}-activated K^+ conductance (Lewis and Hudspeth 1983; Ohmori 1984a), the latter being utilized for frequency tuning by electric resonance. The analysis of current-voltage characteristics showed that the hair cells behave linearly over a wide range of the potential. For instance, the inner hair cells of the guinea pig organ of Corti were linear over a range from −40 to −150 mV, with the membrane potential being −30 to −35 mV and receptor potential −15 to −35 mV (Russell 1983). The reversal potentials for the receptor current of hair cells vary extensively in different objects and range from −9 to +11 mV in the vestibular apparatus of frogs (Corey and Hudspeth 1979a, b), are close to zero in the vestibular apparatus of guinea pigs (Crowford and Fettiplace 1979), or correspond to the values of the endocochlear potential in the inner hair cells of the organ of Corti in guinea pigs (Russell 1983). Hyperpolarization of hair cells causes nonlinearity in the current-voltage characteristics of the receptor. The nonlinearity seems to arise from anomalous and delayed rectification. In the hair cells of the frog vestibular apparatus anomalous rectification develops at a membrane potential below −100 mV and delayed rectification at membrane potentials more positive than about −50 mV. The presence of anomalous and delayed rectification is responsible for a relatively low receptor potential of the hair cells in vertebrates (Corey and Hudspeth 1979a, b).

No electroreceptor studies using microelectrodes have been undertaken so far; however, the presence of ampullary canals up to several centimeters long make it possible to clamp the current and voltage on receptor membranes. The use of different electrophysiological methods enabled detailed study of the receptor mechanisms of electroreceptors. For example, it was established that the response of teleosts to stimulus polarity corresponds to the action of current directly on the presynaptic secretory membrane. If anodal stimulus is applied to the exterior of the receptor, the inner (presynaptic) face of the receptor cell is depolarized. The rate of transmitter release is increased and the postsynaptic fiber fires more rapidly (Roth 1968, 1969; Obara 1976). The literature contains detailed data on the primary processes in the ampullae of Lorenzini of the marine catfish (*Plotosus*) (Obara and Oomura 1973; Akutsu and Obara 1974; Obara 1976; Obara and Sugawara 1984; Sugawara and Obara 1984; Bennett and Obara 1986). The specific activity of the ampullae of Lorenzini was found to manifest itself primarily in the form of ampullary potential which is recorded extracellularly in the ampullary canal and represents the mass response of sensory epithelium. The responses behave linearly for small polarization of either sign (up to ± 200 mV). When the stimulus intensity is increased, the responses to depolarizing anodal stimuli start to become nonlinear and there is a regenerative response induced by the entry of Ca ions through the basal membrane.

The amplitude of the potential is determined by the external Ca^{2+} concentration according to Nernst's equation. The ampullary responses are unaffected by tetrodotoxin. A cathodal stimulus applied externally hyperpolarizes the inner face of the receptor cell and decreases the rate of transmitter release; the nerve fiber fires at a lower rate.

The ampullae of Lorenzini of elasmobranchs respond to the polarity of electric stimuli just the opposite of the receptors of teleosts. As already known, lumen-negative stimuli increase the afferent discharge and lumen-positive stimuli decrease it (Murray 1962; Bennett and Obara 1986). This polarity inversion, compared to the teleost tonic receptors, results from the excitability of the lumenal faces of the receptor cells. It has been suggested that in elasmobranchs the regenerative response arises on the apical surface of the cell, while in teleosts on the basal one (Obara and Bennett 1972; Bennett 1970; Clusin and Bennett 1977a, b; 1979a, b; Obara 1976; Obara and Sugawara 1979; Bennett and Clusin 1979; Bennett and Obara 1986). Therewith it was postulated that the cathodal stimulus applied to the canal pore depolarized the apical membrane, changing its permeability, thereby giving rise to a negative response of sensory epithelium (Fig. 7). The latter causes the opening of Ca channels and the increase in current flowing through the lumenal membrane. This current flows out through the basal membrane, depolarizing it and causing it to release more transmitter. The anodal stimulus, in contrast, hyperpolarizes the apical membrane and closes Ca channels, thus decreasing the inward current through this face, which leads to the hyperpolarization of the basal face and reduced secretion of the transmitter.

According to another hypothesis (Brown and Govardovsky 1983a, b; Brown and Ilyinsky 1984), the unusual response of the Lorenzinian ampullae in elasmobranchs to the polarity of electric stimulation, i.e., an increase of the afferent discharge by lumen-negative stimuli which hyperpolarizes the presynaptic membrane of the receptor cell, is not linked with the spontaneous activity of the

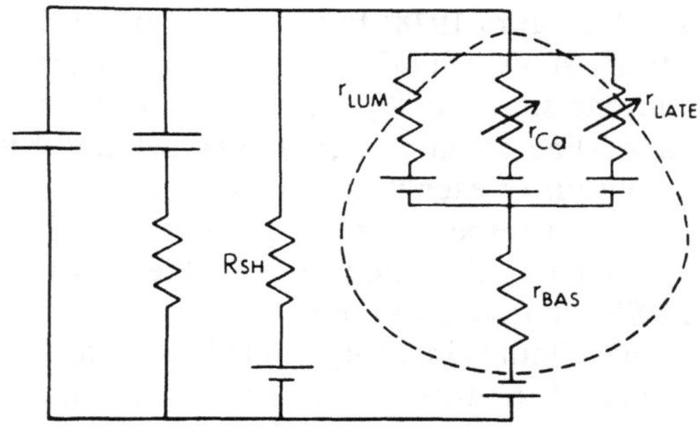

Fig. 7. Equivalent circuit of the ampullary epithelium (Clusin and Bennett 1977b). Active currents arising in the lumenal membranes are represented by a fixed resistor (r_{LUM}) in parallel with two variable resistors. The early calcium current flows through r_{Ca} and the late current through r_{LATE}. The basal faces of the receptor cells are represented by a fixes resistance (r_{BAS}). Most of the leakage current flows through the supporting cell, the marginal zone cells, or the intercellular clefts. These pathways are represented by a single resistor R_{SH}. The batteries E_{Ca}, E_{LUM}, E_{BAS}, and E_{SH} are drawn but not labelled. The ampullary capacity is attributed to cells that are electrically in parallel with the receptor cells, namely the supporting cells and marginal zone cells, whose basal membrane area is large. Some of the ampullary capacitance is in series with a significant fixed resistance

apical membrane of the receptor cell, as Bennett's hypothesis suggests. The authors argue that the action potential of the receptor epithelium (the basic argument in Bennett's hypothesis) is a secondary phenomenon arising only when the stimuli used are too strong and has nothing to do with the normal functioning of the ampullae of Lorenzini. They proposed a model for the mode of action of the electroreceptors of the Lorenzinian ampullae which accounts for their own data and those in the literature. The model presumes the presence of stationary current-voltage characteristics with a section of negative resistance of the apical membrane of a receptor cell.

4 Electrical Responses of the Postsynaptic Membrane

The excitatory postsynaptic potentials (EPSPs) in the afferent fibers of the acousticolateral system were first described in experiments on the goldfish saccular macula (Furukawa and Ishii 1967). In this study it was found that slow generator potentials exist in the nerve fibers along with action potentials. In subsequent studies the generator potential was recorded intracellularly from unmyelinated endings of the lateral-line afferents in the burbot *Lota lota* (Flock et al. 1973a, b). In the absence of mechanical stimulation the afferent fibers exhibit spontaneous EPSPs up to 5 mV. The synaptic delay for evoked EPSPs was 0.9–1.0 ms at 14 °C. Spontaneous generator potentials were also observed in auditory fibers of goldfish (Kyogoku et al. 1986). The sound-evoked generator potentials were produced with a delay of 0.4–0.8 ms.
A detailed analysis of EPSP characteristics was made in a series of studies on the goldfish sacculus (Furukawa and Ishii 1967; Furukawa et al. 1972; Furukawa

and Matsuura 1978; Furukawa et al. 1982). In response to each sound wave the EPSPs show a marked run-down synaptic depression in amplitude (Furukawa and Matsuura 1978). A similar phenomenon has also been observed in the lateral-line organ of *Xenopus* (Flock and Russell 1976). In successive EPSPs the amplitude of each was reduced by a fixed coefficient with respect to the preceeding one, irrespective of sound intensity, indicating that the run-down condition was due to a depletion of available transmitter quanta. In those cases where the EPSPs completely disappeared during continuous sound stimulation, an increase in the stimulus intensity immediately caused EPSP. We regard this phenomenon as one of the forms of sensory adaptation based on transmitter depletion.

A supposition was made that hair cells have many transmitter release sites that vary in sensitivity (Furukawa and Matsuura 1978). If the hair cells were stimulated with a weak sound, the transmitter release would occur in a few sites, but as the sound intensity increased, more and more release sites would be activated. This possibility seems valid since a single afferent fiber receives inputs from many release sites of differing sensitivities. This seems to be also true, for example, for ampullae of Lorenzini of the skate since morphological studies showed that six to eight afferent nerve fibers innervate several thousands of sensory cells.

Apparently the discrete quantal model of neurotransmitter release proposed initially for neuromuscular junction (Del Castillo and Katz 1954) is applicable to many other types of synapses as well. The authors derived that the decrease in EPSP amplitude during continuous sound stimulation and the emergence of responses upon increasing sound intensity were associated with a change in parameter n which characterized a readily available store of transmitter substance, but not of parameter p which characterized the mean probability of release and is independent of the stimulus intensity (Furukawa et al. 1972; Furukawa and Matsuura 1978).

The studies of synaptic transmission in the frog semicircular canals revealed the presence of small, tetrodotoxin-insensitive, depolarizing potentials in the afferents that summate and, after reaching the threshold, lead to the generation of tetrodotoxin-sensitive action potentials (Rossi et al. 1977).

The EPSPs were also recorded in the afferent fibers of electroreceptors in response to electric stimulation (Obara and Bennett 1972; Steinbach 1974; Obara 1976) (Fig. 8). Results showed that depolarizing generator potentials clearly represented EPSPs. Their generation was accompanied by an increase in the conductance of the postsynaptic membrane. Inhibitory stimuli were observed to cause a hyperpolarizing generator potential (IPSPs). Apparently they cannot be regarded as true IPSPs since they are not associated with the release of an inhibitory transmitter. They are, however, a consequence of repolarization of the postsynaptic membrane from its previous state of constant depolarization as a result of a decrease in excitatory transmitter release. During the hyperpolarization of postsynaptic potential no change of membrane conductance was observed.

In the electroreceptors the synaptic processes are characterized by a number of peculiarities. Consideration of the stimulus-response relation infers this. For instance, in the absence of stimulation the electroreceptors show a constant transmitter release, thereby giving rise to tonic impulse activity in the afferent nerve fibers. Stimuli below 1 mV, depending on the polarity, cause either complete

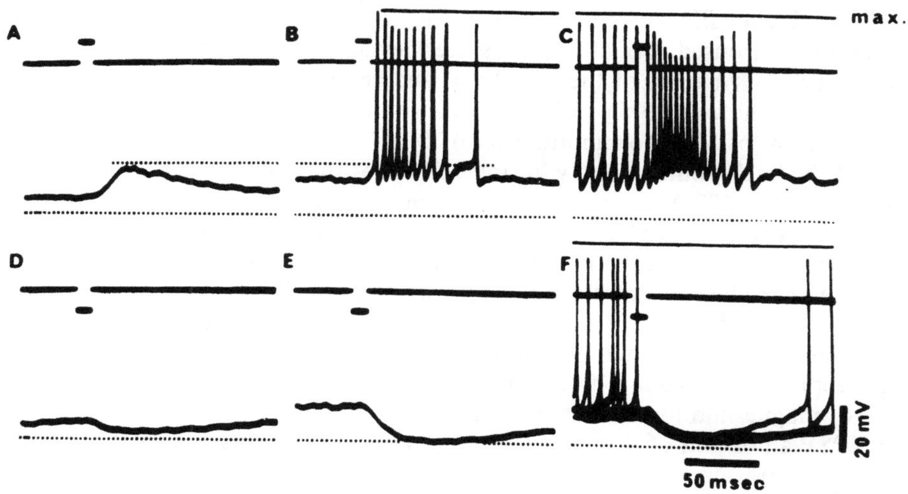

Fig. 8. Intracellular recordings from the afferent terminal (Obara 1976). Repeated short current pulses given to the ampulla induced a gradual membrane fluctuation as well as the depolarizing generator potential (GP), appearing either alone (*A*) or with facilitation of the afferent discharge (*B* and *C*). Note the time course of the subthreshold slow potential and also of the spike peak and the afterpotentials. The *line* marked by *max.* indicates the maximum height of the spike. The *upper dotted line* shows the firing level, the *lower one* the most negative membrane potential. *D–F*, The hyperpolarizing GP with concomitant afferent suppression. Several sweeps are superimposed in *F*. Note the distinct inflection at the onset of GPs of both polarities

inhibition or a sharp increase of spike activity. This undoubtedly coincides with the changes in transmitter release. The input-output relation in the electroreceptors appears to be considerably steeper than, for instance, with a squid giant synapse (Katz and Miledi 1967) or a spinal synapse of lampreys (Martin and Ringham 1975), and is shifted along the voltage axis toward more negative values of the potential, were the transmitter release occurs in the absence of stimulation. Furthermore, a larger, proportional change in output is obtained for a smaller, presynaptic potential.

No miniature postsynaptic potentials characteristic of other receptors of the acousticolateral system have been detected so far in electroreceptors. This is probably due to an extensive convergence of receptor cells on nerve fibers, which makes it impossible to record the contribution of individual synapses to the postsynaptic potential. In addition, it should be noted that the desensitization of postsynaptic receptors in tonic electroreceptors is poorly developed. Desensitization is important for ensuring a constant high receptor sensitivity given continuous transmitter release (Teeter and Bennett 1981; Akoev et al. 1980a).

5 The Effect of Ions on Synaptic Processes

The analysis of the effect that various ions have on the synaptic processes in the receptors of the acousticolateral system affords some insight into the peculiarities of signal transmission from the receptor cells to the afferent fibers. The effect of divalent ions aroused special interest since they are known to play an important role in various synapses of the peripheral and central nervous system.

Perfusion of the guinea pig scala tympani with solutions containing an elevated concentration of Mg^{2+} or an addition of Mn^{2+} or Co^{2+} produced a reversible drop of the background activity in single fibers of the auditory nerve (Robertson and Johnstone 1979). At the same time an increase in the concentration of calcium ions from 2.5 to 8 m*M* did not affect the background activity in the nerve fibers, while a complete removal of these ions led to a drastic decline in activity. An increase in the concentration of Mg^{2+} up to 10 m*M* caused a rise of receptor thresholds, the effect being antagonized by elevated concentration of Ca^{2+}. The background and evoked activities in single nerve fibers of the frog labyrinth were also suppressed in calcium-free solutions or in the presence of high concentrations of magnesium ions (Rossi et al. 1977).

The EPSP amplitude of the goldfish auditory nerve decreased in a solution with a high concentration of Mg^{2+} and increased as the concentration of Ca^{2+} rose (Furukawa and Matsuura 1978). According to these authors, the effects related to Ca^{2+} were caused by a rise of parameters n and p (see above).

There are a number of reports regarding the effects of divalent ions on the synaptic processes in electroreceptors. For example, a removal of calcium ions from a solution or an increase in the concentration of Mg^{2+} blocked the synaptic transmission in the ampullae of Lorenzini of skates (Steinbach 1974; Nier and Henzel 1973; Clusin and Bennett 1977a; Akoev et al. 1980b) and in the tuberous electroreceptors of the weakly electric freshwater mormyrid (Steinbach and Bennett 1971). A rise in the concentration of Mg^{2+} ions produced an increase in the threshold and a decrease in the amplitude of EPSP (Steinbach 1974), whereby Ca^{2+} is the Mg^{2+} antagonist in these processes. However, at a concentration of Mg^{2+} above 10 m*M* the changes in the concentration of Ca^{2+} became ineffective. La^{3+} induced the same effect as Mg^{2+}, only at lower concentrations. A decrease in the concentration of calcium ions in the solution reduced the EPSP amplitude, and in the calcium-free solution with 5 m*M* EGTA the synaptic processes were entirely blocked.

A more detailed study on the effect of divalent ions on synaptic transmission in the skate ampullae of Lorenzini was recently undertaken (Akoev et al. 1980b, 1981). It was found that the increase in the Mg^{2+} concentration in a solution up to 15 – 30 m*M* caused a gradual decrease in the frequency of background activity, followed by a complete suppression of the resting activity. Under these conditions the receptors did not respond to electric and temperature stimuli. Perfusion of the basal membranes with a Ca^{2+}-free solution plus 10 m*M* EGTA affected both evoked and resting activities. In the Ca^{2+}-free solution a three-phase change of receptor excitability was observed.

It is known that, on the one hand, a decrease in the Ca^{2+} concentration causes a reduction in the electrochemical gradient for these ions; this diminishes Ca^{2+} influx to the presynaptic membrane during depolarization and consequentially reduces transmitter release. On the other hand, a decrease in the concentration of calcium ions decreases the electrostatic screening of the fixed negative surface charges near the Ca^{2+} channels of the presynaptic membrane, which causes depolarization across the channels and increases the Ca^{2+} influx and the number of open calcium channels. Such a decrease in concentration of Ca^{2+} leads to an increase of the transmitter release. One may suppose that the phase of decrease in the excitability of ampullae of Lorenzini in a Ca^{2+}-free solution is due to the decrease in the electrochemical gradient for these ions, whereas the phase of excitability increase may be explained by an effective depolarization across the voltage-sensitive Ca^{2+} channels which become sensitive to small changes in membrane potential.

It is interesting to note that changes in the resting activity of the receptors in a Ca^{2+}-free solution occur much later than those in the evoked responses. These facts imply that other mechanisms based on the background and evoked activity of the receptors exist (Bernard 1983). Indeed, there are data which show that the background release of transmitters in various synapses depends on the concentration of calcium ions in the cytoplasm, whereas the mechanism of the evoked activity depends on the external concentration of Ca^{2+}. These two processes are characterized by different temperature sensitivity (Barrett et al. 1978).

Divalent ions play an important role in the synaptic transmission in other electroreceptors as well. For instance, the resting and evoked activity of the electroreceptors in the transparent catfish (*Kryptopterus*) disappeared rapidly in a solution with 15 – 30 m*M* Mg^{2+} (Teeter and Bennett 1981). In a solution with an elevated concentration of Ca^{2+} the thresholds of evoked responses increased, though the level of resting activity did not change. The activity of ampullary electroreceptors of sturgeons was also blocked in the solution with 10 m*M* Co^{2+} or Mg^{2+} (Teeter et al. 1980).

The available data on the role of divalent ions in synaptic transmission in the receptors of the acousticolateral system provide evidence that Ca^{2+} is essential for transmitter release processes. As in the case of other synapses there is every reason to assume that the basal membrane of the receptor cell has potential-dependent calcium channels which trigger transmitter release processes. Indeed, the experiments on the ampullae of Lorenzini in the marine catfish demonstrated the existence of oscillatory responses of receptor cells, which were blocked by perfusing the basal membrane with tetraethylammonium (TEA) and 4-aminopyridine (4-AP) (Sugawara and Obara 1984; Brown et al. 1985). These responses were also blocked by Co^{2+} and Cd^{2+}. The authors believe that the basal (presynaptic) membrane of electroreceptor cells has calcium-activated potassium channels which apparently make the initial phase of receptor adaptation possible. The presence of the calcium-activated potassium conductance is also necessary to ensure steady operation of the receptors with an extremely high sensitivity.

The existence of calcium channels was demonstrated in the vestibular hair cells of the chick (Ohmori 1984a, b). The calcium channels are located on the presynaptic membrane and initiate neurotransmitter release, whereas the transducer channels are located on the apical membrane, as in the case of electroreceptors (Clusin and Bennett 1977a, b; 1979a, b). It was shown that the properties of Ca^{2+} channels in the vestibular hair cells of the chick are similar to those found in the presynaptic axon of the squid giant synapse (Llians et al. 1981).

Numerous investigations show that the hair cells of the acousticolateral system are located in a medium with a high concentration of potassium ions (Ilyinsky 1975). According to Davis' theory, the increased concentration of potassium ions is responsible for a high receptor sensitivity and constant current flow between the apical and basal membranes of the hair cell. This current flows outward through the basal face. There is a depolarization tendency here which accounts for transmitter release during the resting phase (Davis 1965). The activation of a hair cell is accompanied by a change in the resistance of the apical membrane, in the value of current flow, and in the rate of neurotransmitter release. However, the experimental studies in recent years involving intracellular recording from the inner hair cells of the organ of Corti have demonstrated that during activation the uninterrupted flow of current remains constant over a wide range of sound stimulus intensities (Russel and Sellick 1978).

The rise in the concentration of potassium ions in the basal membrane region increases the frequency of the background activity in the hair cells of the organ of Corti (Leng and Comis 1979), the vestibular apparatus (Valli et al. 1977), the lateral line of fish (Hashimoto and Katsuki 1972), and the ampullae of Lorenzini of skates (Akoev et al. 1984a). However, although the background activity of the hair cells in the organ of Corti and the ampullae of Lorenzini in skates rose in a hyperpotassium medium, no changes in the threshold were observed. This contradicts the basic postulate of Davis' theory (Leng and Comis 1979; Akoev et al. 1984a). According to Davis, there is no practical difference in the mechanisms of spontaneous and evoked releases of the transmitter. However, at present it is acknowledged that the spontaneous and the evoked release of the transmitter are based on different mechanisms (Bernard 1983). In the first case the rate of transmitter release is determined by the intracellular concentration of Ca^{2+} and in the second by the entry of Ca^{2+} during depolarization.

6 Pharmacology of Synaptic Transmission

6.1 Glutamate and Aspartate

Lately great interest has been focused on the idea that glutamic and aspartic amino acids (GLU and ASP) possibly play a role in synaptic transmission in the receptors of the acousticolateral system. These substances are known to act as neurotransmitters in a number of synapses in the peripheral and central nervous systems of vertebrates and invertebrates (Gerschenfeld 1973; Krnjevic 1974).

Investigations on neurons of the central nervous system demonstrated the existence of at least three types of postsynaptic receptors for which one of the three amino acid agonists kainate (KA), quisqualate (Q), or *N*-methyl-D-aspartate (NMDA) was potent and selective (McLennan and Lodge 1979; Watkins 1978, 1981; Watkins and Evans 1981; Foster and Fagg 1984).

The role of amino acids in synaptic transmission was most thoroughly studied on hair cells of the organ of Corti. For instance, electrophysiological experiments demonstrated that L-GLU and L-ASP induced into the scala tympani in millimolar concentrations reduced the compound action potential (AP), whereby the cochlear microphonic (CM) was not influenced. Single auditory afferents were activated upon application of these substances, and as a result consequent sound-evoked increases in discharge rates were smaller than normal (Klinke and Oertel 1977; Bobbin 1979; Bobbin and Thompson 1978; Comis and Leng 1979). Godfrey et al. (1976a, b) found high levels of GLU and ASP in both the sensory and the supporting cells of the organ of Corti.

The studies on the receptors of the guinea pig cochlea showed that KA excited the primary afferent nerve fibers and that the effect was due to the influence on the postsynaptic receptors (Bledsoe et al. 1981). Another agonist of excitatory amino acid receptors – Q – had fundamentally the same effect, while NMDA exerted no effect either on the AP or on the CM (Jenison and Bobbin 1983; Jenison et al. 1986). From these experiments it follows that Q is about two orders of magnitude as effective as L-GLU and fourfold as effective as KA. This gave rise to the assumption that Q and KA postsynaptic receptors in mammalian cochlea and NMDA does not.

The results of experiments in which the effect of the antagonists of amino acid receptors was studied also favor this hypothesis (Fex and Martin 1980; Bobbin et al. 1981a). It was shown that the antagonist of NMDA receptors, D-, L-α-aminoadipate (DLAA) produced no effect on the AP and the CM. *Cis*-2,3-piperidine dicarboxylate (PDA) – an antagonist of Q and KA receptors – reduced the AP and had no effect on the CM. Glutamate diethyl ester (GDEE) – antagonist of amino acid receptors (Haldeman and McLennan 1972) – reduced the amplitude of the AP by 80% and that of the CM by 60%. The data presented are consistent with current concepts regarding the existence of the Q-type receptors in the mammalian cochlea (Bobbin et al. 1984).

In several investigations an attempt was made to identify the putative neurotransmitter by means of a perilymph analysis during acoustic stimulation of cochlear hair cells. For example, it was shown that the perilymph of animals exposed to acoustic noise excited the afferent auditory units of nonexposed animals. However, the concentrations of both GLU and ASP did not differ from the levels in perilymph collected during silence (Sewell et al. 1978). These data have also been confirmed by other authors (Medina and Drescher 1981), although their studies involved the analysis of 14 primary amines, including GLU, ASP, taurine, and glycine. The perfusion of the cochlea with artificial perilymph before and after acoustic stimulation revealed no changes in the concentrations of GLU, ASP or other amino acids. Melamed et al. (1982) reported similar results for an analysis of perilymph amines, including γ-aminobutyric acid (GABA).

Interesting data have been obtained in studies on the uptake of a number of labeled amino acids which are putative neurotransmitters in hair cells of the organ of Corti and the vestibular apparatus of gerbils (Ryan and Schwartz 1984). It was shown that proline, alanine, and glycine were slightly accumulated by the inner hair cells, while L-GLU, L- and D-ASP, GABA, histidine, ornithine, arginine, methionine, taurine, and choline showed no preferential accumulation in the hair cells as compared with the supporting cells of the organ of Corti. Of the substances investigated glutamine accumulated predominantly in the hair cells of the organ of Corti and to a greater extent in the inner hair cells than in the outer ones. In contrast to the cochlear hair cells no accumulation of labeled glutamine was observed in the hair cells of the vestibular apparatus. These data indicate that the hair cells of the organ of Corti possess a high-affinity uptake system for this amino acid. It is unlikely that glutamine itself is the neurotransmitter; it is more likely a precursor for the synthesis of the GLU, this reaction being activated by glutaminase. For instance, it is known that in the central nervous system, where the involvement of GLU in synaptic transmission has been experimentally substantiated, GLU is converted into glutamine by the glial cells, thereby terminating GLU action. In this context the existence of the highly specific system of glutamine uptake is quite understandable. It gives rise to a pool of neurotransmitter precursor and, on the other hand, is designated to clean the synaptic cleft.

Recently a detailed investigation was undertaken to study the content of 81 primary amines in the perilymph of the guinea pig inner ear which had been exposed to noise (Drescher et al. 1983). Of all the possible neurotransmitters the following substances were characterized with the greatest concentration: taurine (13.3 μM), GLU (3.6 μM), ASP (0.8 μM), glycine (40 μM), GABA (0.14 μM), and methionine-enkephalin (4.3 μM). The substances were collected from control animals in silence. Sound stimulation was accompanied by an increase in the concentration of GABA, ASP, and a methionine-enkephalin-like component. There was no evidence that the compounds GLU, taurine, and glycine are likely to be afferent neurotransmitters. The authors came to the conclusion that GABA is an afferent transmitter in the hair cells, since its concentration closely correlates with the sound exposure values.

An alternative technique of afferent neurotransmitter release stimulation was applied by Jenison et al. (1985) using solutions with a high concentration of potassium ions. The perfusion of the guinea pig cochlea with a highly concentrated potassium solution (50 mM) revealed a Ca^{2+}-dependent increase in the concentration of GLU, taurine, and a coeluting taurine-GABA fraction. The results obtained in the above study provided the evidence that presynaptic structures of hair cells can release GLU and taurine in a calcium-dependent manner. This conclusion seems questionable since both GLU and taurine could have been released from the synaptic terminals of efferent fibers or supporting cells of the organ of Corti.

In the hair cells of the frog vestibular apparatus GLU and ASP increased the frequency of background activity, as it did in the receptors of the organ of Corti (Annoni et al. 1984). The threshold for activation is 0.1 – 0.2 mM. NMDA is relatively more potent; KA and Q are considerably more potent in their ability to activate (Fig. 9). The authors believe that GLU and its agonists induced a reversible depolarization of vestibular afferents since it persists in a high Mg^{2+} solution. The increase in the frequency of discharges in the afferent fibers persisted throughout the whole period of agonist application, demonstrating the lack of postsynaptic receptor desensitization.

Other authors (Valli et al. 1985) have observed the excitatory effect of L-GLU on the receptors of the frog semicircular canal too. However, their experiments failed to demonstrate the restoration of impulse activity upon perfusion with GLU when the release of a natural transmitter is prevented by a high Mg^{2+} solution. Under these conditions GLU only caused a long-lasting depolarization in the afferent nerve fibers. The authors proposed that GLU had at least two sites of action, a presynaptic and a postsynaptic one. The presynaptic action is responsible for an increased release of the natural transmitter, which in turn produces a substantial increase in the firing frequency in afferent fibers. This GLU effect is antagonized by DAA. On the other hand GLU produced a long-lasting

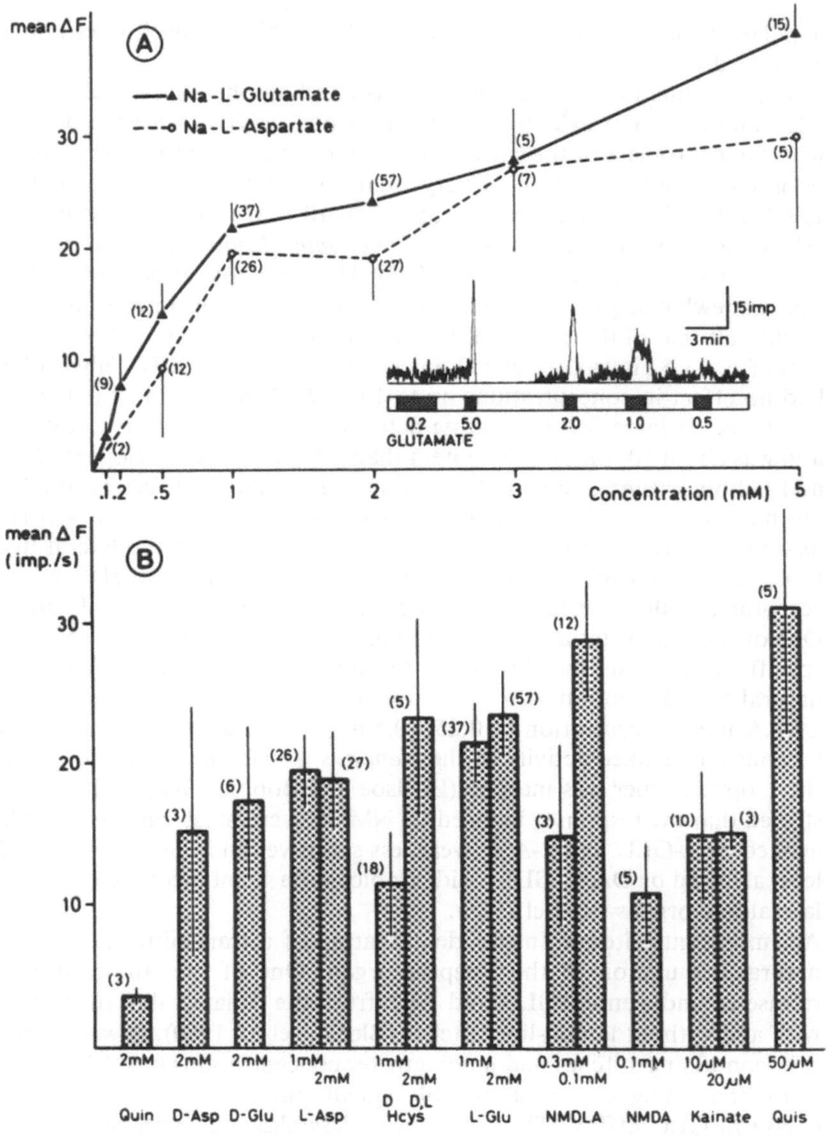

Fig. 9A, B. Dose-response characteristics of acidic amino acids on afferent excitation, recorded extracellularly (Annoni et al. 1984). **A** Dose-response curves to glutamate (*solid triangles*) and aspartate (*open circles*). The numbers in *parentheses* indicate the number of afferents tested at a given concentration. *Vertical bars* represent SEM. *Inset*, Responses of one cell to given concentrations of glutamate (in millimolar concentration). In this case 5 m*M* glutamate produced depolarization inactivation of the afferent. **B** Responses of afferents to given concentration of acidic amino acids. The numbers in *parentheses* represent the number of afferents tested at a given concentration. *Vertical bars* represent SEM. *Δ* F represents the peak frequency attained during drug application minus the resting firing frequency of the axon prior to drug administration. *Hcys*, homocysteic acid; *Glu*, glutamate; *Asp*, aspartate; *NMDA*, *N*-methyl-D-aspartic acid; *Quin*, quinolinic acid; *Kainate*, kainic acid; *Quis*, quisqualic acid

depolarization of postsynaptic afferent fibers which by itself does not cause any afferent dischargers.

The lateral-line mechanoreceptors in fish and amphibians are also activated by GLU and ASP. For example, the spontaneous activity of nerve fibers associated with the neuromast of the catfish lateral line increased upon perfusion with solutions containing L-GLU in a very high concentration up to 50 m*M* (Katsuki 1973). L-GLU in a concentration of $10^{-2}-10^{-3}$ *M* produced a strong excitatory effect on the neuromasts of the *Xenopus laevis* (Bobbin et al. 1981b; Zimmerman 1979; Bledsoe et al. 1983). However, the effect induced by L-ASP was somewhat higher than that of L-GLU, while the potency of D-GLU was about half that of the L form. KA in a concentration 5 – 20 μ*M* exerted an excitatory effect, i.e., it was about 250 times as effective as L-GLU. Dihydrokainic acid had no effect in concentrations up to 10 m*M*. Q was the most potent excitatory substance. It increased the resting activity at a concentration as low as 10^{-5} *M* and was about 10^3 times more potent than L-GLU. Other analogues of L-GLU, L- and D-homocysteate, also induced a strong excitatory effect on the lateral-line mechanoreceptors; the L form was about 10 times as effective as L-GLU. NMDA also influenced receptor activity. Thus, the hair cells of the lateral line, unlike those of the cochlea (except for Q and KA receptors), also have NMDA receptors, while the effects induced by the L and D forms of GLU are similar.

One of the important approaches to neurotransmitter identification is the use of specific antagonists that block the responses of the afferent fibers not only to natural stimulation but also to the application of the putative neurotransmitter. DLAA in a concentration of 0.25 – 0.5 m*M* reduced the frequency of the spontaneous and evoked activity of the *Xenopus* lateral-line sensory organs, whereas the L optic isomer was inactive (Bledsoe and Bobbin 1982). It has been demonstrated that the responses induced by NMDA were most sensitive to DAA, those evoked by L-GLU and L-ASP were less sensitive, and the responses to KA were least affected by DAA. GDEE did not block the spontaneous activity in *Xenopus* lateral-line organs (Russel 1976).

An important criterion in the identification of a transmitter is its release upon natural stimulation of the receptor organ. One of the studies dealt with the release of endogenous GLU and ASP from the isolated skin of *Xenopus laevis* with and without lateral-line stitches (Bledsoe et al. 1980). It was found that the skin containing stitches had much greater concentrations of GLU and ASP than those containing no receptors. Subsequent stimulation caused a significant increase in GLU efflux. The results were less clear cut with respect to the ASP efflux. These results support the idea that GLU may be the afferent transmitter in the lateral-line organs of *Xenopus laevis*.

The afferent transmitter in the electroreceptors still remains to be identified. The most probable substances are L-GLU and ASP. Indeed, a 10^{-5} *M* concentration of L-GLU induced a frequency rise in the spontaneous activity and a decrease in the thresholds in response to electrical stimulation of mormyromasts (Steinbach and Bennett 1971).

Evidence is also available that the ampullary receptors of transparent catfish (*Kryptopterus*) are excited by bath application of L-GLU in a concentration of 1 μ*M*. This process was accompanied by an enhanced sensitivity to electrical stimulation (Teeter and Bennett 1981). This effect may have been due to the summation of the actions of endogenous neurotransmitters and exogenous L-GLU. At high concentrations ($10^{-3} M$) L-GLU provoked a strong excitatory effect followed by the activity block; this is apparently due to an excessive depolarization of the afferent fiber membrane. It is interesting to note that L-GLU restored the receptor resting activity which was blocked by high Mg^{2+} concentration. Thus, it is very probable that L-GLU does act on the postsynaptic membrane.

It has been demonstrated that L-GLU in a concentration of 10^{-9} – $10^{-4} M$ produces an excitatory effect on the ampullary electroreceptors of marine catfish (*Plotosus*) (Higuchi et al. 1980). Moreover, the observed effects persisted in solution with an increased Mg^{2+} concentration (30 m*M*). Iontophoretic application of L-GLU to the postsynaptic membrane of nerve fibers also produced an excitatory effect (Nagai and Obara 1985). However, KA and Q were the most potent, while L-ASP induced only a small long-latency response. D-GLU, D-ASP, L-cysteate, L-cysteine sulfonate, and NMDLA usually produced no effect on the responses of the postsynaptic membrane. At the same time some analogues of L-ASP, D-ASP, L-cysteate and L-cysteine sulfonate induced persistent potentiation of GLU responses. The antagonists of amino acid receptors – GDEE and DLAA – had no effect on either the resting activity or GLU-induced responses.

On the other hand there are data contradicting the involvement of GLU in synaptic processes in the electroreceptors of marine catfish (*Plotosus*). For instance, Joro spider toxin (JSTX), which blocks the postsynaptic potentials and responses to the application of L-GLU, KA, and Q in the lobster neuromuscular synapse (Kawai et al. 1982), blocked the excitatory effects of externally applied L-GLU in the electroreceptors of *Plotosus* (Nagai et al. 1984). However, resting activity and electrically evoked activity were not blocked at the same time. The analysis demonstrated that the electric stimulation of the receptors led to the release of the endogenous neurotransmitter into the solution. This fluid was able to activate other ampullae, however, the authors, using thin-layer chromatography (sensitivity $10^{-7} M$), failed to detect L-GLU in it (Umekita et al. 1980). The results of the study of neurotransmitters in the ampullary electroreceptors of *Plotosus* provide pharmacological evidence against the GLU hypothesis. According to Nagai and Obara (1985) the excitatory effect of GLU and its agonists is probably due to interaction with the extrasynaptic receptors of unmyelinated nerve terminals.

We believe, however, that the results cited above may be explained by the GLU hypothesis as well. One possible mechanism underlying the effect of JSTX may be its ability to block only open GLU channels. In this case its blocking efficacy could depend on a number of open GLU channels. Therefore, use-dependence experiments are necessary to clarify this problem.

The effect of L-GLU and L-ASP on the synaptic processes in the electroreceptors (ampullae of Lorenzini) of skates has been studied in detail (Steinbach 1974; Akoev et al. 1980a, 1981, 1984b, 1987a; Akoev and Adrianov 1981). L-GLU and L-ASP in concentrations of 10^{-4}–$10^{-7} M$ induced depolarization of the postsynaptic membrane and increased the frequency of the background and evoked activity of the receptors in a dose-dependent manner in normal solution (Fig. 10, 11) when synaptic transmission was blocked by a high magnesium solution (Fig. 13). The effect of amino acids was inhibitory in only about 10% of the cases and

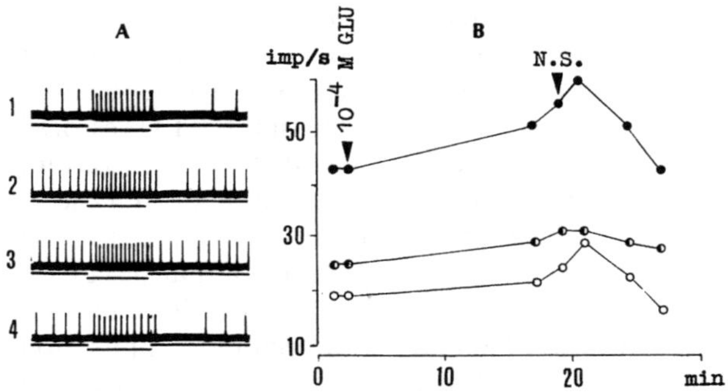

Fig. 10 A, B. Effect of bath application of L-GLU on the background and evoked activity of individual afferent fibers of the ampullae of Lorenzini (unpublished data). **A** *1*, Response to electrical stimulation in normal solution; *2* and *3*, responses in 10^{-6} *M* L-GLU (4 and 6 min respectively); *4*, recovery in normal solution (20 min). **B** Dynamics of frequency changes of background activity (*solid circles*), evoked activity (*open circles*), and the frequency of evoked activity attained during L-GLU application minus the background firing frequency of the ampullae of Lorenzini prior to the drug administration (*solid* and *open circles*). Intensity of electrical pulses, 1.5 μA; stimulus duration, 250 ms

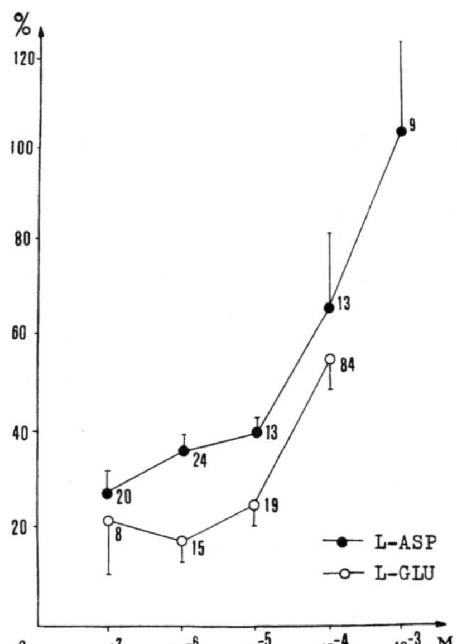

Fig. 11. Dose-response characteristics of acidic amino acids on afferent excitation of the ampullae of Lorenzini (Akoev et al. 1987a). Dose-response curves to L-ASP (*solid circles*) and L-GLU (*open circles*). The *numbers* at the points indicate the number of afferents tested at a given concentration. *Vertical bars* represent SEM. *Ordinate*, The frequency after 3-min application of amino acids, in percent

was observed in receptors with a relatively high level of background activity (usually more than 25/s). The inhibitory effect of L-GLU was also noted in the electroreceptors of *Plotosus* (Higuchi et al. 1980) and may have been due to the pre- and postsynaptic action of exogenous amino acids on the ampullae of Lorenzini. Nevertheless, there is not doubt that L-GLU and L-ASP mainly exert postsynaptic influence on the terminals of afferent fibers innervating the ampullae of Lorenzini.

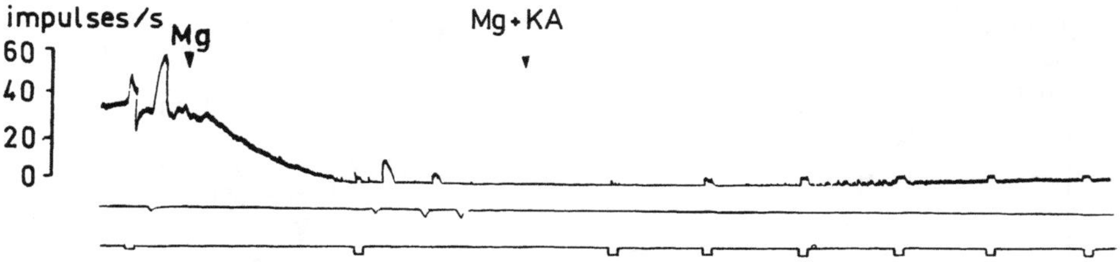

Fig. 12. Effects of 10^{-6} *M* KA on the resting and evoked firing rate of the ampullae of Lorenzini in Mg^{2+}-rich solution (Akoev et al. 1986a). *Upper beam,* Firing frequency, imp/s; *middle beam,* temperature stimulus; *lower beam,* stimulus current. Intensity of applied currents in normal solution was 0.15 and 1.5 μA in high Mg^{2+} solution. Duration of electrical stimuli, 10 s. Intensity (cooling) of applied temperatures: 1.5 °C in normal solution and 3 °C in high Mg^{2+} solution

It is well known that KA is a highly potent amino acid agonist in those synapses where GLU seems to be involved in transmission (Coyle 1983). Experiments on the ampullae of Lorenzini of skate (Akoev et al. 1986a, 1987c) established that the bath application of KA produced significant changes in the resting firing. The effect of KA was clearly correlated with the initial level of background activity. In the units with initial rates from 8 to 20/s KA provoked an increase in activity, followed by a decrease, while in the units with higher levels of resting activity (22 – 40/s) KA caused a reduction in firing frequency. The threshold concentration of KA was as low as 10^{-9} *M*. The essential difference between the effect of KA and that of L-GLU and L-ASP was that KA restored both the resting and evoked activities in a highly concentrated magnesium solution (Fig. 12). This implies that KA may interact with GLU receptors located on the presynaptic membranes of the ampullae of Lorenzini and selectively cause the release of a natural transmitter. Recent studies on brain synaptosomes showed that KA interacts with GLU receptors in the presynaptic membrane and somehow raises the permeability for calcium ions, thereby activating transmitter release processes (Pastuszko et al. 1984). KA stimulates only the release of GLU and ASP, unlike a highly concentrated potassium solution which activates the release of either GABA or other amino acids as well.

Q and NMDA elicited a transient and dose-dependent increase in the firing rate of the ampullae of Lorenzini. The responses of NMDA were abolished by highly concentrated magnesium solution. The threshold concentration for Q was 10^{-8} *M* and for NMDA 10^{-5} *M*. Thus, according to our findings the relative potency of excitatory amino acid agonists is: KA > Q > NMDA.

A wide variety of antagonists of amino acid receptors blocked synaptic transmission in the ampullae of Lorenzini. Among them GDEE, 2-amino-5-phosphonovaleric acid (APV), kynurenic acid (KENYA), PDA, and D-glutamyl glycine (DGG) were significantly more potent in blocking the synapses than DAA, 2-amino-4-phosphonobutyric acid (APB), or glutamic acid dimethylester (GDME). The influence of GDEE (threshold concentration 10^{-6} *M*) supports the conclusion regarding the postsynaptic effects of amino acids (Akoev et al. 1981, 1987b; Akoev and Andrianov 1981). This agent blocked both the background and evoked activity of the ampullae of Lorenzini and antagonized the post-

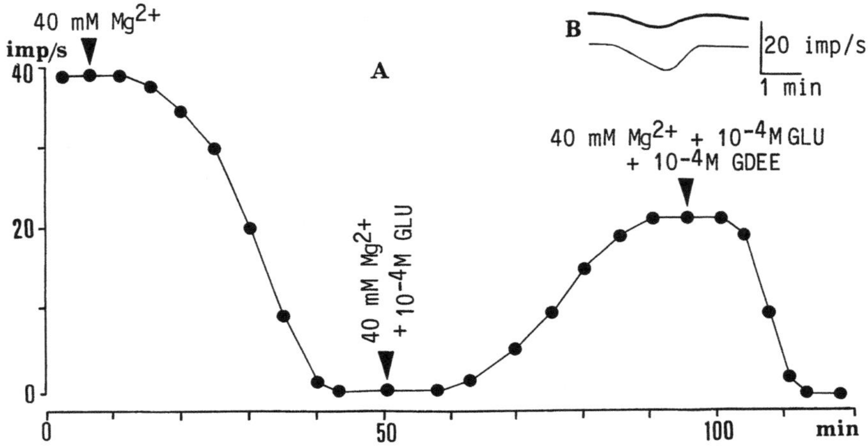

Fig. 13. A Effect of saline containing 40 mM Mg^{2+}, 10^{-4} M L-GLU and 10^{-4} M GDEE on the firing rate of an afferent fiber of the ampullae of Lorenzini (unpublished data). Effect of GDEE on depolarization due to L-GLU. **B** Inverted thermal response of the postsynaptic afferent fiber of the ampullae of Lorenzini in saline containing 40 mM Mg^{+} plus 10^{-4} M L-GLU. *Upper beam*, Firing frequency, imp/s; *lower beam*, temperature stimulus. Intensity of applied temperature (cooling), 3 °C

synaptic amino acid induced depolarization as well (Fig. 13). An example of the actions of PDA is depicted in Fig. 14. This substance blocked the background and evoked activity and also reversibly blocked the postsynaptic amino acid-induced responses. L-GLU was more resistant to blockade than L-ASP. The opposite effect was observed in the case of KENYA.

Recent evidence suggests that several different types of receptors exist for excitatory amino acids in the neurons of the central nervous system where numerous synaptic inputs are received. With respect to the ampullae of Lorenzini, morphological findings indicate that afferent nerve fibers synapse with only one type of sensory cell; hence, there is no reason to believe in the existence of several types of postsynaptic receptors for excitatory amino acids. If this is the case, we may suppose that there is a single receptor channel complex in the postsynaptic membrane of the ampullae of Lorenzini with several separate binding sites for various agonists. Additional experiments are necessary to clarify this problem.

Glutamate binding activity of plasma membranes of the ampullae of Lorenzini was measured by means of a filtration method employing Millipore filters with L-[^{3}H]GLU (Akoev et al. 1986b). It was found that membrane preparations exhibit predominantly one class of binding sites with a dissociation constant (K_d) of 0.2 μM and corresponding binding site densities (B_{max}) of 30 pmol/mg protein. The K_d and B_{max} values determined in these experiments correspond well to the constants determined for the mammalian brain (Foster and Fagg 1984). The results suggest that L-GLU membrane receptors are present on plasma membranes of the ampullae of Lorenzini. These findings are consistent with those regarding L-GLU or a related substance as a putative neurotransmitter for the electroreceptor synapses of the ampullae of Lorenzini.

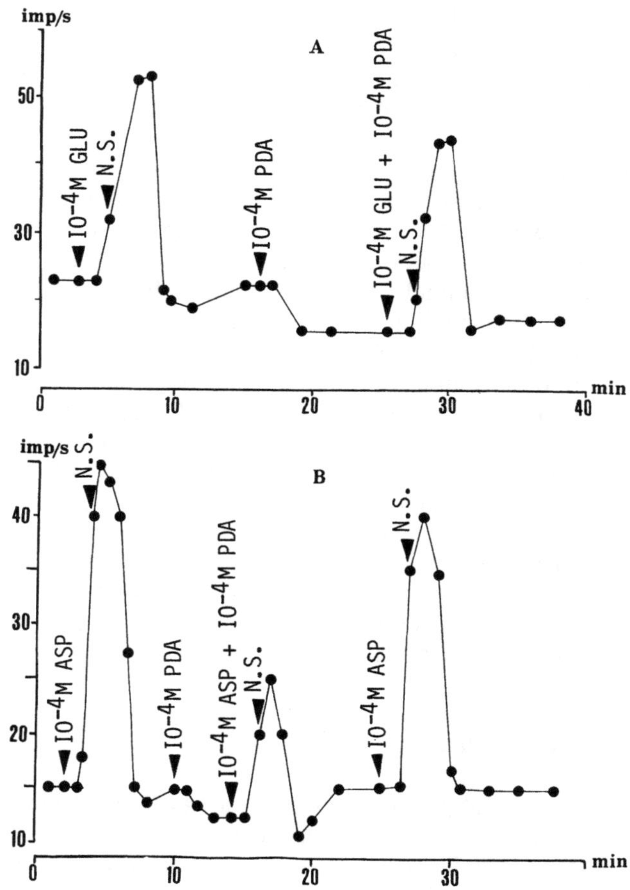

Fig. 14 A, B. Effect of 10^{-4} *M* PDA on depolarization due to L-GLU (**A**) and L-ASP (**B**) in the same afferent fiber of the ampullae of Lorenzini (unpublished data)

6.2 Catecholamines

A possible role of catecholamines as afferent neurotransmitters in the receptors of the acousticolateral system has been studied in a number of investigations. Experiments on single fibers of the auditory nerve of cats (Bobbin and Thompson 1978; Klinke and Evans 1977) and guinea pigs (Comis and Leng 1979) showed that perfusion of the scala tympani with catecholamines in a concentration of $10^{-4}-10^{-5}$ *M* either reduced the frequency of the resting activity in the receptors or produced no effect. The amplitude of the cochlea AP of the guinea pig did not change under the influence of catecholamines (Bobbin and Thompson 1978). α-and β-receptor blocking agents (phentolamine, propranolol) had no effect on the resting and evoked receptor activities. Epinephrine (En) and norepinephrine (NorEn) ($10^{-3}-10^{-7}$ *M*) were equally ineffective with respect to the resting activity of the *Xenopus laevis* neuromast (Russell 1976).
Studies on the ampullae of Lorenzini of skates demonstrated that catecholamines reduced the frequency of the resting activity of the receptors and that adrenoreceptor antagonists did not block the receptor responses to electrical stimuli.

Besides, phentolamine, the α-adrenoreceptor antagonist, blocked En effects (Akoev et al. 1985). No obvious effects were produced by catecholamines on electroreceptors of the marine catfish *Plotosus* (Bennett and Obara 1986). Comis and Leng (1979) reported a depression of resting activity with $3 \times 10^{-4} M$ NorEn.

Histochemical fluorescent methods failed to detect catecholamines in the afferent synapses of hair cells of the organ of Corti of the cat or rabbit (Hilding 1964; Fex et al. 1965; Densert 1974). Application of reserpine or 5,6-hydroxydopamine, substances known to produce a depletion of adrenergic vesicles in the presynaptic terminals, to the afferent endings exerted no significant effects on synaptic transmission and the ultrastructural characteristics of the afferent synapse (Borg et al. 1974).

On the other hand, a number of authors detected changes in the diameter and density of the hair cell synaptic bars of the frog vestibular epithelium and lateral-line organs after treatment with reserpine or monoamine oxidase (Thornhill 1972; Osborne and Thornhill 1972; Monagham 1975). These results seem to support the idea of possible catecholamine involvement in synaptic transmission. However, experiments with radioactive precursors failed to detect the synthesis of NorEn and dopamine in the hair cells of the frog vestibular apparatus and fish lateral line (Flock and Lam 1974). Taken together, these findings do not support the choice of catecholamines as the transmitter substances used by afferent synapses of the acousticolateral system.

6.3 Acetylcholine

The question of possible acetylcholine (ACh) involvement in synaptic transmission in the receptors of the acousticolateral system has long been raised. Earlier publications indicated that ACh can function as an afferent transmitter in the organ of Corti (Gisselsson 1950). This assumption was based on the fact that acetylcholinesterase blocking agents caused an increase in the latency of the AP. However, more detailed investigations revealed that acetylcholinesterase blocking agents did not influence the latency of the AP and the CM (Gannon et al. 1966). The involvement of ACh in the chemical transmission in hair cells of the organ of Corti is also supported by facts such as the presence of ACh in endolymph and an increase in acetylcholinesterase activity and ACh concentration in the endolymph upon sound stimulation of the receptors (Vinnikov and Titova 1964). Other investigators were unable to observe the release of ACh during sound stimulation (Gisselsson 1950; Norris and Guth 1974).

It was assumed that the presence of acetylcholinesterase in the cochlea (Churchill et al. 1956) was usually associated with the terminals of the olivocochlear bundle rather than with afferent synapses (Iurato et al. 1971). Thus, after the section of the crossed and uncrossed olivocochlear bundles the activity of the enzyme associated with ACh synthesis (acetylcholintransferase) was greatly reduced or destroyed (Jasser and Guth 1973). This points to the fact that the structures which remained after degeneration, including hair cells, were unable to synthesize ACh and that the presence of the enzyme was associated with the efferent synapses. This assumption is also supported by the analysis of the distribution of acetylcholintransferase and acetylcholinesterase (Fex and Wenthold 1976).

Flock and Lam (1974) demonstrated by means of labeled precursors of ACh synthesis (methyl-[^{3}H]choline chloride) that those organs (bullfrog basilar papilla, the crista ampullaris of the skate) that have efferent innervation were capable of synthesizing ACh. In constrast, the frog basilar papilla, which has no efferent innervation, was not able to synthesize ACh. Hence, the biochemical evidence supports the hypothesis that the efferents are cholinergically mediated.

Electrophysiological experiments on the organ of Corti indicated that ACh in a concentration 50 µ*M* caused a reduction in the background and evoked activity of single fibers of the auditory nerve (Comis and Leng 1979). Important evidence speaks against the involvement of ACh in the afferent synaptic transmission in the organ of Corti (Galley et al. 1973; Bobbin and Konishi 1974) and in the vestibular labyrinth of the frog (Rossi et al. 1980; Annoni et al. 1984): Cholinolytic agents (atropine, *d*-tubocurarine, and curare) produced no effect on the synaptic processes, while ACh itself did not induce firing in the afferent fibers after the release of the transmitter had been blocked by perfusion with a Mg^{2+}-rich solution. In the mechanoreceptors of the *Xenopus laevis* lateral-line organs ACh and carbachol caused a concentration-dependent increase in the background and evoked activity (Winberry and Bobbin 1983; Bobbin et al. 1985). Atropine, the antagonist of the ACh receptor, blocked the excitatory effects of ACh and carbachol and reduced the frequency of the background activity in the nerve fibers but had no effect on evoked activity. The data obtained indicate that the excitatory effect of ACh on the background activity may be due to the activation of specific cholinergic receptors, but these receptors are likely to be located on the presynaptic membrane.

There are a number of reports concerned with the effect of ACh on synaptic transmission in electroreceptors. It was shown that neither ACh nor *d*-tubo-

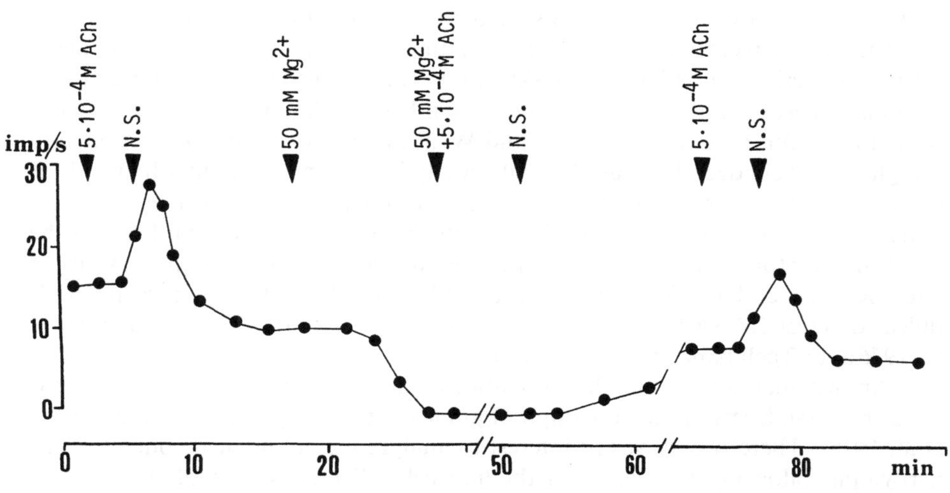

Fig. 15. Effect of saline containing 5×10^{-4} *M* ACh on the resting firing rate of an afferent fiber of the ampullae of Lorenzini in Mg^{2+}-rich solution. (Akoev et al. 1983)

curarine influenced the spontaneous activity of electroreceptors in transparent catfish (Teeter and Bennett 1981) and the marine catfish *Plotosus* (Bennett and Obara 1986). The effect of ACh and cholinolytics on the synaptic processes in the ampullae of Lorenzini of skates has also been investigated (Akoev et al. 1983). ACh exerted a strong excitatory effect on the ampullae of Lorenzini; however, these effects were not related to its action on the postsynaptic membrane, since when the synaptic transmission was blocked with a Mg^{2+}-rich solution, ACh did not induce impulse activity in the afferent fibers (Fig. 15). This points to the fact that ACh has no effect on the permeability of ionic channels of the postsynaptic membrane. Therefore, the observed effects of ACh on the resting activity of the ampullae of Lorenzini can be explained by a presynaptic action of ACh. One may suppose that perfusion with a solution containing ACh results in depolarization of the presynaptic membrane and thus the amount of endogenous transmitter release increases.

The cholinolytic agents atropine and *d*-tubocurarine did not block the synaptic transmission (Akoev et al. 1983) and had no influence on the amplitude of EPSP in the Lorenzinian ampullae (Steinbach 1974). So, ACh is unlikely to be the afferent transmitter in receptors of the acousticolateral system.

6.4 γ-Aminobutyric Acid

Numerous investigations have been devoted to the study of GABA in synaptic transmission in the receptors of the acousticolateral system. The idea of GABA as an acousticolateral sensory transmitter has both opponents and supporters.

On the basis of biochemical analysis, Flock and Lam (1974) suggested that GABA is the neurotransmitter at the afferent synapse. They showed that bullfrog basilar papilla, the lateral-line canal organ of the toadfish, and the crista ampullaris of the semicircular canal of the skate were capable of synthesizing [^{3}H]GABA from [^{3}H]GLU. Moreover, they demonstrated the presence of glutamate decarboxylase activity in the above-mentioned sensory organs. Supporting this idea the authors also found that picrotoxin blocked both resting and evoked activity in the nerve fibers of the crista ampullaris of the skate. Lately it has also been shown that chick vestibular sensory epithelium is capable of synthesizing GABA (Meza 1984; Meza et al. 1982). On the other hand, Fex and Wenthold's studies (1976) revealed very low glutamic acid decarboxylase (GAD) activity in the inner and outer hair cells of the cochlea. This evidence opposes the idea that GABA is a likely afferent transmitter in the organ of Corti. Besides, it was found that GABA synthesis in the cochlear receptors is associated with the efferent synapses rather than the afferent ones (Roberts and Ryan 1971; Schwartz and Ryan 1983). Biochemical methods failed to detect GABA in the inner and outer hair cells of the cochlea (Godfrey et al. 1976a, b; Tachibana and Kuriyama 1974).

It is known that sound stimulation causes an increase in the firing rate of the auditory nerve fibers in the perilymph of guinea pigs or frogs exposed to it (Sewell et al. 1978). Therefore, it was of interest to analyze the amino acid content of the perilymph following stimulation of the hair cells. The activation of the cochlear receptors of the guinea pig induced a significant increase in the concentration of the GABA-like component, ASP, methionine enkephalin, and two unknown derivates labelled "A" and "B" in the perilymph (Drescher et al. 1983). These experiments do not exclude the possibility that GABA is released into the perilymph by cochlear efferent endings. At the same time no increase was observed in the concentration of other putative neurotransmitters (GLU, glycine, taurine) in the perilymph (Drescher et al. 1983; Medina and Drescher 1981). It is relevant to note here that the GLU concentration cannot increase in the perilymph if it is actively eliminated from the synaptic cleft through uptake by the presynaptic endings and that GABA accumulation can be associated with decarboxylation of the released GLU.

The electrophysiological data on the effect of GABA and its antagonists on the receptors of the acousticolateral system are also contradictory. Some authors noted an excitatory effect of GABA on the cat sacculus receptors and that GABA antagonists blocked both the background and evoked activity (Flock and Lam 1974; Felix and Ehrenberger 1977, 1982). Other researchers have generally found that GABA and its antagonists exerted no significant effect on the impulse activity of nerve fibers innervating the receptors of the acousticolateral system; they found no evidence that these effects were inhibitory in nature. For instance, the injection of GABA and bicuculine in concentrations of 10 mM and 1 mM respectively had no effect on the AP and the CM of the guinea pig (Bobbin and Thompson 1978). Picrotoxin, a GABA antagonist, did not block the evoked activity in the auditory nerve or affect the amplitude of the AP of the cat auditory nerve (Desmendt and Monako 1962; Bobbin and Guth 1970; Klinke and Oertel 1977). In a concentration of $10^{-3}-10^{-7}$ M GABA had no influence on the spontaneous activity of the clawed frog neuromasts (Russell 1976). In turn, thiosemicarbazide and 3-mercaptopropionic acid, which are known to inhibit glutamic acid decarboxylase and GABA synthesis, did not change the parameters of the AP of the cat auditory nerve or the ultrastructure of the afferent synapse of the African clawed toad (Desmendt and Monaco 1962; Bobbin and Guth 1970). In experiments with hair cells of the vestibular system of the bullfrog GABA and picrotoxin had no effect on the impulse activity in the afferent fibers (Guth and Norris 1984). When the synaptic transmission was blocked with high Mg^{2+} and low Ca^{2+} concentrations, the addition of GABA to the solution, unlike ASP and GLU, did not change the resting membrane potential of the vestibular afferent fiber in the frog (Annoni et al. 1984). In addition, bicuculine and picrotoxin did not block the synaptic transmission.

A number of studies employing methods which record the impulse activity in the single nerve of the lateral-line receptors have demonstrated only inhibitory effects of GABA. For example, in a concentration 10^{-4} M GABA reduced the frequency of the background activity in the clawed frog lateral-line receptors (Bledsoe et al. 1983). GABA did not change the background activity of the transparent catfish electroreceptors (Teeter and Bennett 1981). On the other hand this substance in a concentration of $10^{-4}-10^{-5}$ M reduced the background activity of the Lorenzinian ampullae of skates and picrotoxin did not block the spontaneous and evoked activity of the receptors (Fig. 16) (Akoev et al. 1985). GABA did not change the amplitude of the EPSP in the ampullae of Lorenzini of skates either (Steinbach 1974).

So far the bulk of the data obtained makes it impossible to consider GABA as a neurotransmitter in the receptors of the acousticolateral system. GABA effects

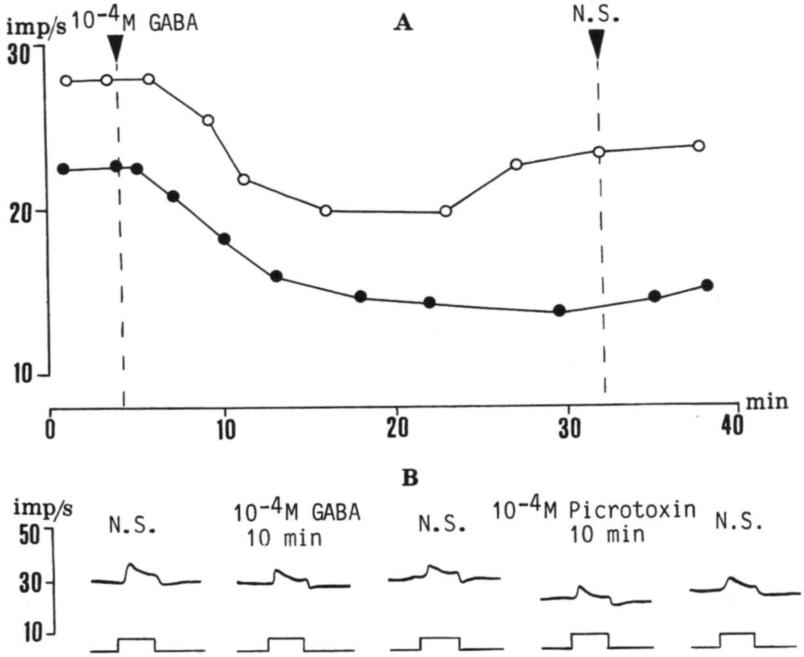

Fig. 16A, B. Effect of GABA (10^{-4} *M*) and picrotoxin (10^{-4} *M*) upon resting (**A**) and evoked (**B**) afferent activity of the ampullae of Lorenzini. (Akoev et al. 1985). **A** Data for two different afferents. **B** *Upper beam*, firing frequency, imp/s; *lower beam*, stimulus current (0.5 μA)

observed at high concentrations seem to be nonspecific, and the existence of postsynaptic GABA receptors is disputable. The most likely explanation of the inhibitory influence of GABA is that it acts on the presynaptic membrane and this effect is associated with a decrease in endogenous neurotransmitter release.

6.5 Histamine, Serotonin, Dopamine, Taurine, Substance P, Adenosine Triphosphate (ATP), Tyramine

Histochemical methods failed to demonstrate the presence of histamine in the organ of Corti (Spoendlin and Lichtensteiger 1966). Histamine in a very high concentration (10 m*M*) induced a 19% reduction of the AP in the auditory nerve (Bobbin and Thompson 1978).

No serotonin synthesis in lateral-line organs was detected by neurochemical methods (Flock and Lam 1974). Histochemical methods also failed to detect serotonin in the organ of Corti (Fex et al. 1965; Borg et al. 1974; Densert 1974). The injection of serotonin antagonists into the cochlea had no effect on its electric activity (Klinke and Oertel 1977). At the same time serotonin reduced the amplitude of the AP in the auditory nerve of the guinea pig, though at rather high concentrations (10 m*M*) (Bobbin and Thompson 1978). On the other hand,

serotonin did not affect the electric activity of the cat cochlea (Klinke and Oertel 1977). According to our data neither serotonin, dopamine nor their antagonists metergoline and fluphenazine influenced the background and evoked activity of skate electroreceptors.
Taurine in a concentration up to 10 m*M* had no effect on the amplitude of the AP of the auditory nerve or the CM of the guinea pig (Bobbin and Thompson 1978). In contrast, in the ampullae of Lorenzini of skates this substance exerted an inhibitory effect in a concentration as low as 10^{-5} *M* (Akoev et al. 1985).
Substance P produced no effect on the electric parameters in the guinea pig cochlea (Norris et al. 1976).
ATP in concentrations of 1 and 10 m*M* reduced the AP in the guinea pig auditory nerve by 38% and 95% respectively (Bobbin and Thompson 1978). The authors attribute these effects to the ability of ATP to chelate calcium.
Octopamine in a 10 m*M* concentration had no effect, while tyramine reduced the amplitude of the AP in the guinea pig auditory nerve (Bobbin and Thompson 1978). The authors believe that tyramine can be regarded as a possible neurotransmitter in cochlea. However, so far there have been no experimental data supporting this point of view.

7 Conclusion

The present review emphasizes that considerable progress toward understanding the mechanisms of synaptic transmission in receptors of the acousticolateral system has been made. Throughout this system the sensory receptors are hair cells. Their fundamental morphology is similar to a marked degree from system to system (auditory, vestibular, lateral-line or electroreceptor) and from species to species. Receptor cells transduce the external stimuli which are mechanical or electrical in nature. They are secondary cells and tonic discharge of spikes in afferent fibers characterizes acousticolateral receptors. Both morphological and physiological lines of evidence indicate that synaptic transmission is chemically mediated in these receptors. Recent studies demonstrated that within basolateral hair cell membranes Ca^{2+} channels are involved in the release of the afferent synaptic transmitter. This Ca^{2+} conductance is partially activated at the resting potential and accounts for the tonic impulse discharge in afferent fibers.
Despite more than a decade of research the identity of the afferent transmitter is not definitely known. The analysis of available experimental data indicates that the most probable neurotransmitters in the receptors of the acousticolateral system are amino acids, i.e., GLU and ASP. However, many aspects of this problem will have to be studied in detail using various methods before their role in synaptic transmission is established. The idea that L-GLU and L-ASP are neurotransmitters invariably implies the existence of specific postsynaptic receptors, though it is not quite clear why L-GLU and L-ASP produced effects at rather high concentrations in a number of synapses. We suspect that at high concentrations L-GLU can interact with nonglutamatergic postsynaptic receptors. The effects of L-GLU and L-ASP may on occasion by associated with the inhibition of the mechanism which induces uptake of the natural neurotransmitter or its precursors. In addition, it is well known that L-GLU exerts an excitatory

effect on various neurones of the central and peripheral nervous system, and in many cases its effects are not primarly associated with synaptic processes. L-GLU can also influence the presynaptic receptors, facilitating the release of the endogenous transmitter. This seems quite reasonable since presynaptic GLU receptors are known to exist (Giorguieff et al. 1977). It should also be pointed out that although the concentration of GLU and ASP is higher in the hair cells than in other areas these amino acids are involved in the Krebs cycle and are obligatory for energy metabolism.

Comparison of the biochemical and pharmacological data about synaptic transmission in various receptors of the acousticolateral system shows that there are significant differences between them. The differences are related to the influence of NMDA and DLAA, which affect the mechanoreceptors in the lateral-line organs but not the receptors of the organ of Corti, the semicircular canal, or electroreceptors. On the other hand, GABA inhibits the activity of the lateral-line mechano- and electroreceptors and has no effect on it in the cochlear receptors, while the influence of ACh is inhibitory in the cochlea and excitatory in the lateral-line organs. Thus, these data indicate that there is little evidence to assume the existence of a single type of postsynaptic receptor among the acousticolateral receptors. It may be supposed that there is only one neurotransmitter throughout the different systems of acousticolateral receptors and that it acts through different subtypes of postsynaptic receptors from system to system. Additional experiments are necessary to clarify this problem.

8 References

Akoev GN, Andrianov GN (1981) The action of divalent ions and drugs on thermal and electrical sensitivity of the ampullae of Lorenzini. In: Szabo T, Czeh G (eds) Sensory physiology of aquatic lower vertebrates. Pergamon, Oxford/Akademiai Kiado, Budapest, pp 57–73 (Advances in physiological science, vol 31)

Akoev GN, Andrianov GN, Volpe NO (1980a) L-Glutamate as a possible neurotransmitter in the ampullae of Lorenzini of the skate. Neurosci Lett 20:307–312

Akoev GN, Andrianov GN, Volpe NO (1980b) Effects of calcium ions on electrical and thermal sensitivity of the ampullae of Lorenzini in the Black sea skates (in Russian). Sechenov Physiol J USSR 66:56–63

Akoev GN, Andrianov GN, Volpe NO (1981) A study of the mechanism of synaptic transmission in the skate ampullae of Lorenzini (in Russian). Neurophysiology (USSR) 13:292–298

Akoev GN, Andrianov GN, Sherman NO (1983) Effects of acetylcholine and cholinolytics on the synaptic transmission in the ampullae of Lorenzini of the skate. Comp Biochem Physiol 74C:95–97

Akoev GN, Andrianov GN, Sherman NO (1984a) Effects of potassium ions on the impulse activity of the ampullae of Lorenzini of the skate. Neurosci Lett 46:341–345

Akoev GN, Andrianov GN, Sherman NO (1984b) Effects of glutamic and aspartic amino acids on the impulse activity of the ampullae of Lorenzini in skates (in Russian). Sechenov Physiol J USSR 70:322–325

Akoev GN, Andrianov GN, Sherman NO (1985) Effects of GABA, taurine and catecholamines on synaptic processes in the ampullae of Lorenzini of the Black sea skates (in Russian). Sechenov Physiol J USSR 71:466–471

Akoev GN, Andrianov GN, Sherman NO (1986a) Action of kainic acid on the synaptic transmission in electroreceptors (ampullae of Lorenzini) of skate (in Russian). Neurophysiology (USSR) 18:147–153

Akoev GN, Andrianov GN, Jakovleva MF, Kuznetsov VI, Moiseeva AB, Ryzhova IV, Sherman NO (1986b) Studies on the mechanism of synaptic transmission in electroreceptors (ampullae of Lorenzini) in skates (in Russian). Proc of IX Meeting on Evol Physiol, Leningrad, p 8

Akoev GN, Andrianov GN, Sherman NO (1987a) Effects of aspartic acid in the ampullary electroreceptors of skates (in Russian). Neurophysiology (USSR) 19:61–67

Akoev GN, Andrianov GN, Sherman NO (1987b) Analysis of the effects of glutamic acid diethylester on synaptic processes in the ampullae of Lorenzini of skates (in Russian). Neurophysiology (USSR) 19:132–138

Akoev GN, Andrianov GN, Szabo T (1987c) Effects of kainic acid on synaptic transmission in electroreceptors (the ampullae of Lorenzini) of skates. Comp Biochem Physiol 88[A]:61–63

Akutsu Y, Obara S (1974) Calcium-dependent receptor potential of the electroreceptor of marine catfish. Proc Jpn Acad 50:247–251

Andrianov GN, Ilyinsky OB (1983) Lateral-line organs (in Russian). In: Kreps EM (ed) Evolutional physiology. Nauka, Leningrad, pp 110–160 (Handbook of physiology, vol 2)

Andrianov GN, Brown HR, Ilyinski OB (1979) A review: physiology of the fish electric receptors (in Russian). Sechenov Physiology J USSR 65:785–800

Annoni JM, Cochran SL, Precht W (1984) Pharmacology of the vestibular hair cell-afferent fiber synapse in the frog. J Neurosci 4:2106–2116

Barets A, Szabo T (1962) Appareil synaptique des cellules sensorielles de l'ampoule de Lorenzini chez la torpille, *Torpedo marmorata*. J Microsc Paris 1:47–54

Barrett EF, Barrett JN, Botz D, Chang DB, Mahaffey D (1978) Temperature-sensitive aspects of evoked and spontaneous transmitter release at the frog neuromuscular junction. J Physiol 279:253–273

Bennett MVL (1970) Comparative physiology: electric organs. Annu Rev Physiol 32:471–528

Bennett MVL, Clusin WT (1979) Transduction in electroreceptors: origins of sensitivity. In: Cone RA, Dowling JE (eds) Membrane transduction mechanisms. Raven, New York, pp 91–116

Bennett MVL, Obara S (1986) Ionic mechanisms and pharmacology of electroreceptors. In: Bullock TH, Heiligenberg W (eds) Electroreception. Wiley, New York, pp 157–181

Benshalom G, Flock Å (1977) Calcium-induced density in synaptic vesicles of afferent and efferent synapses on hair cells in the lateral line organ. Brain Res 121:173–178

Benshalom G, Flock Å (1980) Ultrastructure of synapses in the lateral line canal organ. Acta Otolaryngol (Stockh) 90:161–174

Bernard C (1983) Action of streptomycin and calcium on the apical membrane of hair cells of the frog isolated semicircular canal. Acta Otolaryngol (Stockh) 96:21–30

Bledsoe SC Jr, Bobbin RP (1982) Effects of D-α-aminoadipate on excitation of afferent fibers in the lateral line of *Xenopus laevis*. Neurosci Lett 32:315–320

Bledsoe SC Jr, Bobbin RP, Thalman R, Thalman I (1980) Stimulus-induced release of endogenous amino acids from skins containing the lateral line organ of *Xenopus laevis*. Exp Brain Res 40:97–101

Bledsoe SC Jr, Bobbin RP, Chihal DM (1981) Kainic acid: an evaluation of its action on cochlear potentials. Hear Res 4:109–120

Bledsoe SC Jr, Chihal DM, Bobbin RP, Morgan DN (1983) Comparative actions of glutamate and related substances on the lateral line *Xenopus laevis*. Comp Biochem Physiol [C]75:199–206

Bobbin RP (1979) Glutamate and aspartate mimic the afferent transmitter in the cochlea. Exp Brain Res 134:389–393

Bobbin RP, Guth PS (1970) Evidence that gamma-aminobutyric acid is not the inhibitory transmitter at the crossed olivocochlear nerve-hair cell junction. Neuropharmacology 9:567–574

Bobbin RP, Konishi T (1974) Action of cholinergic and anticholinergic drugs at the crossed olive-cochlear bundle-hair cell junction. Acta Otolaryngol (Stockh) 77: 56–65

Bobbin RP, Thompson MN (1978) Effects of putative transmitters on afferent cochlear transmission. Ann Otol Rhinol Laryngol 87:185–190

Bobbin RP, Bledsoe SC Jr, Chihal DM (1981a) Effects of various excitatory amino-acid antagonists on guinea pig cochlear potentials. Assoc for Res in Otolaryngol, St Petersburg Beach, Florida, p 27

Bobbin RP, Bledsoe SC Jr, Chihal DM, Morgan DN (1981b) Comparative action of glutamate and related substances on the *Xenopus laevis* lateral line. Comp Biochem Physiol [C]69:145–147

Bobbin RP, Bledsoe SC Jr, Jenison GL (1984) Neurotransmitters of the cochlea and lateral line organs. In: Berlin CJ (ed) Hearing sciences: recent advances. College Hill Press, San Diego, pp 159–180

Bobbin RP, Bledsoe SC Jr, Winberry S, Ceasar G, Jenison GL (1985) Comparative actions of GABA and acetylcholine on the *Xenopus laevis* lateral line. Comp Biochem Physiol [C]80:313–318

Borg E, Densert O, Flock A (1974) Synaptic vesicles in the cochlea. Acta Otolaryngol (Stockh) 78:321–332

Brown HR, Govardovsky VI (1983a) Receptor mechanisms of the electroreceptors (ampullae of Lorenzini) in skates (in Russian). Neurophysiology (USSR) 15:178–185

Brown HR, Govardovsky VI (1983b) Electrical model of Lorenzinian ampullae receptor (in Russian). Neurophysiology (USSR) 15:235–241

Brown HR, Ilyinsky OB (1984) Physiology of electroreceptors. Nauka, Leningrad

Brown HR, Govardovsky VI, Cherepnov VL (1985) Action of calcium and potassium channel blockers on changes in transepithelial potential and spike responses of Lorenzinian ampullae in the Black sea skate (in Russian). Neurophysiology (USSR) 17:652–660

Churchill JA, Schuhknecht HF, Doran R (1956) Acetylcholinesterase activity in the cochlea. Laryngoscope 66:1–15

Clusin WT, Bennett MVL (1977a) Calcium-activated conductance in skate electroreceptors. Current clamp experiments. J Gen Physiol 69:121–143

Clusin WT, Bennett MVL (1977b) Calcium-activated conductance in skate electroreceptors. Voltage clamp experiments. J Gen Physiol 69:145–182

Clusin WT, Bennett MVL (1979a) The oscillatory responses of skate electroreceptors to small voltage stimuli. J Gen Physiol 73:683–702

Clusin WT, Bennett MVL (1979b) The ionic basis of oscillatory responses of skate electroreceptors. J Gen Physiol 73:703–723

Comis SD, Leng G (1979) Action of putative neurotransmitters in the guinea pig cochlea. Exp Brain Res 36:119–128

Corey DP, Hudspeth AJ (1979a) Response latency of vertebrate hair cells. Biophys J 26:499–506

Corey DP, Hudspeth AJ (1979b) Ionic basis of the receptor potential in a vertebrate hair cell. Nature 281:675–677

Cornford ME, Barajas L (1980) Cupular secretion by *Xenopus laevis* line organs: autoradiographic evidence for incorporation of ^{3}H-glucose and ^{35}S-sulfate. Am J Anat 158:115–133

Crowford AC, Fettiplace R (1979) Reversal of hair cell responses by current. J Physiol (Lond) 295:66P

Dallos P, Santos-Sacchi J, Flock Å (1982) Intracellular recordings from cochlear outer hair cells. Science 218:582–584

Davis H (1965) A model for transducer action in the cochlea. Cold Spring Harbor Symp Quant Biol 30:181–190

Del Castillo J, Katz B (1954) Changes in endplate activity produced by presynaptic polarization. J Physiol (Lond) 124:586–604

Densert O (1974) Adrenergic innervation in the rabbit cochlea. Acta Otolaryngol 78:345–356

Derbin C (1970) Effects de la section du nerf latéral sur les jonctions sensorineurales des ampoules de Lorenzini de la torpille, *Torpedo marmorata*. J Microsc (Paris) 9:119–126

Desmendt JE, Monaco P (1962) The pharmacology of a centrifugal inhibitory pathway in the cat's acoustic system. Proc 1st Int Pharmacol Meet, vol 8:pp183–188

Drescher MJ, Drescher DG, Medina JE (1983) Effects of sound stimulation at several levels on concentrations of primary amines, including neurotransmitter candidates in the perilymph of the guinea pig inner ear. J Neurochem 41:309–320

Fayre D, Sans A (1979) Morphological changes in afferent vestibular hair cell synapses during postnatal development of the cat. J Neurocytol 8:765–775

Felix D, Ehrenberger K (1977) The action of GABA and acetylcholine in the labyrinth of the cat. In: Portmann M, Aran JM (eds) Inner ear biology. Inserm, Paris, pp 147–157

Felix D, Ehrenberger K (1982) The action of putative neurotransmitter substances in the cat labyrinth. Acta Otolaryngol (Stockh) 93:101–105

Fettiplace R, Crowford AC (1978) The coding of sound pressure and frequency in cochlear hair cells of the terrapin. Proc R Soc Lond [Biol] 203:209–218

Fex J, Martin MR (1980) Lack of effect of D,L-α-aminoadipate, an excitatory amino acid antagonist, on cat auditory nerve responses to sound. Neuropharmacology 19:809–811

Fex J, Wenthold R (1976) Choline acetyltransferase, glutamate decarboxylase and tyrosine hydroxylase in the cochlea and cochlear nucleus of the guinea pig. Brain Res 109:575–585

Fex J, Fuxe K, Lennerstrand G (1965) Absence of monoamines in olivocochlear fibers in cat. Acta Physiol Scand 64:259–262

Flock Å (1965) Electron microscopic and electrophysiological studies on the lateral line canal organ. Acta Otolaryngol [Suppl] (Stockh) 199:1–90

Flock Å (1971) Sensory transduction in hair cells. In: Loewenstein WR (ed) Principles of receptor physiology. Springer, Berlin Heidelberg New York, pp 396–414 (Handbook of sensory physiology, vol 1)

Flock Å (1974) Information transfer at the synapse between hair cell and sensory nerve fibers. Rheinisch-Westfael Akad Wiss 53:347–375

Flock Å, Lam DMK (1974) Neurotransmitter synthesis in inner ear and lateral line organ. Nature 249:142–144

Flock Å, Russell IJ (1976) Inhibition of efferent fibres: action on hair cells and afferent synaptic transmission in the lateral line canal organ of burbot, *Lota lota*. J Physiol (Lond) 257:45–62

Flock Å, Wersäll J (1962) A study of orientation of sensory hairs of the receptor cells in the lateral line organ of a fish with special reference to the function of the receptors. J Cell Biol 15:19–27

Flock Å, Jørgensen JM, Russell IJ (1973a) Passive electrical properties of hair cells and supporting cells in the lateral line canal organ. Acta Otolaryngol (Stockh) 76:190–198

Flock Å, Jørgensen JM, Russell IJ (1973b) The physiology of individual hair cells and their synapses. In: Møller A (ed) Basic mechanisms of hearing. Academic, New York, pp 273–306

Foster AC, Fagg GE (1984) Acidic amino acid binding sites in mammalian neuronal membranes: their characteristics and relationship to synaptic receptors. Brain Res Rev 7:103–164

Frischkopf LF (1973) Effects of stimulation on synaptic body size in auditory hair cells in the frog. Conference on eighth nerve system in vertebrates other than mammals, Moffet Field, California

Furukawa T, Ishii Y (1967) Neurophysiological studies on hearing in goldfish. J Neurophysiol 30:1377–1403

Furukawa T, Matsuura S (1978) Adaptive rundown of excitatory postsynaptic potentials at synapses between hair cells and eighth nerve fibres in the goldfish. J Physiol (Lond) 276:193–209

Furukawa T, Ishii Y, Matsuura S (1972) Synaptic delay and time course of postsynaptic potentials at the junction between hair cells and eighth nerve fibres in the goldfish. Jpn J Physiol 22:617–635

Furukawa T, Kuno M, Matsuura S (1982) Quantal analysis of a decremental response of hair cell-afferent fibre synapses in the goldfish sacculus. J Physiol (Lond) 322:181–195

Galley N, Klinke R, Oertel W, Pause M, Storch W-H (1973) The effect of intracochlearly administered acetylcholine blocking agents on the efferent synapses of the cochlea. Brain Res 64:55–63

Gannon RP, Laszo CA, Moscovitch DN (1966) The effect of physostigmine on the latency of the cochlear potentials. Acta Otolaryngol (Stockh) 61:536–546

Gerschenfeld HM (1973) Chemical transmission in invertebrates' central systems and neuromuscular junction. Physiol Rev 53:1–120

Giorguieff MF, Kemel ML, Glowinski J (1977) Presynaptic effect of L-glutamic acid on the release of dopamine in rat striatal slices. Neurosci Lett 6:73–77

Gisselsson L (1950) Experimental investigation into the problem of humoral transmission in the cochlea. Acta Otolaryngol [Suppl] (Stockh) 82:3–78

Gleisner L, Flock Å, Wersäll J (1973) The ultrastructure of the afferent synapse of hair cells in the frog labyrinth. Acta Otolaryngol (Stockh) 76:199–207

Godfrey DA, Caster JA, Berger SJ, Matschinsky FM (1976a) Level of putative transmitter amino acids in the guinea pig cochlea. J Histochem Cytochem 24:468–470

Godfrey DA, Krzanovski JJ, Matschinsky FM (1976b) Activities of enzymes of the cholinergic system in the guinea pig cochlea. J Histochem Cytochem 24:470–472

Guth PS, Melamed B (1982) Neurotransmission in the auditory system: a primer for pharmacologists. Annu Rev Pharmacol Toxicol 22:383–412

Guth SL, Norris CH (1984) Pharmacology of the isolated semicircular canal: effect of GABA and picrotoxin. Exp Brain Res 56:72–78

Guth PS, Norris CH, Bobbin RP (1976) The pharmacology of transmission in the peripheral auditory system. Pharmacol Rev 28:95–125

Guth PS, Sewell WF, Tachibana M (1981) The pharmacology of the cochlear afferents and cochlear nucleus. In: Brown RD (ed) Pharmacology of hearing. Wiley, New York, pp 99–136

Haldeman S, McLennan H (1972) The antagonistic action of glutamic acid diethylesters towards amino acid-induced and synaptic excitation of central neurons. Brain Res 45:393–400

Hama K (1965) Some observations of the fine structure of the lateral line organs and the Japanese sea eel *Lyncorymba nystromi*. J Cell Biol 24:193–210

Hama K, Saito K (1977) Fine structure of the afferent synapse of the hair cells in the saccular macula of the goldfish, with special reference to the anastomosing tubules. J Neurocytol 6:361–373

Harris GG, Frischkopf LS, Flock Å (1970) Receptor potentials from hair cells of the lateral line. Science 167:76–79

Hashimoto T, Katsuki Y (1972) Enhancement of the mechanosensitivity of hair cells of the lateral line by environmental potassium ions. J Acoust Soc Am 52:553 – 557

Higuchi T, Nagai T, Umekita SI, Obara S (1980) The afferent neurotransmitter in the ampullary electroreceptors: L-glutamate mimics the natural transmitter. Neurosci Lett [Suppl] 4:S7

Hilding DA (1964) Catecholamine activity in the inner ear. J Acoust Soc Am 36: 1016 – 1021

Hudspeth AJ (1982) Extracellular current flow and the site of transduction by vertebrate hair cells. J Neurosci 2:1 – 10

Hudspeth AJ (1983) Mechanoelectrical transduction by the hair cells in the acousticolateralis sensory system. Annu Rev Neurosci 6:187 – 215

Hudspeth AJ (1986) The ionic channels of vertebrate hair cells. Hear Res 22:21 – 27

Hudspeth AJ, Corey DP (1977) Sensitivity, polarity and conduction change in the response of vertebrate hair cells to controlled mechanical stimuli. Proc Natl Acad Sci USA 74:2407 – 2411

Hudspeth AJ, Jacobs R (1979) Stereocilia mediate transduction in vertebrate hair cells. Proc Natl Acad Sci USA 76:1506 – 1509

Ilyinsky OB (1975) Physiology of sensory systems. Physiology of mechanoreceptors (in Russian). Nauka, Leningrad (Handbook of physiology, vol III)

Ilyinsky OB, Chekmasova NM, Chereptov VL (1974) The structure of ampullary organs of lateral-line systems of some elasmobranchs (in Russian). Proc Acad Sci USSR 216:1198 – 1201

Iurato S, Luciano L, Pannese E, Reale E (1971) Histochemical localization of the acetylcholinesterase (AChE) activity in the inner ear. Acta Otolaryngol [Suppl] (Stockh) 279:6 – 50

Jasser A, Guth PS (1973) The synthesis of acetylcholine by the olivocochlear bundle. J Neurochem 20:45 – 53

Jenison GL, Bobbin RP (1983) Effects of quisqualate on guinea pig cochlear potentials. Assoc for Res in Otolaryngol, St Petersburg Beach, FL, January 23 – 27

Jenison GL, Bobbin RP, Thalman R (1985) Potassium-induced release of endogenous amino acids in the guinea pig cochlea. J Neurochem 44:1845 – 1853

Jenison GL, Winbery S, Bobbin RP (1986) Comparative actions of quisqualate and *N*-methyl-D-aspartate, excitatory amino acid agonists, on guinea pig cochlear potentials. Comp Biochem Physiol [C]84:385 – 389

Jørgensen I, Flock Å (1973) The ultrastructure of lateral line sense organs in the adult salamander *Amblistoma mexicanum.* J Neurophysiol 2:135 – 142

Jørgensen I, Flock Å, Wersäll J (1972) The Lorenzinian ampullae of *Polyodon spathula.* Z Zellforsch Mikrosk Anat 130:362 – 377

Katsuki Y (1973) The ionic receptive mechanism in acoustico-lateralis system: In: Møller AR, Boston P (eds) Basic mechanisms of hearing. Academic, New York, pp 307 – 334

Katz B, Miledi R (1967) A study of synaptic transmission in the absence of nerve impulses. J Physiol (Lond) 192:407 – 436

Kawai N, Niwa A, Abe T (1982) Spider venom contains specific receptors blocker of glutaminergic synapses. Brain Res 247:169 – 171

Klinke R (1981) Neurotransmitters in the cochlea and the cochlear nucleus. Acta Otolaryngol (Stockh) 91:541 – 554

Klinke R, Evans EF (1977) Evidence that catecholamines are not the afferent transmitter in the cochlea. Exp Brain Res 28:315 – 324

Klinke R, Oertel W (1977) Amino acids – putative afferent transmitter in the cochlea? Exp Brain Res 30:145 – 148

Krnjevic K (1974) Chemical nature of synaptic transmission in vertebrates. Physiol Rev 54:418 – 541

Kyogoku S, Matsuura S, Kuno M (1986) Generator potentials and spike initiation in auditory fibers of goldfish. J Neurophysiol 55:244 – 255

Leng G, Comis S (1979) Effect of increased potassium in scala tympani on auditory nerve sensitivity. Experientia 35:767 – 768

Lewis RS, Hudspeth AJ (1983) Voltage- and ion-dependent conductance in solitary vertebrate hair cells. Nature 304:538 – 541

Llinas R, Steiberg IX, Walton K (1981) Presynaptic calcium currents in squid giant synapse. Biophys J 33:289 – 322

Martin AR, Ringham GL (1975) Synaptic transfer at a vertebrate central nervous system synapse. J Physiol (Lond) 251:409 – 426

McLennan JN, Lodge D (1979) The antagonism of amino acid-induced excitation of spinal neurons in the cat. Brain Res 169:83 – 90

Medina JE, Drescher DG (1981) The amino acid content of perilymph and cerebrospinal fluid from guinea pigs and the effect of noise on the amino-acid composition of perilymph. Neuroscience 6:505 – 509

Melamed B, Norris C, Bryant G, Guth PS (1982) Amino acid content of guinea pig perilymph collected under conditions of quiet or sound stimulation. Hear Res 7:13 – 18

Meza G (1984) Some characteristics of glutamic acid decarboxylase of chick ampullary cristae. J Neurochem 43:634 – 639

Meza G, Carabez A, Ruiz M (1982) GABA synthesis in isolated vestibulary tissue of chick inner ear. Brain Res 241:157 – 161

Monagham P (1975) Ultrastructural and pharmacological studies on the afferent synapse of lateral line sensory cells of the African clawed toad, *Xenopus laevis*. Cell Tissue Res 163:239 – 247

Mullinger AM (1964) The fine structure of ampullary receptors in *Amiurus*. Proc R Soc Lond [Biol] 160:345 – 359

Mullinger AM (1969) The organization of ampullary sense organs in the electric fish, *Gymnarchus niloticus*. Tissue Cell 1:31 – 52

Mulroy MJ, Altman DW, Weiss TF, Peake WT (1974) Intracellular electric responses to sound in a vertebrate cochlea. Nature 249:482 – 485

Murray RW (1962) The response of the ampullae of Lorenzini of elasmobranchs to electrical stimulation. J Exp Biol 39:119 – 129

Nagai T, Obara S (1985) Afferent facilitation induced by ionophoretic application of acidic amino acids in the ampullary electroreceptors of *Plotosus*. J Neurophysiol 53:853 – 867

Nagai T, Obara S, Kawai N (1984) Differential blocking effects of spider toxin on synaptic and glutamate responses in the afferent synapse of the acoustico-lateralis receptors of *Plotosus*. Brain Res 300:183 – 187

Nier K, Henzel H (1973) Effects of calcium on afferent responses of the ampullae of Lorenzini. Pflugers Arch 338:281 – 287

Norris CH, Guth PS (1974) The release of acetylcholine (ACh) by the crossed olivocochlear (COCB) bundle. Acta Otolaryngol (Stockh) 77:318 – 326

Norris CH, Guth PS, Hastings RC (1976) The infusion of picrotoxin, bicuculine, aminooxyacetic acid, and several putative transmitters on afferent transmission. Fed Proc 35:621

Obara S (1976) Mechanism of electroreception in ampullae of Lorenzini of the marine catfish *Plotosus*. In: Reuben JP, Purpura DP, Bennett MVL, Lendall ER (eds) Electrobiology of nerve, synapse and muscle. Raven, New York, pp 129 – 147

Obara S, Bennett MVL (1972) Mode of operation of ampullae of Lorenzini of the skate, *Raja*. J Gen Physiol 60:534 – 557

Obara S, Oomura Y (1973) Disfacilitation as the basis for sensory suppression in a specialized lateralis receptors of the marine catfish. Proc Jpn Acad 49:213 – 217

Obara S, Sugawara Y (1979) Contribution of Ca to the electroreceptor mechanism in *Plotosus* ampullae. J Physiol (Paris) 75:335 – 340
Obara S, Sugawara Y (1984) Electroreceptor mechanisms in teleost and non-teleost fishes. In: Bolis L, Keynes RD, Maddrell SHP (eds) Comparative physiology of sensory systems. Cambridge University Press, Cambridge, pp 509 – 523
Ohmori N (1984a) Studies of ionic currents in the isolated vestibular hair cells of the chick. J Physiol (Lond) 350:561 – 581
Ohmori N (1984b) Mechanoelectrical transducer has discrete conductance in the chick vestibular hair cell. Proc Natl Acad Sci USA 81:1888 – 1891
Ohmori N (1985) Mechano-electrical transduction currents in isolated vestibular hair cells of the chick. J Physiol (Lond) 359:189 – 217
Osborne MP (1977) Role of vesicles with some observations on vertebrate sensory cells. In: Gottrell GA, Usherwood PNR (eds) Synapses. Academic, New York, pp 40 – 63
Osborne MP, Thornhill RA (1972) The effect of monoamine depleting drugs upon the synaptic bars in the inner ear of the bullfrog (*Rana catesbeina*). Z Zellforsch Mikrosk Anat 127:347 – 355
Pastuszko A, Wilson DF, Erecinska M (1984) Effect of kainic acid in rat brain synaptosomes: the involvement of calcium. J Neurochem 43:747 – 754
Roberts BL, Ryan K (1971) The fine structure of the lateral-line sense organs of dogfish. Proc R Soc Lond [Biol] 179:157 – 168
Robertson D, Johnstone BM (1979) Effects of divalent cations on spontaneous and evoked activity of single mammalian auditory neurons. Pflugers Arch 380:7 – 12
Rossi ML, Valli P, Casella C (1977) Post-synaptic potentials recorded from afferent nerve fibers of the posterior semicircular canal in the frog. Brain Res 135:67 – 75
Rossi ML, Prigioni J, Valli P, Casella C (1980) Activation of the efferent system in the isolated frog labyrinth: effect on the afferent EPSPs and spike discharge recorded from single fibers of the posterior nerve. Brain Res 185:125 – 137
Roth A (1968) Electroreception in the catfish *Amiurus nebulosus*. Z Vergl Physiol 61: 196 – 202
Roth A (1969) Electrische Sinnesorgane beim Zwergwels *Ictalurus nebulosus* (*Amiurus nebulosus*). Z Vergl Physiol 65:368 – 388
Russell IJ (1976) Amphibian lateral line receptors. In: Llinas R, Precht W (eds) Frog neurobiology. Springer, Berlin Heidelberg New York, pp 543 – 549
Russell IJ (1983) Origin of the receptor potential in inner hair cells of the mammalian cochlea – evidence for Davis theory. Nature 301:334 – 336
Russell IJ, Sellick PM (1978) Intracellular studies of hair cells in the mammalian cochlea. J Physiol (Lond) 284:261 – 290
Russell IJ, Richardson GP, Cody AR (1986) Mechanosensitivity of mammalian auditory hair cells in vitro. Nature 321:517 – 519
Ryan AF, Schwartz IR (1984) Preferential glutamine uptake by cochlear hair cells: implication for the afferent cochlear transmitter. Brain Res 290:376 – 379
Saito K (1980) Fine structure of the sensory epithelium of the guinea pig organ of Corti: afferent and efferent synapses of hair cells. J Ultrastruct Mol Struct Res 71:222 – 232
Sand O (1975) Effects of different ionic environments on the mechano-sensitivity of lateral line organs in the mudpuppy. J Comp Physiol 102[A]:27 – 42
Schwartz IR, Ryan AF (1983) Differential labelling of sensory cells and neural populations in the organ of Corti following amino acid incubations. Hear Res 9:185 – 200
Sejnowski TJ, Yodlowski ML (1982) A freeze-fracture study of the skate electroreceptor. J Neurocytol 11:897 – 912
Sewell WF, Norris CH, Tachibana M, Guth PS (1978) Detection of an auditory nerve activating substance. Science 202:910 – 922

Shelton PMJ (1971) The structure and function of lateral-line system in larval *Xenopus laevis*. J Exp Zool 178:211 – 231
Smith CA, Sjöstrand FS (1961) Structure of nerve endings on the external hair cells of the guinea pig cochlea by serial sections. J Ultrastruct Res 5:523 – 556
Spoendlin H (1972) Innervation densities of the cochlea. Acta Otolaryngol (Stockh) 73:235 – 248
Spoendlin H, Lichtensteiger W (1966) The adrenergic innervation of the labyrinth. Acta Otolaryngol (Stockh) 61:423 – 434
Steinbach AB (1974) Transmission from receptor cells to afferent nerve fibers. In: Bennett MVL (ed) Synaptic transmission and interneuronal communication. Raven, New York, pp 105 – 140
Steinbach AB, Bennett MVL (1971) Effect of divalent ions and drugs on synaptic transmission in phasic electroreceptors in mormyrid fish. J Gen Physiol 58:580 – 598
Sugawara Y, Obara S (1984) Damped oscillation in the ampullary electroreceptors of *Plotosus* involves Ca-activated transient K conductance in the basal membrane of receptor cells. Brain Res 302:171 – 175
Szabo T (1965) Sense organs of the lateral line system in some electric fish of the *Gymnotidae, Mormyridae,* and *Gymnarchidae*. J Morphol 117:229 – 249
Szabo T (1972) Ultrastructural evidence for mechanoreceptor function of the ampullae of Lorenzini. J Microsc (Paris) 14:343 – 350
Szabo T (1974) Anatomy of the specialized lateral line organs of electroreception. In: Fessard A (ed) Electroreceptors and other specialized receptors in lower vertebrates. Springer, Berlin Heidelberg New York, pp 13 – 59 (Handbook of sensory physiology, vol III/3)
Szabo T, Wersäll J (1970) Ultrastructure of an electroreceptor (mormyromast) in a mormyrid fish *Gnathonemus petersii*. J Ultrastruct Res 30:473 – 490
Szamier RB, Wachtel AW (1970) Special cutaneous receptor organs of fish. VI. Ampullary and tuberous organs of *Hypopomus*. J Ultrastruct Res 30:450 – 471
Tachibana M, Kuriyama K (1974) Gamma-aminobutyric acid in the lower auditory pathway of the guinea pig. Brain Res 69:370 – 374
Teeter JH, Bennett MVL (1981) Synaptic transmission in the ampullary electroreceptors of the transparent catfish, *Kryptopterus*. J Comp Physiol 142[A]:371 – 377
Teeter JH, Szamier RB, Bennett MVL (1980) Ampullary electroreceptors in the sturgeon *Scaphirhynchus platorynchus* (Rafinesque). J Comp Physiol 138[A]:213 – 223
Thornhill RA (1972) The effect of catecholamine precursors and related drugs on the morphology of the synaptic bars in the vestibular epithelia of the frog *Rana temporaria*. Comp Gen Pharmacol 3:89 – 97
Toyoshima K, Shimamura A (1982) Comparative study of ultrastructures of the lateral-line organs and palatal taste organs in the African clawed toad, *Xenopus laevis*. Anat Rec 201:371 – 381
Umekita S-I, Matsumoto Y, Abe T, Obara S (1980) The afferent neurotransmitter in the ampullary electroreceptors: stimulus-dependent release experiments refute the transmitter role of L-glutamate. Neurosci Lett [Suppl] 4:S7
Valli P, Zucca G, Casella C (1977) The importance of potassium in the function of frog semicircular canals. Acta Otolaryngol (Stockh) 84:344 – 351
Valli P, Zucca G, Prigioni I, Botta L, Casella C, Guth P (1985) The effect of glutamate on the frog semicircular canal. Brain Res 330:1 – 9
Vinnikov YaA, Titova LK (1964) The organ of Corti. Consultants Bureau, New York
Wagner HJ (1973) Darkness-induced reduction on the number of synaptic ribbons in fish retina. Nature 246:53 – 55
Waltman B (1966) Electrical properties and fine structure of the ampullary canals of Lorenzini. Acta Physiol Scand [Suppl] 264:1 – 60

Watkins JC (1978) Excitatory amino acids. In: McGeer EG, Olney JW, McGear PLM (eds) Kainic acid as a tool in neurobiology. Raven, New York, pp 37 – 69

Watkins JC (1981) Pharmacology of excitatory amino acid receptors. In: Roberts PJ, Storm-Mathisen J, Johnston GAR (eds) Glutamate: transmitter in the central nervous system. Wiley, New York, pp 1 – 24

Watkins JC, Evans RH (1981) Excitatory amino acid transmitters. Annu Rev Pharmacol Toxicol 21:165 – 204

Wersäll J, Flock Å, Lundquist PG (1965) Structural basis for directional sensitivity in cochlear and vestibular sensory receptors. Cold Spring Harbor Symp Quant Biol 30:115 – 132

Wickham MG (1973) Large inclusion in receptor cells of the amphibian lateral line. Am J Anat 138:465 – 480

Winberry SL, Bobbin RP (1983) Actions of acetylcholine and carbachol on the spontaneous activity of afferent fibers in the *Xenopus laevis* lateral line (abstr). 6th Meet Assoc Res Otolaryngol, Florida, p 47

Yamada Y, Hama K (1972) Fine structure of the lateral-line organ of the common eel, *Anguilla japonica*. Z Zellforsch Mikrask Anat 124:454 – 464

Zakon H (1986) The electroreceptive periphery. In: Bullock TH, Heiligenberg W (eds) Electroreception. Wiley, New York, pp 103 – 156

Zimmerman D McG (1979) Onset of neuronal function in the lateral line. Nature 282:82 – 84

Experimental Models of Sensorineural Hearing Loss – Effects of Noise and Ototoxic Drugs on Hearing

J. Syka*

Institute of Experimental Medicine, Czechoslovak Academy of Sciences, Lidových milicí 61, Prague 2, 120 00, Czechoslovakia

* The author wishes to thank Dr. Donald Robertson and Prof. Charles Edwards for valuable comments on an earlier version of the manuscript.

1 Introduction

Sensorineural hearing loss, i.e., the loss of hearing which is caused by pathological processes in the cochlea and consequently in the auditory pathway is the subject of many recent clinical and experimental studies. Increasing noise levels in the environment as well as the increasing occurrence of ototoxic side effects of drugs significantly increase the attention which is paid to the problem of sensorineural hearing loss. Although the experimental studies on the mechanisms of the noise-induced and drug-induced hearing loss can be traced back into the last century, only the last 40 years have seen significant improvement in the understanding of cochlear pathology and pathophysiology. Undoubtedly the progress in pathological studies would not have been possible without the great progress in understanding cochlear ultrastructure and function. New techniques have made possible the investigation of changes in very subtle structures of the auditory receptors, such as stereocilia, and the correlation of these microstructural properties with the responses of single auditory nerve fibers. New insights into the pathological mechanisms have been achieved using intracellular recordings from hair cells. Combined measurements of ionic concentrations in cochlear fluids together with stria-produced potentials have given another dimension to the complex picture of sensorineural hearing loss. Structural and functional changes as a consequence of exposure to noise have also been found in the central part of the auditory pathway. These changes have been traced using not only anatomical and electrophysiological methods, but also with modern biochemical techniques. Specific changes have been observed in young developing animals which were exposed to either noise or ototoxic drugs.
Although the mode of insult differs for acoustic trauma and drug ototoxicity, the end results are usually similar, i.e., sensory-cell degeneration and subsequent sensorineural hearing loss. Therefore, Kiang et al. (1976) used both noise and the ototoxic antibiotic kanamycin in their pioneering study to produce characteristic changes in cochlear structure and function. Their approach served for a long time as an experimental model of sensorineural hearing loss. Most recent investigations have demonstrated subtle micromechanic and metabolic changes in the membrane of hair cells including stereocilia resulting from noise exposure and ototoxic drug treatment. It can be expected that in the near future more precise information about molecular mechanisms of the induced hearing loss will appear.

This review attempts to cover the different approaches used in the investigation of noise- and drug-induced hearing loss in experimental animals. Where necessary, the results of related psychophysical studies are included as well.

2 Effects of Noise

2.1 Morphological Changes After Noise Exposure

2.1.1 Hair-Cell Damage

Discrete morphological changes in the organ of Corti after overstimulation with intense sound were first described 80 years ago (Witmaack 1907; Yoshii 1909). Numerous studies since have described in detail the damage resulting from noise exposure and the results may be found in many recent reviews (Henderson et al. 1976; Hamernik et al. 1982; Schmiedt 1984; Slepecky 1986; Saunders et al. 1985; Lim 1986). This review will mention briefly alterations found in hair cells and surrounding cells of the organ of Corti when exposed to noise. The details of the morphological changes will be related to the functional changes.

Cells injured by acoustic trauma show several changes, including pycnosis of the nucleus, ruptured vacuoles, and swelling (Liberman and Kiang 1978; Liberman and Mulroy 1982). The endoplasmatic reticulum of the outer hair cells (OHCs) may show vacuolation and vesiculation (Lim and Dunn 1979) and swelling of mitochondria may occur (Lim and Melnick 1971; Spoendlin 1971; Ward and Duvall 1971). In the region below the basal body the number of lysosomes in the OHC increases (Engström and Ades 1960; Beagley 1965b; Lim and Melnick 1971; Angelborg and Engström 1973; Slepecky et al. 1982). With more serious damage the cuticular plate appears to "soften" (Lim and Melnick 1971) and bulge outwards into the subtectorial space (Lindeman and Bredberg 1972). The rupture of a cell may eventually result in expulsion of the cellular content into endolymphatic or perilymphatic space. The ruptured cells are replaced with scar tissue, which is formed from phalangeal plates of the Deiter's cells (Engström et al. 1970; Bohne 1976a; Hawkins and Johnsson 1976; Bohne and Rabbit 1983). When the hair cell degenerates, small holes sometimes appear in the reticular lamina. The holes may represent a pathway through which endolymph can enter the fluid spaces of the organ of Corti. According to Bohne (1976b) the presence of endolymph (or excessive potassium ions) in the fluid spaces of the organ of Corti may result in swelling of uninjured cells and nerve fibers and eventually in rupture. Bohne (1976b) observed similar changes after perfusion of scala tympani in chinchilla with artificial endolymph.

2.1.2 Stereocilia Damage

Studies using the scanning electron microscope (first used in 1969 by Lim and Lane) have clearly demonstrated that the first subtle changes after acoustic overstimulation may occur in the stereocilia of hair cells (for review see, e.g., Saunders et al. 1985; Slepecky 1986). Hunter-Duvar (1977) and Hunter-Duvar et al. (1982) have described in different species (chinchilla, squirrel monkey, guinea pig, cat) disruption and bending of stereocilia of inner (IHCs) and outer hair cells (OHCs) after loud sound stimulation. A 1-kHz stimulus at 90 dB SPL for 3 h bent and disrupted the stereocilia of the first row of OHCs. At higher intensity bending and disruption were found in stereocilia of IHCs, and still further increase in the exposure was accompanied by pathological changes of stereocilia in the second outer row of hair cells (OHC2) and then in the third outer row of hair cells (OHC3). The initial bending of stereocilia was followed by fusion and later by autolytic processes.

A similar sequence of changes in stereocilia was found in acute experiments in the guinea pig made by Robertson and Johnstone (1980). Guinea-pig cochleas were examined for structural damage using the scanning electron microscope immediately after exposure to pure tones with intensities from 96 to 126 dB SPL. Again, the first observable abnormalities consisted of collapse or fusion of the stereocilia of OHC1 (Fig. 1). With increasing sound intensity the OHC1 damage spread in the basal direction and to a lesser extent in the apical direction. The stereocilia of IHCs showed a mild separation and derangement of the tallest stereocilia, which proceeded until the stereocilia completely disappeared. The symmetrical spread of damage in the IHCs around the point of maximum effect was in contrast to the asymmetrical spread stereociliary damage to OHCs. The damage to the second and third rows of OHCs occurred at quite high intensities of exposure and was frequently associated with severe disruption of the whole organ of Corti.

Similar sequences of the spread of damage to stereocilia have been observed in chinchillas exposed to impulse noise (Slepecky et al. 1982). The stereocilia in the first row of OHCs showed significant damage. Stereocilia were no longer erect and in some cases were fused or missing. Changes in stereocilia structure were correlated with the extent of permanent threshold shifts in these chinchillas, whereas other alterations in the cell ultrastructure could not be related to shifts in hearing threshold.

In contrast to the above-mentioned data, the results of Engström (1983) point to a different sequence of stereociliary damage after noise exposure in the rabbit. Broadband noise (2 – 7 kHz) at 115 dB SPL for 15 or 30 min was used in this study. The first changes, consisting of fusion, fractionation, and inclination of stereocilia, were always observed in the IHC stereocilia. The damage to IHCs was correlated with loss of auditory sensitivity. When animals were allowed to survive for several weeks after noise exposure, typically giant stereocilia were found. One possible explanation for the occurrence of the principal damage to IHC stereocilia would be the difference in species, but Engström (1983) also found very similar damage to stereocilia predominantly on IHCs in rats and guinea pigs.

Immediately after exposure to noise, floppy stereocilia (Hunter-Duvar 1977; Lim 1980; Erlandsson et al. 1980) and stereocilia where the cross bridges between actin filaments have been broken (Tilney et al. 1982) or the actin filaments are depolymerized (Tilney et al. 1982) are present; however, they disappear several days after the noise exposure. These changes may be related to the functionally observed temporary threshold shift (TTS) and the subsequent recovery from TTS.

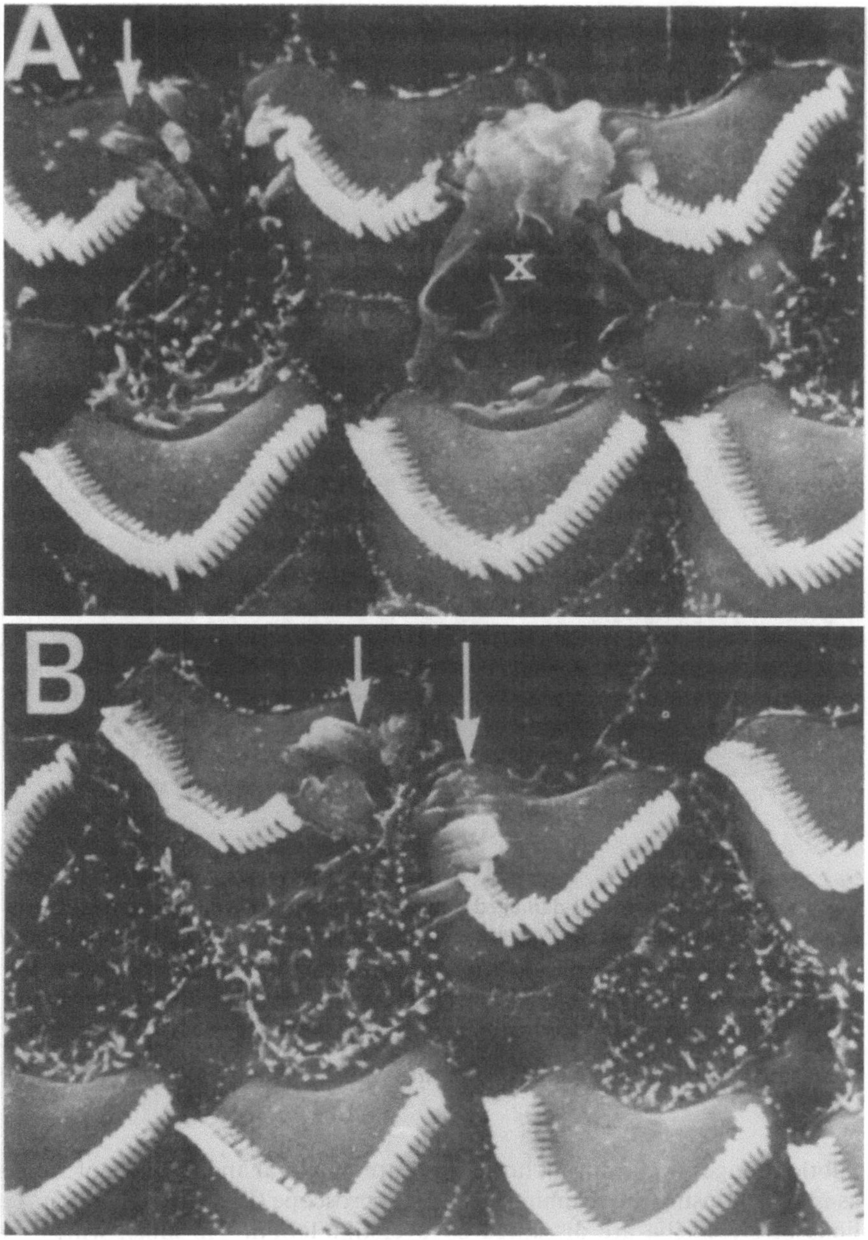

Fig. 1A – B. Typical appearence of OHC abnormalities immediately after exposure to 10 kHz, for 1 h at 106 dB SPL. **A** Region showing protrusions of reticular lamina *X* as well as stereocilia collapse (*arrow*). **B** Region showing stereocilia collapse and fusion (*arrows*). Plate width 1 μm. (Robertson et al. 1980)

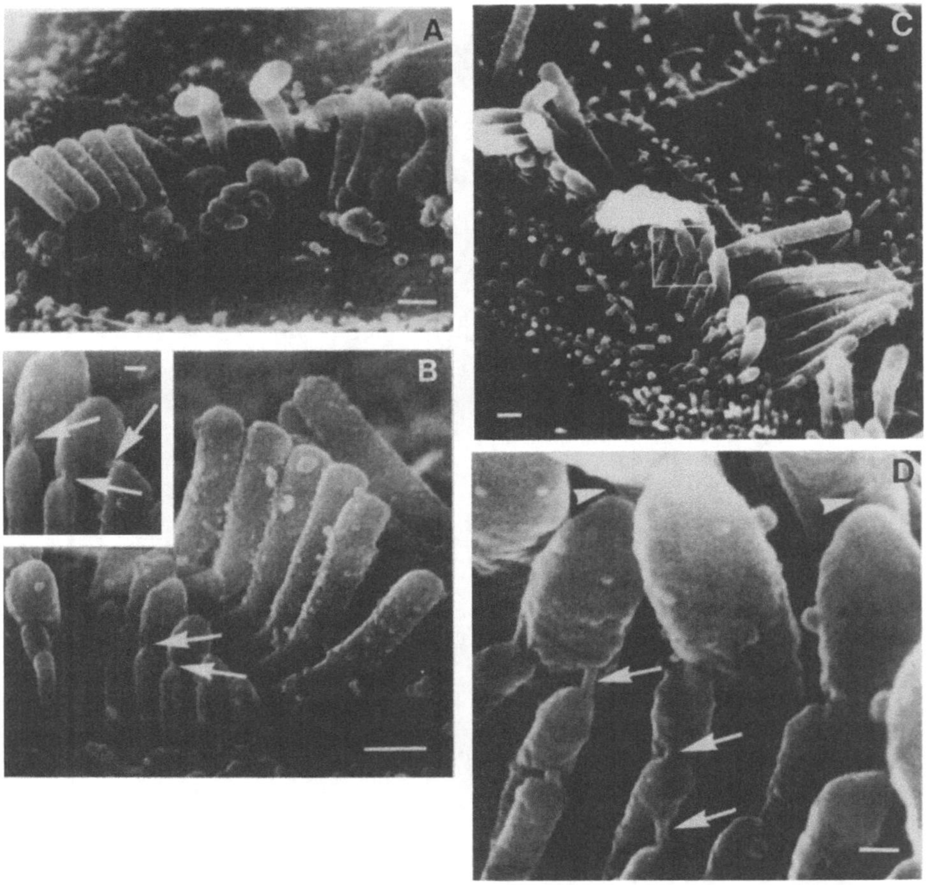

Fig. 2A–D. IHC in sound-damaged cochlea. **A** The stereocilia of the tallest row have flexed at the point at which the stereocilia of the next row originally abutted. *Scale bar*, 500 nm. **B** The tallest stereocilia have separated from the shorter stereocilia. Tip links between the rows of shorter stereocilia are intact (*arrows*). See also tip links in the *inset*. **C** Disruption of the tallest stereocilia in IHCs. High power view of stereocilia in central square = **D**, showing intact tip links among the shorter stereocilia (*arrows*), while links to the taller stereocilia are missing (*arrowheads*). *Scale bar*, 100 nm. (Pickles et al. 1987)

Subtle changes occur during noise exposure in lateral and tip links of stereocilia, as is evident using the scanning electron microscope (Pickles et al. 1987). Lateral links of stereocilia are disrupted by overstimulation. Tip links, which may be associated with sensory transduction, also disappear (Fig. 2). In OHCs they disappear in all rows of stereocilia, whereas in IHCs they are most vulnerable in the tallest rows of stereocilia. Significant changes in the structure of the tallest outer row of IHC stereocilia after noise exposure are also evident using the transmission electron microscope (Liberman 1987). The shorter row can remain ultrastructurally normal even when the tall row is completely missing. The most common stereocilia pathology in chronic acoustic trauma are changes in their root-

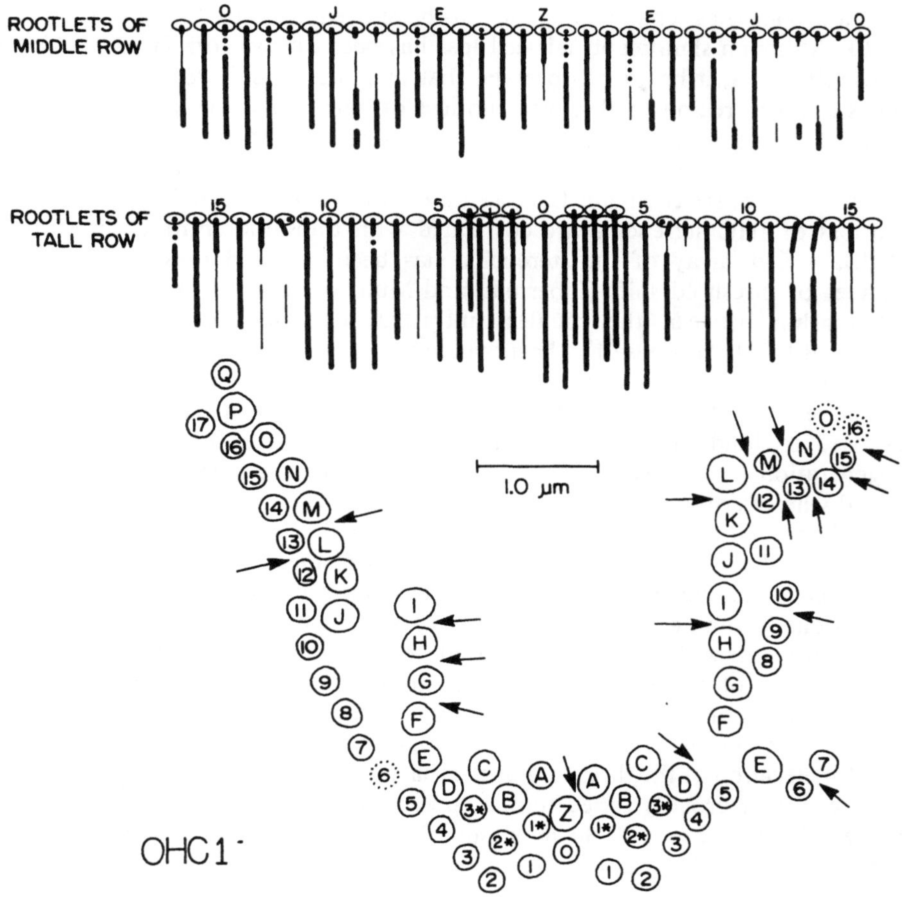

Fig. 3. Condition of the intracuticular rootlets (*top*) and the degree of disarray of the stereocilia (*bottom*) for one first-row OHC from a region with 40 dB PTS. *Dotted thick lines*, fractured rootlets; *thin lines*, attenuated rootlets; *missing lines*, missing rootlets; *bent lines*, displaced rootlets. The *arrows* in the bottom tracing are positioned between adjacent stereocilia for which both members of the pair show rootlet damage. (Liberman 1987)

lets (Liberman 1987). Fracture, attenuation, and complete loss of the stereociliary rootlets are characteristic for both IHCs and OHCs, particularly for the first row of OHCs (Fig. 3). In addition, ectopic supracuticular cytoplasm appears in hair cells with fused stereocilia.

2.2 Functional Correlates of Hair-Cell Damage

The first attempts to correlate changes in functional properties of hair cells (measured at their output as responses of auditory nerve fibers, single neurons of the spiral ganglion, or the N1 component of the auditory nerve action potential) with morphological changes of hair cells have been made by Kiang et al. (1970, 1976). Although significant hair-cell loss has been always correlated with large threshold shifts, it has been observed that in some cases 40 to 50 dB threshold shifts occur in regions with almost no loss of hair cells (Liberman and Mulroy 1982). New insight into this problem came when scanning electron microscopy was adapted for evaluation of the pathology of the organ of Corti, particularly for the study of stereocilia of IHCs (Hunter-Duvar 1977). Significant fusion and the loss of IHC stereocilia are correlated with threshold shifts of more than 60 dB, and disarray of the stereocilia has been reported present with lesser degrees of threshold shift (Liberman and Mulroy 1982). In the guinea pig the relationship between the cochlear ultrastructure and the neural threshold indicated by N1 response has been studied in acute experiments carried out by Robertson and Johnstone (1980) and in animals surviving many days after noise exposure by Robertson (1982). In the chronic experiments the animals were exposed to loud pure tones from 107 to 129 dB SPL for 1 h. The actual degeneration of hair cells was quite sparse, but abnormalities in stereocilia were found whenever the N1 thresholds were elevated. The most damaged stereocilia were in the first row of OHCs, and the damage in this region correlated well with notches in the N1 threshold curve. Significant changes in parameters of spiral ganglion-cell tuning curves were also present in this region. The classification of alterations in tuning curves was similar to that used by Liberman and Mulroy (1982): (a) units with no response to sound at the highest sound pressure used, (b) total absence of a sharply tuned portion on the tuning curve, giving it a broad shape, (c) distortion of the tuning curve shape manifested as a partial dissociation of the low frequency tail and high frequency tip (W shape of curve), and (d) elevation of the threshold of the sharply tuned tip. The unresponsive cells and cells with broadly tuned tuning curves (types a and b) were found in the guinea pig in regions with the maximum threshold shift and also with maximum damage to hair cells and their stereocilia. The W-shaped tuning curves were not combined with the hypersensitivity of the tail regions, in contrast to observations in the cat (Kiang et al. 1976; Liberman and Mulroy 1982).

Subtle changes in IHC and OHC stereocilia in cats with permanent noise-induced shifts and their relationship to parameters of auditory nerve-fiber tuning curves have been described (Liberman and Dodds 1984b). This was made possible by the labeling of selected fibers with intracellularly injected horseradish peroxidase. It was concluded that a normal tuning curve tip requires that the stereocilia on both the IHCs and OHCs are normal. Selective damage to the OHCs (i.e., to their stereocilia) is associated with elevation of the tips and hypersensitivity of tuning curve tails. Destruction of the OHCs in a region where the IHCs appear normal is associated with flat or bowl-shaped tuning curves without a tip. When both OHCs and IHCs show destruction, both tips and tails of the curve are elevated (Fig. 4). The first row of OHCs makes a greater contribution to the tuning curve tips than do the OHCs in the other two rows. In regions with complete disarray of stereocilia in the first row of OHCs the tuning curve tips were typically elevated, even though the stereocilia of second and third row OHCs were completely normal.

In a related study, individual fibers in cats with permanent noise-induced threshold shift were labeled to estimate the shift in characteristic frequency (CF)

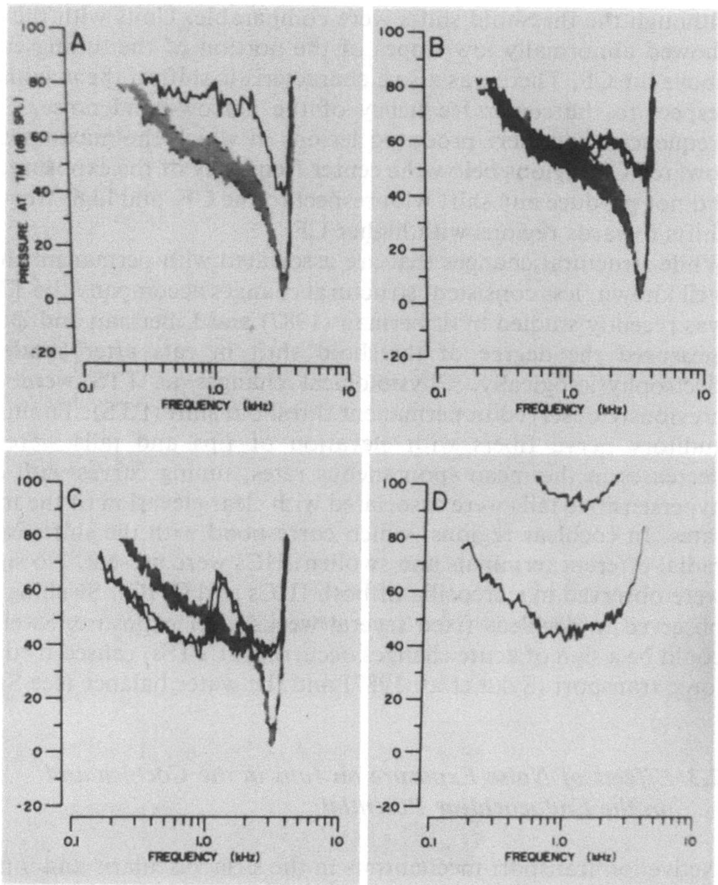

Fig. 4A–D. Different types of tuning curve abnormalities observed in cats with permanent threshold shifts. One or more abnormal tuning curves are compared with a group of normal tuning curves (*stippled*) of similar characteristic frequency (CF). **A, B** Type V tuning curves with hyposensitive tips and tails. **C** Type W tuning curve with hypersensitivity of the tail. **D** Bowl-shaped tuning curve without the tip. (Liberman and Mulroy 1982)

of abnormal tuning curves. The problem, which is related to pitch shift in psychophysical studies, was studied after acute noise exposure by Robertson et al. (1980). In chronic animals the estimation of the CF shift was not accurate unless the "original" CF could be estimated on the basis of cochlear location of the labelled fiber. Narrow-band noise stimuli with center frequencies from 1.5 to 9 kHz and intensities ranging from 109 to 121 dB were used in this study, with survival times from 35 to 78 days. The results showed that at least for units whose CF was originally above 1 kHz the CF shifts down as the threshold rises. The largest CF shifts were observed in units contacting IHCs within 1 mm of cochlear areas where the entire organ of Corti was destroyed. For units connected with IHCs at greater distances from the damaged area the CF shifts were smaller, although the threshold shifts were comparable. Units with the largest CF shifts showed abnormally low slopes of the portion of the tuning curve immediately above the CF. There was also a characteristic shift of the maximum damage with respect to the center frequency of the narrow-band noise. Exposures at low frequencies routinely produced lesions in which the maximum damage shifted towards CF regions below the center frequency of the exposure, mid frequencies did not produce any shift with respect to the CF, and high frequencies produced shifts towards regions with higher CF.

While structural changes that are associated with permanent threshold shift are well known, less consistent structural changes accompany the TTS. This problem was recently studied by Liberman (1987) and Liberman and Dodds (1987), who measured the degree of threshold shift in cats after subtle noise exposure electrophysiologically. Physiological changes at TTS were similar to those previously observed in permanent threshold shift (PTS). Tuning curves of single auditory nerve fibers with elevation of tips and tails were associated with decreases in the mean spontaneous rates, tuning curves with elevated tips and hypersensitive tails were associated with clear elevation of the mean spontaneous rates. In cochlear regions, which correspond with the shifts in the CF, swollen radial efferent terminals and swollen IHCs were present. No significant changes were observed in stereocilia of both IHCs and OHCs. Swelling of IHCs was not observed in cochleas fixed several weeks after exposure. Swelling of the IHCs could be a sign of acute changes occurring at TTSs, caused by disturbances in the ionic transport (Syka et al. 1987) and the water balance (see Sect. 2.4.1).

2.3 Effects of Noise Exposure on Ions in the Cochlea and on the Endocochlear Potential

Active ion transport mechanisms in the stria vascularis and ionic permeabilities of the endolymph-perilymph barrier play an important role in the maintenance of the unique endolymph electrolyte concentration gradients and the endocochlear potential (Smith et al. 1954; Bosher 1979; Konishi et al. 1979; Melichar and Syka 1977, 1987a, b; Bosher 1980a; Konishi and Salt 1980). The endocochlear potential (EP) is thought to consist of two components: positive and negative. The positive component of the EP (approximately +120 mV) is assumed to be produced by electrogenic K^+ transport, localized in the stria vascularis, whereas the negative potential is considered to be a modified K^+ diffusion potential dependent mainly upon the concentration gradient of K^+ between endolymph and perilymph. The effects of different noise exposures on the concentrations of ions in different compartments of the cochlea and the EP in guinea pig have been investigated (Konishi et al. 1979, 1982; Konishi and Salt 1980; Salt and Konishi 1979; Melichar et al. 1980; Syka et al. 1981; Syka et al. 1987). The changes in the cochlea were dependent on the intensity and duration of the noise, but in all cases the noise changed the ionic permeability of the endolymph-perilymph barrier and, specifically the permeability of the organ of Corti. At lower intensities (less than 100 dB for 20 min) there is no change in the EP and in the concentrations of ions in the cochlea (Salt and Konishi 1979). When the noise level is increased to 105 dB SPL (broad-band noise) a characteristic change

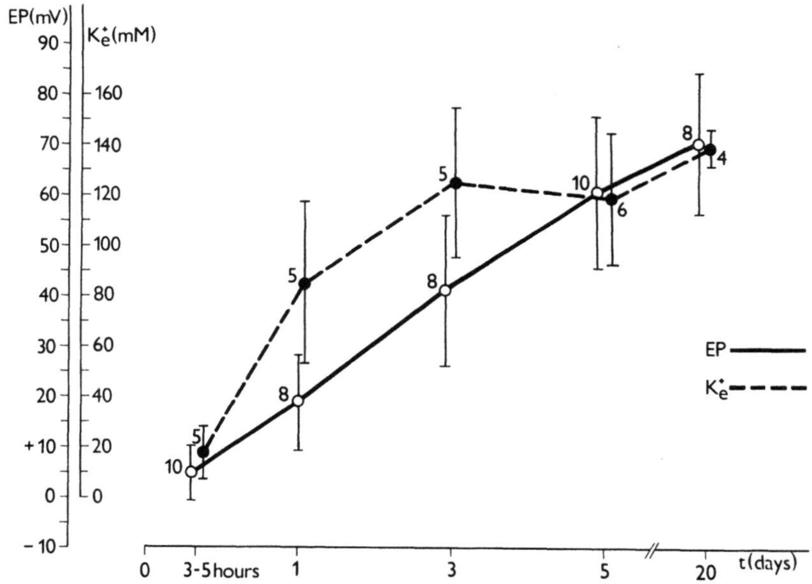

Fig. 5. Recovery of the endocochlear potential (EP) and K^+ in endolymph after noise exposure (*ordinate*). *Abscissa*, Time after exposure. *Numerals* at each point indicate the number of experiments. (Melichar et al. 1980)

in the EP occurs, leading to a postexposure increase of the EP. Recently this change has been found also with short-lasting exposure (30 s) to a pure tone at 105 dB SPL; the change is correlated with changes in K^+ concentration in the cortilymph and in extracellular DC receptor potentials (Syka et al. 1987). An increase in the intensity of the noise to 115 dB SPL (Salt and Konishi 1979) further increases the EP, along with an increase in the K^+ concentration. However, repeated exposure to 115 dB SPL octave-band noise centered at 350 Hz decreases the EP and K^+, which apparently is a sign of traumatic changes in the organ of Corti, leading finally to a breakdown of the endolymph-perilymph barrier and possibly also to a failure of the pump mechanisms in the stria vascularis. In guinea pigs after noise exposure to 142 dB one-third octave-band noise centered at 1 kHz for 1 h there is a similar profound decrease of the EP and K^+ concentration (Melichar et al. 1980; Syka et al. 1981). The EP and K^+ in endolymph, which initially decreased to 5.7 mV and 18.9 mmol/l respectively, returned to normal values in 5 – 10 days (Fig. 5). Histological examination of the cochlear partitions soon after acoustic trauma revealed various degrees of pathological changes, from ruptures in the reticular lamina to complete detachment of the organ of Corti and perforations of the basilar membrane. Several days after the exposure, flat epithelial cells had replaced the degenerated cells of the organ of Corti. An abnormally increased EP was observed in some animals 25 days after exposure. In the second turn, where the noise produced maximum damage, the anoxic negative component of the EP was less negative (Fig. 6). Because the concentration of K^+ in endolymph and perilymph attained essentially normal values at that time, this finding supported the hypothesis that the

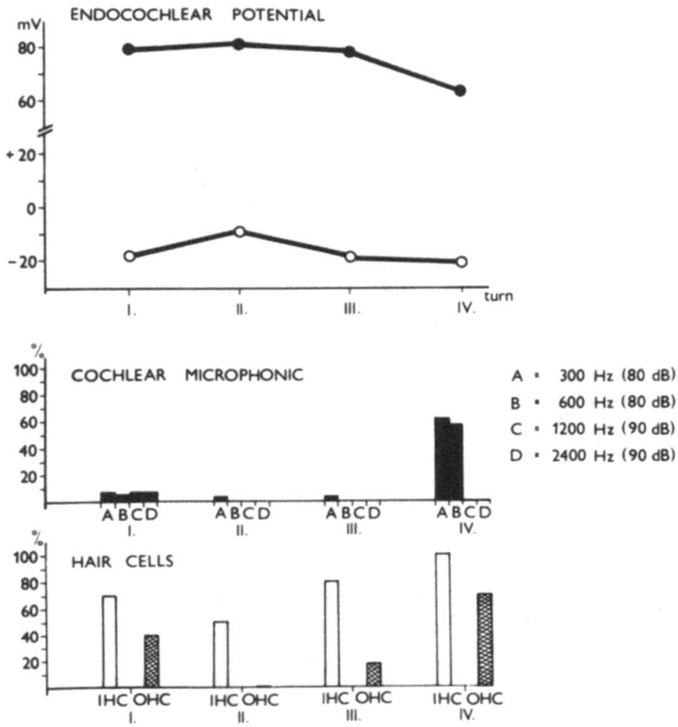

Fig. 6. EP, anoxic negative EP, distribution of IHCs and OHCs, and relative amplitudes of cochlear microphonics (CM) to 300, 600, 1200, and 2400 Hz tones (expressed as percentage of amplitudes found in control nonexposed animals) in four cochlear turns in a guinea pig 20 days after noise exposure. (Syka et al. 1981)

negative component of the EP is dependent not only on the K^+ diffusion potential, but also on the functional state of the organ of Corti.

The decreased permeability to ions of the noise-damaged cochlea has been shown using other techniques (Konishi 1979, 1982; Konishi and Salt 1980). In animals exposed to broad-band noise at levels between 95 and 115 dB for 7 days, the significant increase in the EP was accompanied by an increase in K^+ and Cl^- concentration in endolymph. The concentrations in the perilymph were not changed. The uptake of radioactive ^{43}K from perilymph to endolymph was decreased. The noise does not seem to alter the permeability to K^+ of Reissner's membrane, but significantly decreases the K^+ permeability of the barrier between the scala media and the scala tympani. In addition, it has been found that the rate of decline of the electrochemical potential difference for K^+ across the endolymph-perilymph barrier during anoxia is reduced in noise-exposed animals when compared to normal guinea pigs (Konishi and Salt 1980). In agreement with this is the finding that the value of the anoxic negative potential in noise-exposed animals is more positive (Konishi and Salt 1980; Melichar et al. 1980; Syka et al. 1981).

Although the decrease in the permeability to ions of the endolymph-perilymph barrier is unequivocal, this does not necessarily exclude participation of pathological changes in the stria vascularis in the observed changes in the transport of ions and the maintenance of their concentrations in the cochlea. Several authors have demonstrated morphological changes in the stria vascularis after high levels of noise exposure, which were characterized by vasoconstriction, strial edema, temporary disappearence of intermediate cells, and abnormal vessel transport (Hawkins 1971; Lipscomb and Roettger 1973; Duvall et al. 1974; Lim 1976). The stria vascularis may become detached from the spiral ligament after excessive noise exposure (Úlehlová 1983). The time course of the morphological changes in the stria after the noise exposure, described by Duvall et al. (1974), corresponds roughly with the time course of the EP and K^+ changes after acoustic trauma (Melichar et al. 1980), including the recovery to normal values over 5 – 20 days postexposure. It is not clear at present whether the observed increase in the EP after overstimulation with loud sound can be attributed also to changes in pump mechanisms of the stria vascularis. In guinea pigs exposed to noise 20 days before the recording, higher intracellular potentials of marginal cells were observed (Melichar and Syka 1987a). While a positive intracellular potential of about +10 mV was measured in marginal cells in vitro in non-exposed animals, +17 mV was found in noise-exposed animals.

In summary, there is strong evidence that acoustic overstimulation disturbs ionic transport mechanisms in the cochlea. At lower intensities, which produce TTSs, the effects occur at the level of ion channels, especially potassium channels, which are blocked. The resulting reduction of potassium standing currents is indicated by changes in the potassium within the cortilymph and by changes of the intracellularly as well as the extracellularly recorded DC receptor potentials. The endocochlear potential increases, the K^+ and Cl^- concentrations in endolymph are higher, and the Na^+ concentration is lower than before exposure. Further increases in the noise intensity result in mechanical damage to the cochlear partitions, including holes in the organ of Corti and basilar membrane, and the communication between endolymph and perilymph spaces results in a decrease of K^+ in scala media. The breakdown of the function of the stria vascularis is signaled by a significant decrease in the positive EP. In about 5 – 20 days after exposure to noise, the values of the EP and K^+ concentration in endolymph return to normal.

2.4 Threshold Shifts

Noise exposure produces characteristic changes in the hearing threshold which are dependent on the parameters of the noise. Short-term exposure with relatively low intensities (e.g., 100 dB SPL sound for 1 min) results in a hearing loss which is only temporary and the threshold recovers after some time (TTS). Exposure to high intensities even for a short period produces damage in the cochlea which is permanent and the threshold shifts are therefore permanent (PTS). If the noise exposure is of long duration (hours or days) and low intensity, the threshold shift first increases reaching after some time an asymptote (asymptotic threshold shift, ATS) and this shift then stabilizes. It depends on the parameters of the noise whether the threshold after cessation of the long-term exposure fully recovers or whether some residual PTS remains. Many aspects of the different threshold shifts have been discussed in extensive reviews (Henderson et al. 1976; Hamernik et al. 1982). Several additional topics which have been studied in animal experiments in recent years will now be considered.

2.4.1 Possible Mechanisms of the Temporary Threshold Shifts

The primary source of mechanical input to the cochlear hair cells during acoustic stimulation is the relative shear between the tectorial membrane and their apical parts produced by basilar membrane motion. The displacement of the stereocilia modulates their transducer conductances. Stimulation by loud sound may cause changes in the mechanical properties of the stereocilia and consequently in conductances of the hair cells. Indeed, several studies have indicated that intense acoustic stimulation changes the micromechanical properties of the sensory hair bundle at the apical pole of the hair cells (Miller et al. 1985; Saunders et al. 1986a). There are many indications that after acoustic overstimulation manifold morphological changes in stereocilia appear, which to some extent may disappear during several days or weeks after exposure.

The development of several new methods for measurement of micromechanical and electrical parameters of the responses of hair cells to acoustic stimuli have opened new perspectives in the investigation of basic mechanisms of TTSs. The stiffness of hair cells has been measured directly under visual control (Strelioff and Flock 1984). The stiffness was defined as the force required to produce a linear 1-μm deflection of the hair-bundle tip. A 1- to 2-μm diameter quartz-glass fiber probe was used to measure the stiffness of stereocilia from isolated coils of the guinea-pig cochlea. The effects of acoustic overstimulation on the stiffness of hair-cell stereocilia have been measured with a slightly modified technique (Saunders et al. 1986a, b; Saunders and Flock 1986). The stereocilia were stimulated by a 100-Hz oscillating water jet delivered through a glass micropipette, rather than by a quartz fiber probe. Overstimulation, which was related to threshold stimulation as detected by visual control, led to a decrease in stiffness which eventually occurred in all sensory hair bundles of all hair-cell rows. The OHC1 stereocilia showed the greatest loss of stiffness, those of OHC2 less, and those of OHC3 and IHCs least of all. The differences in susceptibility of stereocilia to overstimulation were more marked for cells in turns 2 and 3. Practically all threshold shifts as indicated by stiffness losses recovered in the course of several minutes. The recovery process was related to the amount of the threshold shift, i.e., the longest times for recovery of the stiffness to prestimulation levels were observed in OHC1. While the correlation between these results from in vitro studies and those from in vivo studies (Robertson and Johnstone 1980; Liberman and Mulroy 1982) showing the susceptibility of different hair cells to acoustic overstimulation is tempting, there are several limitations to a straightforward comparison. The cells in vitro are not in contact with the tectorial membrane, and their viability is limited. There are differences in stereociliary morphology between hair cells of different rows and cochlear locations (Iurato 1961; Lim 1980; Wright 1981, 1984). Generally, the stereocilia on OHC1 are the shortest, while those on OHC3 and IHC are longest. As Saunders and Flock (1986) have pointed out, it takes about 15 dB more stimulation to produce a threshold response in OHC1 stereocilia than in OHC3 stereocilia. If the stimulus delivered to the sensory hair bundle is normalized for the height of stereocilia then the amount of threshold shift among hair-cell rows would be practically the same. Saunders and Flock (1986) have suggested that the loss of stereociliary stiffness after acoustic overstimulation is due either to injury in the rootlet region of the hair bundle (Tilney et al. 1982; Liberman and Dodds 1987) and or to broken interconnecting links between stereocilia (Pickles et al. 1984; Comis et al. 1985). In further experiments Saunders et al. (1986b) have demonstrated that active cellular processes control and limit the changes in stereociliary micromechanics. The threshold shift after block of metabolism by cooling or by poisoning with cyanide did not reach a plateau, as in normal control cells, and no postexposure recovery of threshold shift was seen in stereocilia bundles of metabolically blocked cells. Changes in stereocilia stiffness after long-term noise exposure have been reported by Canlon et al. (1987). IHC stereocilia showed after noise exposure (1 kHz pure tone at 105 dB SPL for 72 h) a decrease in stiffness while the OHC stereocilia remained unaltered. This effect occurred after 1 h exposure and remained constant with exposures up to 7 days. After 1 month recovery from noise exposure (with a decrease of the threshold shift from 50 dB to 30 dB) normal threshold values of stereocilia movement were observed.

Whatever the mechanisms of the loss of stereociliary stiffness, these results have serious implications for the function of IHCs and OHCs. Stereocilia and apical membranes of hair cells are places where transduction channels for ions, basically for potassium, are localized. Because of the difference in potassium concentration across the membranes of hair cells and the presence of a membrane potential of about 120–160 mV, the receptor current is apparently carried by potassium ions travelling down their electrical gradient into the hair cells. By implication potassium ions have to leave the hair cells through the basolateral membrane along their chemical gradient. It has been shown by Russell (1983) and Ashmore and Meech (1986) that the hair-cell membrane is highly permeable to potassium. Assuming that after entering the hair cells through the apical membranes, potassium ions will be driven through the basolateral membrane to the extracellular space, it may be expected that potassium will accumulate within the extracellular space of the organ of Corti. This accumulation of potassium in the extracellular space will depolarize the hair cells. Indeed, Cody and Russell (1985) have shown that the OHC membrane potential depolarizes during acoustic overstimulation. The changes in the potassium concentration in the organ of Corti in guinea pig during loud sound stimulation were examined by Syka et al. (1987). The K^+ concentrations were measured with double-barrel K^+ selective microelectrodes, which were inserted into the organ of Corti in a manner similar to that described by Russell and Sellick (1978) for intracellular recording from hair cells. The resting K^+ level in cortilymph was found to be slightly higher (by about 0.4 mmol/l) than in perilymph. During stimulation with moderate sound intensities the K^+ concentration increased and after cessation of stimulation (10–30 s duration) decreased to prestimulation level. The maximum increase in the K^+ level

(3 mmol/l) during stimulation was observed in the vicinity of the IHCs. During sound stimulation loud enough to produce temporary threshold shift (105 dB SPL for 30 s), however, a different pattern of K^+ concentration change appeared. After an initial increase, the K^+ level started to decrease during stimulation and after cessation of the stimulus it decreased to K^+ levels lower than the normal resting level. Subsequent stimulation produced a smaller increase in K^+ concentration without the undershoot after stimulation. It may be assumed from the results that loud sound stimulation results in reduction of potassium standing currents through hair cells. It seems to be possible that processes which lead to damage to the micromechanics of stereocilia as described by Saunders and Flock (1986) are responsible for blocking potassium transduction channels in hair-cell membranes and consequently result in reduction of potassium standing currents.

Undoubtedly threshold shifts are accompanied by changes in neurotransmitters both in the cochlea and the central auditory pathway. The increasing amount of information on the distribution and role of neurotransmitters in the auditory system is in contrast with virtually no available data about their changes during and after noise exposure. Drescher et al. (1983) have shown an increase in the concentration of met-enkephalin-like component and GABA-like component in perilymph after noise exposure to 115 dB SPL. These results have recently been supported by Eybalin et al. (1987) who observed a significant decrease in endogenous met-enkephalin levels in the cochlear tissue after exposure to white noise (110 dB SPL) for 1 h. Data on changes in neurotransmitters within the cochlea may significantly contribute to the understanding of the mechanisms of TTSs.

2.4.2 Modulation of Temporary Threshold Shifts

In experimental animals, the amount of TTS, which is essentially constant for constant noise exposure, may be changed under certain conditions. Probably most impressive is the finding that monaural TTS produced by an intense pure tone is decreased if an acoustic stimulus of the same frequency is presented simultaneously to the other ear (Cody and Johnstone 1982a). The TTS in guinea pigs evoked by a 10 kHz pure tone at 107 dB SPL for 1 min is decreased by about 30% by simultaneous exposure of the other ear to a 10 kHz pure tone at 85 dB SPL. When the other ear is stimulated with a different frequency, the reduction in TTS as measured by the reduction in CAP does not occur. Also when the animal is injected intraperitoneally with strychnine, the reduction in TTS caused by simultaneous binaural stimulation no longer occurs. It is therefore suggested that acoustic activation of auditory efferents is responsible for the reduced ipsilateral sensitivity loss.

Other experimental modulations of TTS lead to less unambiguous results. Simultaneous presentation of a second tone (10.5 or 2 kHz) with the original tone (16 kHz) produces significantly less threshold shift than does the 16-kHz tone alone (Cody and Johnstone 1982b). The reduced threshold shift disappears as the intensity of the second tone is decreased. The cochlear nonlinearity obtained with two-tone exposure is similar to the nonlinearities of the "two-tone suppression" type. However, in related psychophysical experiments, Smoorenburg

(1980) was not able to demonstrate an effect of TTS on two-tone suppression in man. His TTS was not large (less than 10 dB), for obvious reasons, and therefore any straightforward comparisons are not possible.

In a similar study, Rajan and Johnstone (1983) analysed residual effects in monaural TTS caused by pure-tone exposures in guinea pigs. The reduction in TTS to high-frequency exposure was measured when the exposure was preceded by stimulation with the same tone but at lower intensity. As a rule, the thresholds for the first, lower intensity exposure, must recover completely to reduce the TTS produced by a second stimulation. The protective effect of the prior exposure was not affected by intraperitoneal injections of strychnine which were supposed to abolish the activity of olivocochlear efferents. The animals were also paralyzed with a muscle relaxant to remove the influence of the middle ear muscles. However, the experimental data of Rajan and Johnstone (1983) do not find their counterpart in psychophysical studies in man, which generally demonstrate either no effect of priming or even increased effects of second exposure (Trittipoe 1959; Ward 1960). However, there was no control of the middle ear reflexes in the psychophysical studies.

2.4.3 Asymptotic Threshold Shift

Long exposure to noise results in a limited growth of the threshold shift, which after some time reaches an asymptote and does not increase further (Kryter 1963). For exposures to noise centered at about 1800 Hz, TTS will be asymptotic after 240 – 480 min of exposure. For exposures below about 1800 Hz, durations in excess of 480 min are needed to produce asymptotic TTS (Kryter et al. 1966). The concept of asymptotic threshold shift was later extensively studied in the Central Institute for the Deaf in St. Louis. Mills et al. (1970) exposed one subject for 1 or 2 days to an octave-band noise centered at 500 Hz at 81 or 92 dB and observed characteristic asymptotic TTS. The threshold at 750 Hz increased for 8 – 12 h and then remained unchanged. Threshold shifts were proportional to exposure intensity. The time constant of temporal integration at 750 Hz was reduced, there was delayed recruitment of loudness, the amplitudes of Békésy tracings were reduced, but frequency discrimination was not affected. Recovery from TTS required 3 – 6 days (Mills et al. 1970). The data were later confirmed by Melnick and Maves (1974) on 10 subjects exposed to low frequency octave-band noise at 90 dB for a period of 24 h. The asymptote was attained by the 12th h. In another study using exposures to 4 kHz for 24 h, the maximum TTS occurred at 4 and 6 kHz (Melnick 1977). An increase from the 80 dB SPL exposure to 85 dB SPL resulted in a proportional increase in TTS, while the time to attain the asymptote was not essentially different for those two intensities. Miller (1974) summarized the data on human TTS growth and decay with respect to the intensity of the exposure and plotted curves for hypothetical growth and recovery of TTS using data from subjects exposed to a band of noise centered at 4 kHz. The effects of exposure to noise of long duration on experimental animals, mainly chinchillas, have also been studied. The changes in TTS parameters produced data essentially similar to that obtained in human volunteers. In chinchillas the TTS produced by prolonged monaural exposures to an

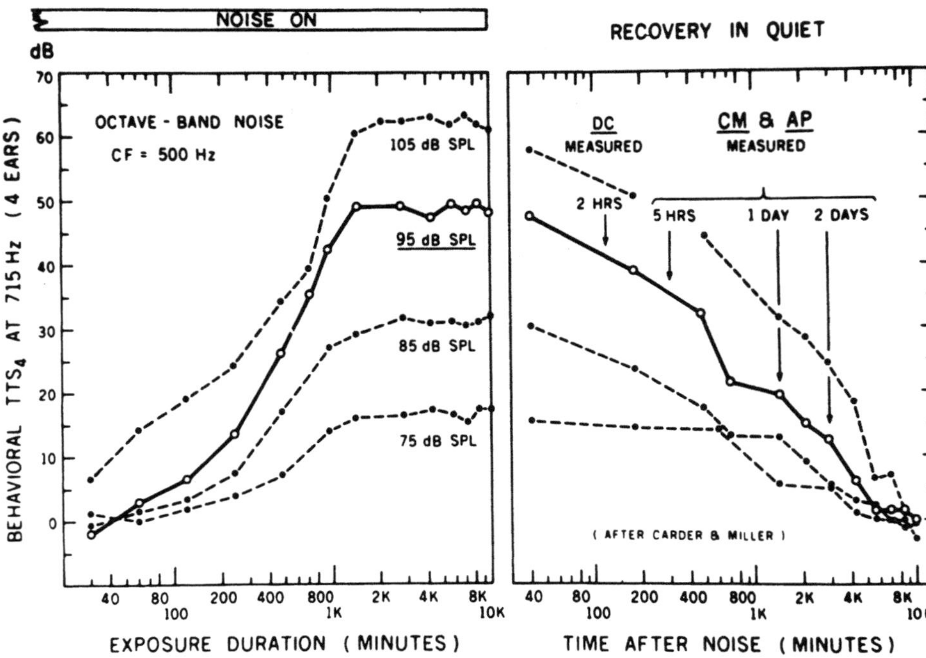

Fig. 7. Growth and recovery of temporary threshold shifts during exposure to an octave band of noise (CF = 500 Hz) at four different sound pressure levels. (Benitez et al. 1972)

octave-band noise centered at 0.5 kHz reached an asymptote after about 24 h of exposure (Fig. 7). The TTS at asymptote, in the frequency region of the maximum effect, increased by 1.6 dB for every 1 dB increase in the noise level above 65 dB. At asymptote, the threshold was elevated to a predictable absolute sound pressure level, regardless of the animal's original threshold (Mills 1973). The TTS reverted from its asymptotic value to zero in 3 to 6 days. The results were also confirmed with octave-band noise centered at 4 kHz (Carder and Miller 1971, 1972; Mills and Talo 1972). Hearing thresholds in all experiments were estimated with behavioral audiometry, on the basis of instrumental avoidance conditioning.

The auditory thresholds estimated with a behavioral conditioning procedure in the chinchilla have been compared with the thresholds of single auditory nerve fibers and of neurons in the cochlear nucleus at their characteristic frequency (CF) (Salvi et al. 1978, 1983). In both cases prolonged exposures to noise were used (3.5 – 5 days), with the aim of producing asymptotic threshold shifts. The threshold elevations in these experiments were between 60 and 75 dB. The behavioral thresholds and these at the CF of the most sensitive units were within 10 – 15 dB of each other. The threshold elevations were flat when the CF of the noise was 500 Hz, whereas only a higher frequency threshold loss was observed when the CF was 4 kHz (Salvi et al. 1983). Because the histological control always revealed discrete pathological changes of hair cells and other cells in the organ of Corti in these animals, it may be expected that a significant permanent threshold shift developed. Of particular interest is whether the concept of the asymptotic threshold shift is valid also for very long-lasting exposures. In animals exposed to moderate levels of noise for 90 days it was found that beyond the 15th day of exposure, there was a 9 – 19 dB increase in thresholds over the initial asymptotic level (Mills 1976; Moody et al. 1976). The question was carefully studied later by Bohne and Clark (1982) who exposed chinchillas to octave-band noise centered at 0.5 kHz at 95 dB SPL for up to 108 days. Exposures of short durations (up to 9 days) produced scattered losses of OHCs in the low-frequency region of the cochlea. The threshold shifts were temporary and averaged 40 dB for low-frequency tones and 43 dB for high-frequency tones. No permanent threshold shifts were present. With longer exposures lesions started to appear in the high-frequency region of the cochlea which comprised both OHCs and IHCs as well as supporting cells. With increasing duration of exposure the damage in both the low-frequency and the high-frequency regions continued to increase. In contrast, the overall threshold shift, estimated by a behavioral conditioning procedure, remained practically constant. Thus, the asymptotic threshold shift for 108 days of exposure was confirmed. In contrast, the permanent threshold shift (PTS) increased systematically as the duration of exposure was lengthened. In the case of animals exposed for 108 days the PTS function was within 10 dB of the level of asymptotic shift. It is not quite clear, therefore, whether any asymptote exists for the permanent hearing loss, given a constant intensity of exposure.

Experimental models are helpful in understanding the effects of noise on man particularly when the parameters of the noise exposure are similar to those occurring in normal everyday life. Typical exposures of workers in industry are intermittent, whereas the majority of data in experimental animals have been obtained from continuous exposures. When comparing chinchillas exposed to continuous or interrupted noise (4 kHz octave-band noise at 86 dB SPL), the incidence and average size of the lesion in the basal turn is reduced in all cases receiving interrupted exposures (Bohne et al. 1987). In another series of experiments chinchillas were exposed to intermittent exposures at 0.5 kHz, 95 dB SPL either 6 h per day for 36 days or 15 min/h for 144 days (Clark et al. 1987, Sinex et al. 1987). Significantly less hearing loss (estimated behaviorally) and cochlear damage (estimated histologically) was observed for intermittent schemes in comparison with continuous exposures of equal energy. Behaviorally estimated thresholds, AP thresholds, and auditory nerve fiber thresholds initially increased by about 40 dB and later recovered by as much as 30 dB of sensitivity even though the exposure was continued. Threshold shifts produced by noise on interrupted schedules did not produce asymptotic threshold shifts.

In contrast to the extensive literature on the physiology and pathology of the guinea-pig auditory system, the problem of asymptotic threshold shift in the guinea pig has been studied only by Syka and Popelář (1980). This is probably due mainly to the fact that the guinea pig fails in many routine conditioning procedures. Syka and Popelář (1980) adopted a conditioning procedure based on the positive reinforcement paradigm described by Petersen et al. (1977). Auditory thresholds were assessed before, during, and after exposure to one-third octave-band noise centered at 2 kHz at 100 dB SPL for 5 days. The results were comparable with those observed in chinchillas under identical conditions. The

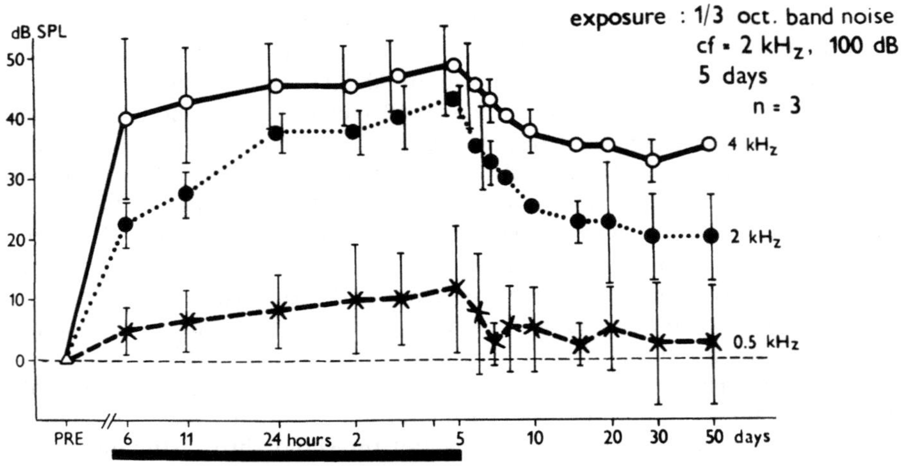

Fig. 8. Growth and decay of threshold shifts at 0.5, 2, and 4 kHz during and after the noise exposure. (Syka and Popelář 1980)

threshold shift at 4 kHz reached 40 dB after 6 h of exposure and recovered only slightly during the 5 days after exposure; however, a 35-dB permanent threshold shift remained in 4 kHz region (Fig. 8). For the 0.5-kHz region, which was below the exposure frequency, the TTS amounted 10 dB by the 5th day and completely recovered after exposure. While the amount of asymptotic threshold shift was similar to that for chinchillas, the permanent threshold shift was larger and also the asymptote was attained earlier in the guinea pig (at least for the 4 kHz). The parameters of exposure were not completely identical in all experiments and therefore any definitive conclusions on interspecies differences in noise susceptibility would be premature.

Asymptotic threshold shift has also been found in monkeys. In the genus *Macaca* the characteristic asymptotic curve of the threshold shift is attained after approximately 10 days' exposure (Moody et al. 1976). In the squirrel monkey the increase in TTS during 4 days of exposure to 0.5- or 4-kHz octave-band noise at different intensity levels was examined and a plateau, expected to occur in the case of asymptotic threshold shift, was not seen (Nielsen 1982). It seems that in the monkey the growth of the threshold shift is slower than in the chinchilla and guinea pig, and, surprisingly, also slower than in man. Further experiments will be necessary to explain these discrepancies.

2.5 Frequency Discrimination and Noise Exposure

2.5.1 Psychophysical Studies

There is much evidence that noise-induced hearing loss is accompanied by abnormalities in frequency discrimination. First a reduction in frequency resolution is consistently found in such measures as critical bands, auditory filter shape, or psychophysical tuning curves. Furthermore, for a short period of time after noise exposure there are temporary pitch shifts (diplacusis) which gradually decline. In addition the just noticeable difference (jnd) for frequency may change after noise exposure.

The currently most popular method for measuring frequency-resolving power is the measurement of psychophysical tuning curves. The estimation of frequency resolution is given as a power of the masking produced by the masker on audibility of a pure tone – a probe. The forward-masking paradigm is used in this case, i.e., the masker precedes the probe. The advantage of this method is that the psychophysical tuning curves are essentially similar to the tuning curves for individual auditory neurons. Practically all studies have concluded that hearing loss is accompanied by reduced frequency resolution. For example, Wightman (1982) characterizes the changes as a broadening of the psychophysical tuning curves (PTCs), which occurs not only in the region of the loss of pure-tone sensitivity (with threshold increases larger than 40 dB), but also in regions of normal pure-tone sensitivity. Of course, the broadening of PTCs was more obvious in the region with the increase in threshold. However, in some subjects completely normal PTCs were found in regions with large threshold shifts. Similar findings have also been published by Bonding (1979a) and by De Boer and Bouwmeester (1974), although in this case critical bandwidths were measured.

Auditory filter shapes have been found to be different in subjects with cochlear impairments from those estimated in normal hearing subjects (Glasberg and Moore 1986). The lower branch of the filter is much less steep in the impaired ear in comparison with the normal. Deficits in discrimination at hearing-loss frequencies were also manifested by more gradual slopes of psychoacoustic functions for frequency discrimination (Nelson and Freyman 1986). Moore and Glasberg (1986) compared frequency selectivity in stimultaneous and forward masking of subjects with unilateral cochlear impairments. PTCs for normal ears were sharper in forward masking than in simultaneous masking, but for the impaired ears the PTCs were similar for both types of masking. The results demonstrated that suppression is reduced but not completely absent in cases of moderate cochlear hearing loss. Surprisingly, the frequency discrimination in the presence of low-frequency masking noise deteriorates to the same extent in hearing-impaired and in normal listeners (Turner 1987). In both groups similar increases were found in frequency difference limen in the presence of noise.

The changes in pitch after noise exposure, i.e., the temporary pitch shifts, are closely related to the changes in frequency discrimination. In 1950 Davis et al. showed that exposure to intense tonal stimuli (120 dB or more for 32 or 64 min) had no effect on test frequencies close to or below the exposure frequency. However, large shifts in pitch approximately one-half octave or more above the frequency of the exposure tone were observed. As noted by McFadden and Plattsmier (1982), Davis et al. (1950) found little, if any, effect of pitch matching from exposure to a broad-band noise. Other studies in this field have disagreed with respect to the direction of the pitch shifts. For example, Ward (1963) observed downward shifts of between 2.5% and 6.5% for frequencies of 150 Hz and 500 Hz, following exposure to high-intensity clicks, and Békésy (1963) found nearly symmetrical shifts downward and upwards away from a low-frequency exposure tone. In their review, McFadden and Plattsmier (1982) suggest that the differences in the results may arise from the differences of the exposures parameters: while Davis et al. (1950) exposed subjects for a relatively long time, the exposures of Békésy (1963) were only for 2 min and 94 dB SPL. Apparently downward pitch shifts disappear after longer exposures.

Alterations in frequency discrimination have also been reviewed by McFadden and Plattsmier (1982). The differences in exposure parameters result in data which are more or less quantitative in showing, for example, changes in the range of 10 – 15 Hz in Δf at 1000 Hz, but only at low sensation levels (Brandt 1967). Generally, the data published for subjects with cochlear impairment demonstrate abnormally large jnd's for frequency (see McFadden and Wightman 1983).

2.5.2 Frequency Tuning Curves in Animals

Changes in threshold tuning of the auditory nerve in cats with experimental hearing loss induced by ototoxic antibiotics were first described by Kiang et al. (1970). Data have accumulated since then showing specific changes in tuning properties of auditory nerve fibers, spiral ganglion cells, and even hair cells after exposure to noise. In cats, after exposure to narrow-band and broad-band stimuli from 100 dB to 115 dB SPL, the tuning curves for auditory units which were connected with the impaired cochlear regions usually displayed elevation of the tips (Kiang et al. 1976). However, in some cases the sensitivity at the low-frequency part – or tail part – of the tuning curve, was increased and this has been called hypersensitivity. The changes found by Kiang et al. (1976) were permanent, since they were measured many days or months after exposure. The focus in this study was the description of global changes in characteristic frequencies of many auditory nerve fibers in relation to the extent of damage in the cochlea. Their plots of unit thresholds at CF were correlated with patterns of hair cell losses in the cochlea: significant hair cell loss was always accompanied by highly abnormal thresholds.

The effects of noise exposure on cochlear and auditory nerve structure and function have been analysed in greater detail by Liberman and Mulroy (1982). Cats were exposed to narrow-band stimuli centered at frequencies between 0.5 and 15 kHz for 1 – 2 h, at intensities from 100 to 125 dB SPL. Measurements of auditory nerve function and subsequent histological preparation were performed 30 – 620 days after exposure. In contrast to the situation in normal animals, many auditory nerve fibers were unresponsive to sound, although some showed spontaneous activity. Some of the units which responded to sound showed several abnormalities in tuning curves (Fig. 4). Some of them preserved more sharply tuned tips and flatter tails (they were described as "V-shaped" curves); others had broadly tuned tips and tails, with possible hypersensitivity of the tail (called "W-shaped" curves), while still others were broadly tuned or flat, consisting only of a tail. Liberman and Mulroy (1982) have found a characteristic distribution of the abnormal tuning curves with respect to the region of threshold shift where narrow band noise was effective: broadly tuned flat tuning curves were distributed within the region of the threshold shift. The V shape tuning curves were found mostly at the low-frequency border of the threshold shift, and those with W shape abnormalities were observed predominantly at the high-frequency edge of the abnormal region.

In the guinea pig, changes in the tuning properties of single auditory neurons (in the spiral ganglion) have been studied by Cody and Johnstone (1980) after repeated exposures to continuous pure tones at 100 dB SPL. Tips of tuning curves were elevated, and the characteristic frequencies had shifted to lower frequencies. Consistent with this was a decrease in the slope of the high-frequency cut-off of the tuning curve. In some cases the authors observed also hypersensitivity of the tuning-curve tails, however, the hypersensitivity disappeared after repeated stimulations.

2.5.3 Auditory Nerve Action-Potential Tuning Curves After Noise Exposure

Auditory nerve action-potential (AP) tuning curves are sensitive indicators of the impairment of cochlear frequency sensitivity. They may be measured in both experimental animals and human subjects (during transtympanic electrocochleography). Two procedures, both based on masking of the action potential evoked with a probe tone, are used: in the simultaneous masking procedure the masker tone and the test tone are presented simultaneously, while in the forward-masking procedure the masker precedes the test tone. In normal subjects the simultaneous masking procedure gives less sharply tuned AP tuning curves than does forward masking. It is supposed (Dallos and Cheatham 1976, 1977; Harris 1979) that two-tone suppression, a nonlinearity of cochlear origin, distorts the sharp tuning of individual auditory nerve fibers which contribute to the AP.

Generally, measurements of AP tuning curves in pathological cochleas have shown tuning curves which are broader than those found in normal cochleas. Human AP tuning curves in pathological cochleas using simultaneous masking were first reported by Eggermont (1977). Harrison et al. (1981 a) have measured AP tuning curves with simultaneous and forward masking in normal human subjects, in patients with sensorineural hearing loss, and in guinea pigs after treatment with ototoxic antibiotics. Broad AP tuning curves with low Q_{10} values were found in patients with sensorineural hearing loss; the tuning had deteriorated by a factor of two to three. The AP tuning curves obtained with forward masking in these patients were as broad as those measured with simultaneous masking.

In the cat, AP tuning curves before and after noise exposure (broad-band noise at 105 dB SPL) were reported by Van Heusden and Smoorenburg (1981 b). After noise exposure the AP tuning curves (estimated with forward masking) were wider. The broadening was not a simple function of the increased threshold, because in normal animals the AP tuning curves measured at the higher SPL were narrower than those found at the same absolute SPL after exposure.

2.6 Intensity Functions

Suprathreshold intensity discrimination is frequently examined in psychophysical and clinical studies, where loudness recruitment is usually used as an indicator of sensorineural deafness. Several recent reviews of the problem are available (McFadden and Plattsmier 1982; McFadden and Wightman 1983). It should be remembered that the first psychophysical experiments concerning loudness perception in noise-exposed subjects were performed by Davis et al. (1950), and their results were essential for further research. They used the alternate binaural loudness balance method and demonstrated a temporary recruitment-like effect on loudness perception. This temporary loudness shift is similar to TTS, in that the maximum effect occurs in a frequency region one-half to one octave above the exposure frequency.

Correlates of the loudness shift are found also in many animal experiments, although the real mechanism of recruitment is still unknown. Van Heusden and Smoorenburg (1981 a) observed steeper slopes in the amplitude versus intensity functions of the eighth nerve action potential in cats exposed to noise. Syka and Popelář (1980) found steeper psychophysical functions in guinea pigs exposed to third-octave band noise centered at 2 kHz. The slopes of functions at frequencies higher than 2 kHz were increased.

By far the most extensive data about intensity functions are those on auditory nerve fibers. In normal auditory nerve fibers, the slopes of discharge rate-vs-intensity functions to pure tone stimuli are frequency dependent. In normal cat cochlear fibers, for stimulus frequencies below the CF, the slope is steep compared with those above the CF (Sachs and Abbas 1974; Evans 1974). In hypoxia this relationship changes so that intensity functions measured at various frequencies above and below CF have similar slopes (Evans 1974). In noise-exposed animals similar changes in the steepness of the intensity function after exposure have been reported by Harrison (1981). The rate functions at the characteristic frequency became abnormally steep and comparable to those of the low-frequency tail region of the normal cochlear fibers. Harrison (1981) extended the finding also to AP suppression-vs-intensity functions, which he measured both in guinea pigs and in human subjects. In both cases the AP suppression curves indicated a loss in frequency dependency of the rate function slopes and the predicted steep rate functions at CF.

The frequency-dependent nonlinearity, which is present in normal animals in rate-intensity functions, is a consequence of the nonlinearity present in the amplitude intensity functions of the DC receptor potentials in IHCs (Russell and Sellick 1978; Patuzzi and Sellick 1983). In guinea pigs with TTS evoked with loud sound stimulation (105 dB SPL) the extracellularly recorded DC receptor potentials in the organ of Corti (and similarly the potassium-concentration functions in cortilymph) are steeper for the CF and for frequencies higher than CF (Syka et al. 1987). Thus, the mechanism of the frequency-dependent nonlinearity in intensity functions must already be present at the level of the organ of Corti.

A model for loudness recruitment, i.e., for an explanation of abnormally steep intensity functions in sensorineural hearing loss has been proposed by Kiang et al. (1970) and Evans (1976). The model is based on the number of active units in the population of auditory nerve fibers. In animals showing pathological changes it is supposed that there exists an intensity range over which a larger than normal percentage of units is activated with each increasing level of stimulus. The activation is more effective over tails of tuning curves than over their tips (Fig. 9). Salvi et al. (1983) have successfully tested this hypothesis in auditory nerve fibers of chinchillas exposed to noise.

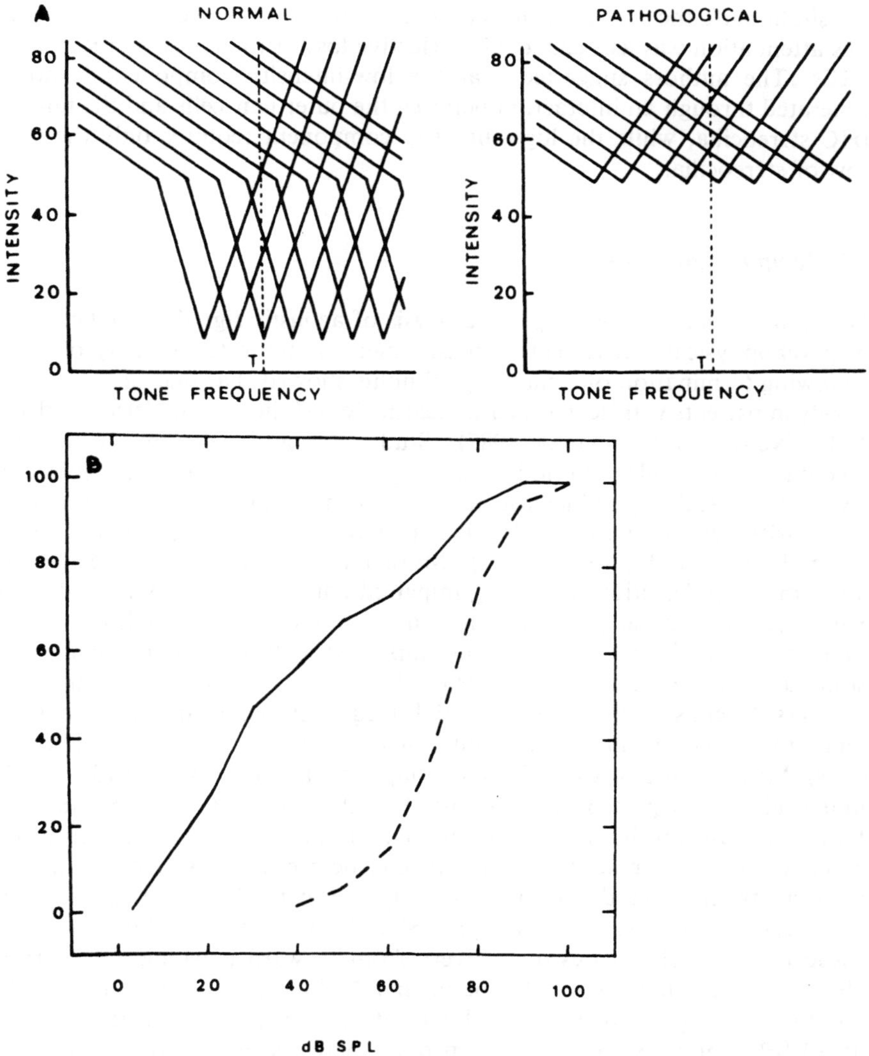

Fig. 9. A Idealized tuning curves (*solid line*) in normal and pathological animals according to Evans (1976). Units begin to respond to the tone (*dashed line*) at the intensity where the dashed line cuts across the tuning curve. **B** Percentage of units in the normal (*solid line*) and noise-exposed (*dashed line*) chinchillas that were activated as the intensity of a 2 kHz tone was increased. (Salvi et al. 1983)

Another aspect of the rate-intensity function of auditory nerve fibers, its nonmonotonic behavior at high intensity levels, has been studied by Liberman and Kiang (1984). The rate and phase-level functions of normal auditory nerve fibers may be separated into a low-intensity component and high-intensity component; the separation is indicated by a dip in the rate function and a simultaneous sudden shift in the phase function. In cats with a noise-induced threshold shift the low-intensity component was attenuated or abolished, and this attenuation was associated with selective loss of the tallest stereocilia of the IHCs. The authors suggested that the low-intensity component is normally generated through an interaction between the outer hair cells and the tall row of IHC stereocilia, while the high-intensity component requires only the shortest row of stereocilia.

2.7 Temporal Analysis

Different aspects of the temporal analysis of acoustic signals have been studied in psychophysical experiments. Measurement of the rate of decay of masking following termination of a masking stimulus showed that masking decays more slowly in patients with sensorineural hearing loss (Elliot 1975; Nelson and Turner 1980; Nelson and Freyman 1987). Studies of gap detection, which means recognition of a silent period in a long-duration stimulus, were less conclusive. Both a tendency to increase and quasinormal values have been found in patients with sensorineural hearing loss (Jesteadt et al. 1976; Irwin et al. 1981; Fitzgibbons and Wightman 1982). More recent data support the concept of abnormal gap detection in hearing-impaired patients. Fitzgibbons and Gordon-Salant (1987a) found gap resolution in impaired listeners significantly poorer than in normal listeners. Resolution improved with broad-band signals at high stimulus levels. Similarly, as in normal listeners, the gap detection in impaired listeners depends on the content of high-frequency signals in the noise. Optimal temporal resolution in hearing impairment requires minimum sensation levels (SLs). Most listeners resolve threshold gaps at minimum levels of 25 – 35 dB SL, independent of degree of hearing loss (Fitzgibbons and Gordon-Salant 1987b). However, gap thresholds were usually larger for impaired ears, even when the comparisons were made at equal SLs (Glasberg et al. 1987). Also, the rate of recovery from forward masking is usually slower for impaired ears, even when the maskers are presented at equal SLs. Larger gap thresholds tend to be associated with slow rates of recovery from forward masking (Glasberg et al. 1987). In electrophysiological experiments Arehole et al. (1987) measured time constants fitting the forward-masking data in chinchillas exposed to 2 kHz at 85 dB SPL for 5 – 8 days. Time constants increased by up to a factor of three after the exposure.

Auditory summation is the phenomenon most frequently studied with temporal analysis. It is mostly examined as the relationship between the threshold of a tone and its duration. In normal subjects the threshold to a long-duration tone is approximately 12 dB lower than that to a short-duration tone. In noise-exposed subjects the difference decreases (Jerger 1955; Wright 1968; Mills et al. 1970). Mills et al. (1970) have found this decrease in the difference at the frequency where the maximum TTS was found after exposure to noise centered at 500 Hz.

In the chinchilla a change in temporal integration after noise exposure has been reported by Henderson (1969). However, Syka and Popelář (1980) were not able to demonstrate any significant change in temporal integration after noise exposure in guinea pigs. Salvi et al. (1983) have suggested that reduction in temporal summation may be observed in the pattern of poststimulation histograms of auditory nerve fibers in noise-exposed chinchillas. However, no significant differences in the decay of poststimulation histograms with respect to normal animals have been found.

2.8 Acoustic Emissions and Cochlear Nonlinearities

Micromechanical processes in the organ of Corti are generally regarded as a source of several types of acoustic emissions, as well as such cochlear nonlinearities as the quadratic and cubic difference tones and two-tone suppression. It is important to know how these phenomena are influenced by noise exposure. In patients with sensorineural hearing loss of different origin, the cochlear echo is not present when the hearing loss exceeds 30 dB (Kemp 1978). Normal cochlear echoes can be observed as long as the hearing loss does not exceed about 15 dB (Rutten 1980). Cochlear echoes were also characteristically suppressed after noise exposure which resulted in TTS (115 dB SPL of 1250 Hz for 5 min; Kemp 1982). High-frequency components of the echo were particularly affected by the exposure. The recovery of the echo was similar to that of TTS: 1 day after exposure the echo was identical to that before exposure. TTS in the subject at one-half octave above the exposure frequency was around 20 dB, whereas the magnitude of the cochlear emissions decreased by about 10 dB. In experimental animals cochlear echoes were most easily recorded in primates (Anderson and Kemp 1979). They can be reduced or abolished by noise exposure or by treatment with ototoxic drugs.

Spontaneous emission is another type of acoustic emission which has been recorded in humans (Kemp 1981; Zurek 1981) and in some mammalian species such as chinchillas (Zurek and Clark 1981), guinea pigs (Evans et al. 1981), and dogs (Ruggero et al. 1983). Though the spontaneous emissions in the guinea pig was not affected by paralysis of the middle ear muscles, it was abolished under hypoxia, and it returned after restoration of a normal oxygen supply. It was suppressed by external tones of the appropriate frequency and intensity, and it responded to changes in middle-air pressure in the same way as do spontaneous emissions in humans. Spontaneous otoacoustic emissions were originally believed to be an objective correlate of tinnitus, but that idea has not been confirmed (Wilson 1986). For example, Zurek (1981) tested 32 normal hearing subjects and found spontaneous emissions in 22 ears from 16 of the subjects. None of these subjects was aware of tinnitus. Wilson and Sutton (1981) tested six tinnitus sufferers but found no corresponding spontaneous emissions.

The third type of acoustic emissions which can be measured in the ear canal are acoustic distortion products or cochlear nonlinearities. In contrast to cochlear echo and spontaneous emissions, they were found in many species of experimental animals. The acoustic distortion products are only 25 – 40 dB below the level of the primary tones and are easier to record. Usually, two-tone acoustic distortion products are recorded in the activity of the auditory nerve fibers. Siegel et al. (1982) produced PTS in the chinchilla by means of noise and showed that two-tone distortion products were abolished when the primary tones were associated with the region of injury in the cochlea. If the primaries were in the

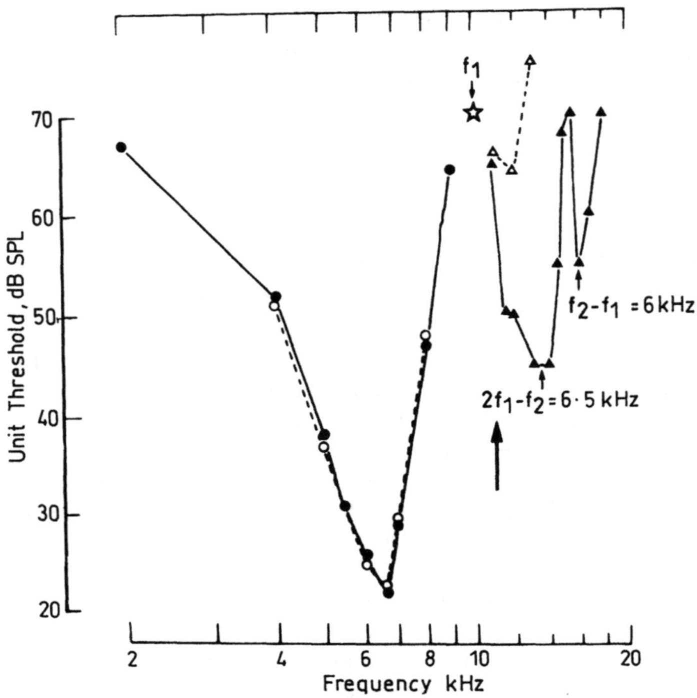

Fig. 10. Effect of exposure (1 min, 110 dB, 11 kHz) on $2f_1-f_2$ and f_2-f_1 response in a single auditory neuron. *Closed circles*, tuning curve before exposure; *open circles*, tuning curve after exposure; *closed triangles*, f_2 intensity required for threshold response before exposure; *open triangles*, f_2 intensity required after exposure; f_1 intensity indicated by *star*; *bold arrow* denotes exposure frequency. (Robertson and Johnstone 1981)

nondamaged region, the neural correlates of distortion products were normal. The generation of distortion products in the guinea pig may also be altered by TTS produced by a tone above the CF when the tone causes TTS in the region of primaries (Robertson and Johnstone 1981; Fig. 10).

In patients with sensorineural hearing loss, acoustic distortion products were found to be reduced or abolished (Smoorenburg 1972). Also TTS was combined with a decrease in the generation of combination tones (2f1 – f2), but another typical cochlear nonlinearity, two-tone suppression, was not changed (Smoorenburg 1980). This finding is not surprising in the context of other data, which also show a different influence of noise on two-tone suppression. Two-tone suppression is reduced or absent in damaged regions of the cochlea (Robertson 1976; Schmiedt and Zwislocki 1980; Salvi et al. 1982). However, some disagreement has been observed in details of the reduction of two-tone suppression. While Salvi et al. (1982) have reported the reduction to be correlated with the amount of threshold shift, less correlation data have been reported by Schmiedt and Zwislocki (1980). The amount of reduction in two-tone suppression has been shown to be correlated with change in the slope of the high-frequency leg of the tuning curve (Schmiedt et al. 1980). In contrast to this, Robertson and Johnstone

(1981) have demonstrated that two-tone suppression may be fatigued without alterations in the sharp tuning of a fiber. In spite of these small discrepancies, virtually all data suggest that both acoustic emissions and cochlear nonlinearities are vulnerable and susceptible to noise damage.

2.9 Changes in the Structure and Function of the Auditory Pathway After Noise Exposure

2.9.1 Morphological Changes

Evidently the main pathological effects of noise are the changes which can be observed in the inner ear. In contrast to the sensitivity of the inner ear receptors, the normal circuitry of the auditory pathway is relatively robust and apparently more resistant to acoustic overstimulation. We may expect pathological changes in the central part of the auditory pathway particularly with long-lasting exposures. Also, noise impulses, which stimulate receptors before the protective contraction of middle ear muscles occurs, may result in pathology of the central auditory pathway.

From the morphological point of view, the question of the extent to which degeneration of the receptors in the inner ear is accompanied by morphological changes in the auditory nerve is interesting. There are several studies which demonstrate the occurrence of pathological changes in dendritic nerve endings. Swelling of dendrites was described first by Beagley (1965 a). The dendritic nerve endings of guinea pigs exposed to octave-band noise at 117 dB SPL for 24 h show swollen mitochondria (Lim and Melnick 1971). Reversible swelling of the dendrites after 1 h exposure to wide-band noise at 130 dB SPL has been described by Spoendlin (1971; Fig. 11). In contrast, Slepecky et al. (1981, 1982) did not see any changes in the synaptic bar, synaptic cleft, and receptor surface on the dendritic bouton of the IHC in chinchilla after impulse-noise exposure. Dendritic swelling after acoustic injuries is present in the cat and guinea pig (Liberman and Mulroy 1982; Robertson 1983; Goulios and Robertson 1983). In guinea pigs exposed to a 110 dB SPL tone, swelling of dendrites beneath IHCs was clearly visible after 22 min of noise, and the swelling was reversible (Robertson 1983). Most important from the functional point of view was the finding that the occurrence of swelling corresponded with the loss of sensitivity in the flat tail part of the tuning curves of those auditory neurons whose CF was approximately 1/2 octave higher than the exposure frequency. In terms of the place frequency map for the guinea-pig cochlea, 1/2 octave higher in frequency corresponds to approximately 1 mm in the basal direction with respect to the representation of the exposure tone frequency and this was the frequency region where the dendritic swelling was greatest. The times for the occurrence of dendritic swelling and the loss of "tail" sensitivity in tuning curves corresponded well. Dendritic swelling is well known in retina; the authors assume that it is caused by a high concentration of transmitter substance released into the extracellular space.

Degeneration of nerve cells of the spiral ganglion after exposure of guinea pigs to noise at 100 dB SPL for up to 30 days has been reported (Rossi et al., 1976). In animals with shorter duration exposures, a reduction in cell body mitochondria

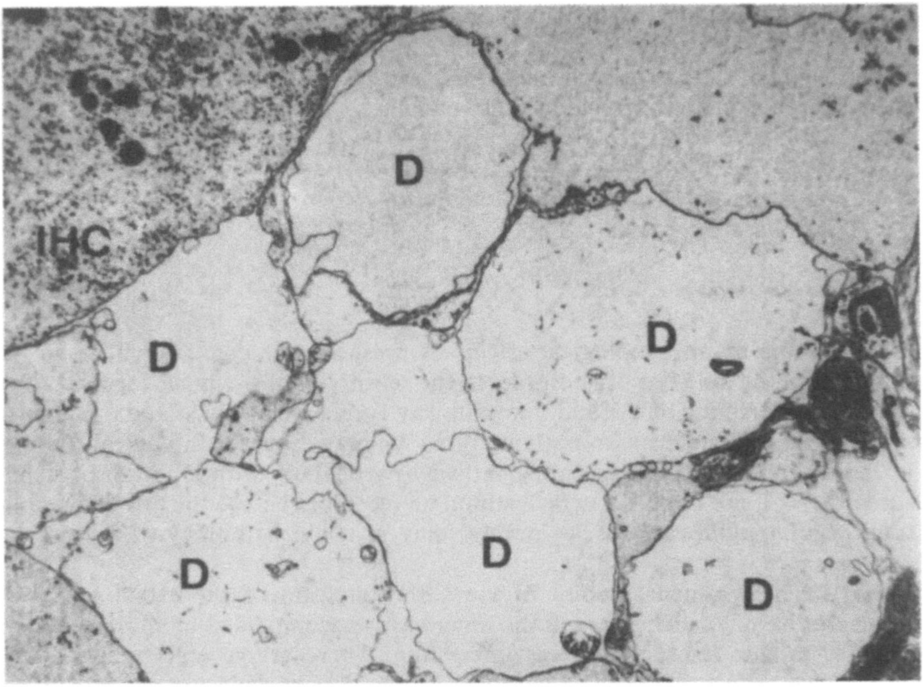

Fig. 11. Swollen afferent dendrites (*D*) to the IHCs immediately after 1 h 130 dB wide-band noise exposure. (Spoendlin 1971)

was observed, while the endoplasmatic reticulum and ribosomes appeared normal. In animals exposed for 30 days, many cells showed an increased number of ribosomes, there were changes in the shape of the nucleus, and many cells of the spiral ganglion had degenerated. Interestingly, nonmyelinated type II cells appeared normal. Spoendlin (1975) has shown that the total loss of the OHC has little effect on the population of spiral ganglion cells, whereas IHC loss is combined with pronounced degeneration in the spiral ganglion. Degeneration of the unmyelinated portion of the auditory nerve fibers is complete in a few days, whereas degeneration of the ganglion cells and their axonal projections to the cochlear nucleus takes several months (Spoendlin 1975; Wright 1976; Morest 1982).

In cats exposed to narrow-band noise centered at 3 kHz for 8 h at 135 dB SPL and examined at different survival times (up to 120 days), both IHCs and OHCs were found degenerated, especially in the middle turn of the cochlea (Morest 1982). In contrast, the spiral ganglion cells were found to be intact even after 120 days. Both dorsal and ventral cochlear nuclei showed axonal degeneration. A band of degenerated axons was present in the posteroventral cochlear nucleus, in the area of the octopus cells. In the superior olivary complex, degeneration was present in the medial nucleus of the trapezoid body, in the lateral nucleus of the trapezoid body, and in periolivary nuclei, while no degeneration was observed in the inferior colliculus. In the chinchilla (also Morest and Bohne 1983) the pattern of degeneration is similar, though not identical. The parameters of exposure were different: octave-band noise at 4 kHz with SPL of 108 dB was used for 1 h 45 min. The survival times were shorter (21 to 32 days). Again, no degeneration was observed in the spiral ganglion, whereas degenerated fibers were traced in the cochlear nuclei and the superior olivary complex. In addition, in the chinchilla degeneration was present in the inferior colliculus. The degeneration in the cochlear nucleus was differentiated according to the diameter of fibers of the auditory nerve. Fine degenerated fibers which were supposed to be connected with OHCs were found mainly in the dorsal cochlear nucleus, whereas coarse fibers connected with the regions of predominant IHC loss were found mainly in the ventral cochlear nucleus.

2.9.2 Threshold Shifts at Different Levels of the Auditory Pathway

Until approximately 1963 the feeling was unanimous that auditory fatigue or temporary threshold shift is of cochlear origin. However, around this time several studies appeared which suggested that not all the changes evoked by noise exposure were of cochlear origin, or in other words that there may be specific changes in the central part of the auditory pathway accompanying noise exposure. Wernick and Tobias (1963) reported a central factor of auditory fatigue in humans. The TTS, from low or high level sound exposures, was found to be greater when the subject was given a mental task than when the subject was mentally relaxed. The problem of the central effect of noise was further approached from a different point of view by Starr and Livingston (1963). They measured the spontaneous electrical activity in different central auditory nuclei, as well as in the eighth nerve in the cat with implanted electrodes. The average activity elicited during the noise (80 dB SPL for 2 h) progressively decreased at higher levels in the central auditory pathway. In the cortex, the ongoing activity actually decreased below the baseline value during the noise exposure. The recovery of spontaneous activity after exposure was also slower in the higher auditory centers.

In an extension of these studies to the single unit level, Starr (1965) found postexposure fatigue of spontaneous activity in all units of the cochlear nucleus in the cat. The duration of the suppression of the spontaneous activity was often related to the duration of the exposure. Benitez et al. (1972) studied the effects of the noise exposure (octave-band noise centered at 500 Hz at 95 dB SPL for several days) on the CM, endocochlear potential, and evoked responses from the brainstem nuclei (recorded between a subcutaneously placed electrode over the tentorium and a reference electrode in the mouth). The endocochlear potential was unchanged, whereas there were shifts in the CM and evoked responses from the brainstem nuclei comparable with behavioral changes in the TTS. AP decreased, however, more than would be expected by behaviorally estimated TTS. Although the results of this study are clearly presented, some technical problems make the interpretation less clear. CM responses to tone pips were compared with AP responses to wide-band clicks and in addition it is not certain which auditory nuclei were responsible for the centrally recorded component.

The peripheral and central components of the auditory fatigue or TTS have been compared by Babighian et al. (1975). In the kangaroo rat, cochlear microphonics, CAP, and evoked responses (as well as single unit responses) to short-lasting (2 – 3 min) tonal stimuli at 85 dB SPL were measured. Suppression of collicular responses was always 20 – 30 dB higher than the suppression of CM and CAP (if present at all). Similar differences in suppression of CM and single-unit responses in the inferior colliculus have also been observed in the guinea pig (Popelář et al. 1978). Salvi et al. (1975) have demonstrated that the loss in sensitivity after noise exposure increases progressively from peripheral to central auditory sites. Chinchillas were exposed to tones (frequency between 0.4 and 8.0 kHz) 8 min in duration at 95 dB SPL. The responses of cochlear nucleus neurons showed losses of about 6 dB; the losses of inferior colliculus neurons were on the average 6 dB greater than those of cochlear nucleus neurons. The auditory evoked responses recorded from the tentorium (cf. also Benitez et al. 1972) disclosed shifts similar to those of the inferior colliculus neurons. Apparently the differences in losses among individual auditory nuclei after exposure to noise are characteristic only for short-term exposure, because after exposures lasting several days the behavioral thresholds and the thresholds at CF of the most sensitive units in the cochlear nucleus were within 10 – 15 dB of each other (Fig. 12; Salvi et al. 1978). Lonsbury-Martin and Martin (1981) compared behaviorally estimated threshold shifts in awake rhesus monkeys with threshold shifts of neuronal responses from the cochlear nucleus and inferior colliculus after 3 min 100 dB tone exposure. The neuronal response shifts amounted to 20 dB, whereas the shifts estimated behaviorally usually did not exceed 5 dB. In some units, however, there were more complex postexposure changes involving increases in discharge rates to more intense tone bursts in comparison with the preexposure responses. In their recent study Lonsbury-Martin et al. (1987) monitored behavioral and single-unit thresholds in monkeys for several months when the animals were exposed to mild noise intensities which produced TTS. After 18 months significant permanent high frequency losses developed in these animals, which were revealed in both behavioral tests and single-unit recordings. In one animal, which was exposed for only 6 months, threshold shifts were detectable only in single-unit recordings, whereas behaviorally estimated thresholds were not significantly shifted.

Changes in evoked activity along the auditory pathway after white-noise exposure of 115 dB SPL for 30 min were reported by Syka and Popelář (1982). The CM isopotential curves recorded at the round window were shifted after the noise by about 1 dB in the range of 0.5 – 8 kHz. The amplitude-intensity function of the click-evoked auditory nerve action potential decreased by about 8 dB, that of the inferior colliculus response by 21 dB, and the amplitude-intensity function of the auditory cortex evoked potential decreased by about 6 dB. These results suggest that in the guinea pig as in the kangaroo rat (Babighian et al. 1975) and the chinchilla (Salvi et al. 1978), the individual auditory nuclei are impaired by noise to different extents and that the impairment does not increase linearly in the pathway up to the auditory cortex. In a related study, Popelář and Syka (1982) have monitored spontaneous and evoked activity of single units in the inferior colliculus using the same parameters of noise exposure. The threshold shifts were comparable with those observed in the evoked potentials of the

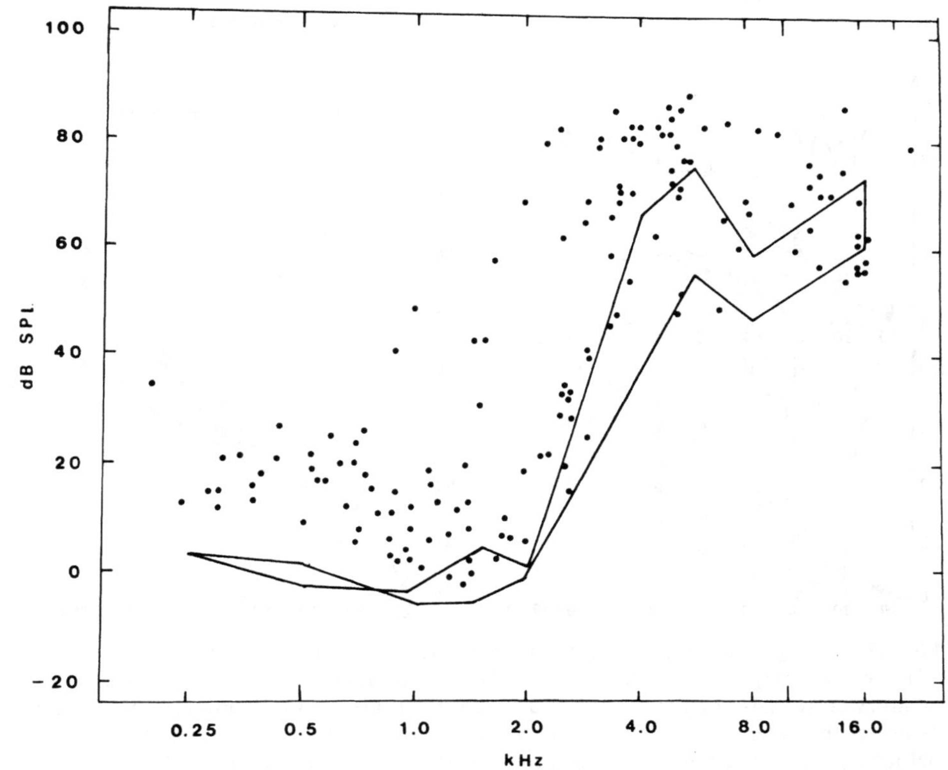

Fig. 12. Mean behavioral thresholds (*solid lines*) and the single neuron thresholds (*filled circles*) in chinchillas exposed to an 86 dB SPL octave-band noise centered at 4.0 kHz for 3.5 – 5 days. (Salvi et al. 1978)

inferior colliculus. The shapes of the poststimulus histograms and the slopes of the spike intensity curves were not changed after exposure. The total number of spikes during the response was, however, reduced and the reduction was in proportion to the threshold elevation. In contrast to Starr (1965), who found that exposure of one ear to noise increased thresholds in olivary complex units for click from the unexposed ear, monaural noise exposure in the guinea pig had no effect on the neuronal activity evoked by stimulation of the nonexposed ear.

So far all studies have demonstrated that in addition to cochlear effects, noise exposure may result in specific changes in nuclei of the auditory pathway, with the greatest suppression observed in the inferior colliculus. Practically all experiments, with the exception of those by Lonsbury-Martin and Martin (1981) were performed in anesthetized animals. Starr and Livingston (1963) have pointed out that anesthesia significantly influences the electrical activity in the auditory system. The electrical activity in the cat's auditory nuclei was suppressed after injection of pentobarbital before, during, and after noise exposure. The influence of anesthesia on changes in auditory function elicited with noise has been demonstrated also by other authors (e.g., Henry 1980). Therefore, Popelář et al. (1987) measured changes in evoked potentials along the auditory

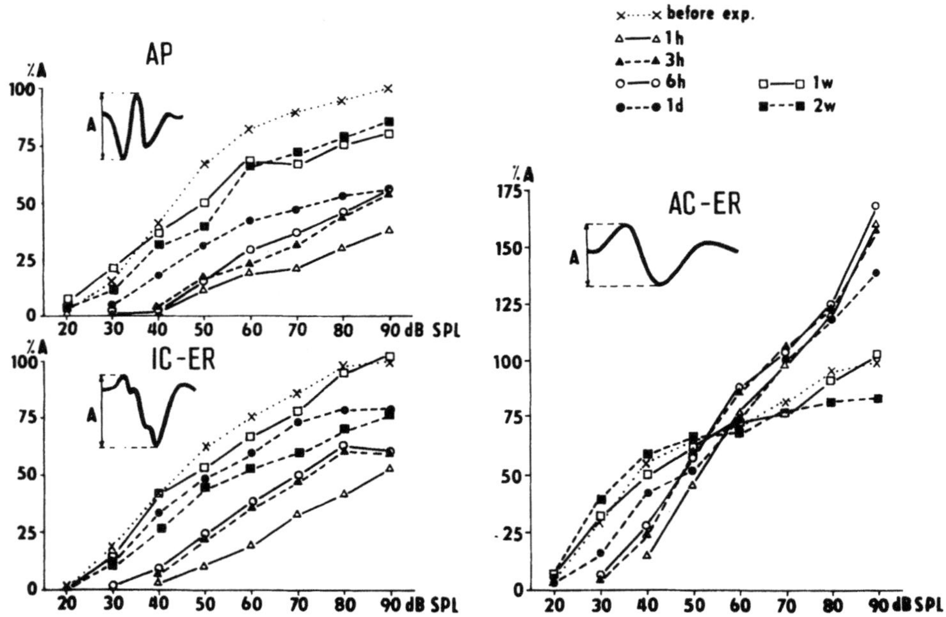

Fig. 13. Average amplitude-intensity functions for the auditory nerve action potential (*AP*), evoked responses in the inferior colliculus (*IC-ER*) and evoked responses in the auditory cortex (*AC-ER*) measured at different times after noise exposure. The *ordinate* expresses the potential amplitude in percentage of the amplitude measured at 90 dB SPL before exposure. The *inset* shows the principle of amplitude measurement. (Popelář et al. 1987)

pathway in awake unrestrained guinea pigs with implanted electrodes. Application of 120 dB SPL white noise for 1 h resulted in practically equal threshold shifts (approximately 30 dB) in the auditory nerve, inferior colliculus, and the auditory cortex (judged on the basis of shifts in thresholds of evoked responses; Fig. 13). The authors conclude that anesthesia caused the large shifts which were previously observed in the amplitude-intensity functions in the inferior colliculus (Syka and Popelář 1982). In control experiments, the animals with implanted electrodes were anesthetized with urethane: the shifts in the inferior colliculus evoked responses were again greater in comparison with those in the auditory nerve and auditory cortex. Further, in unanesthetized animals there were signs of hypersensitivity immediately after noise exposure, as evidenced by an amplitude enhancement of the auditory cortical evoked responses (Popelář et al. 1987). The enhancement was present for 24 h after exposure and was not observed when the animals were anesthetized with urethane while exposed to noise.

2.9.3 Latency Changes After Noise Exposure

It is generally accepted that a tone-burst-elicited compound action potential (CAP) at threshold reflects the activity of a small group of unit fibers. As sound pressure of a tone burst increases, there is an increase in the number of active unit fibers, where characteristic frequency is mostly of a higher frequency than that of the stimulus (Özdamar and Dallos 1976). Therefore the N_1 latency with increasing burst intensity decreases because the fibers of higher CF have shorter response latencies. Several studies have been aimed at examining the effect of noise exposure on N_1 latency. Salvi et al. (1979), who used octave-band noise (5 days, center 0.5 kHz at 95 dB SPL) in the chinchilla found shortened click-elicited N_1 latency and single auditory nerve fibers latency after exposure. The threshold latency of the single-unit response and CAP was a function of absolute SPL in both normal and noise-treated animals. This finding corresponds with the hypothesis proposed by Wang and Dallos (1972) that in the guinea pig the latency of the CAP as a function of SPL remains constant independently of the pattern and degree of hair-cell loss in the cochlea. The hypothesis was confirmed in guinea pigs treated with kanamycin (Wang and Dallos 1972). In contrast to this, Van Heusden and Smoorenburg (1981 b) have reported CAP latency values at threshold before and after inducement of the trauma to be not significantly different (i.e., the latency increased with respect to absolute SPL). In their experiments cats were exposed to broad-band noise at approximately 105 dB SPL for half an hour. The greatest threshold shifts were found between 2 and 6 kHz. It was suggested that shifts in latency (with respect to equal SPL before and after exposure) result from the shift of the excitation area to an area that responds to lower frequencies. Prolongation of N_1 threshold latencies was also observed after acute pure-tone trauma (4 kHz at 95 dB SPL for 10 min), i.e., N_1 latencies at threshold were similar before and after the exposure (Sikora and Morizono 1983). However, the N_1 latency increased at and below the trauma frequency and decreased at frequencies higher than 4 kHz. While the exposures of Van Heusden and Smoorenburg (1981 b) and Sikora and Morizono (1983) resulted in TTS, the data of Salvi et al. (1979) and Wang and Dallos (1972) were from animals with significant PTS. This may be the main reason for the discrepancy in the N_1 latency function with respect to SPL reported in these studies.

Characteristic changes in the latency distribution of the first spike after a 2-min exposure to 95 dB SPL tones were observed in neurons of the cochlear nucleus in the rhesus monkey (Lonsbury-Martin and Martin 1981). Units which decreased their driven activity after exposure displayed corresponding increases in latencies, whereas those which increased driven activity decreased their latencies of the first spike. Latency distribution changes with respect to the threshold acquired after exposure (white noise at 115 dB SPL for 30 min) were observed in single units of the inferior colliculus in the guinea pig (Popelář and Syka 1982). The latencies for both monotonic and nonmonotonic units were functions of the neural threshold, i.e., the postexposure latencies were longer than the response latencies at the same absolute SPL recorded before the exposure.

2.9.4 Spontaneous Activity

Spontaneous activity of auditory nerve fibers and neurons in the auditory pathway after noise exposure as a rule decreases (Kiang et al. 1976; Salvi et al. 1978). However, after exposure to short-lasting tones of moderate intensity, an increase in the spontaneous activity was observed in units of the cochlear nucleus in the rhesus monkey (Lonsbury-Martin and Martin 1981). Only high spontaneous-rate neurons changed their spontaneous activity (increasing or decreasing), whereas low spontaneous-rate units were relatively unaffected. Also, in the inferior colliculus neurons were described with increased spontaneous discharge rates after noise exposure (Salvi 1976).

Long-lasting exposure (5 days) to a relatively mild intensity (95 dB SPL octave-band noise centered at 500 Hz) did not result in any significant changes in the spontaneous activity of auditory nerve fibers in chinchillas (Salvi et al. 1983). This finding is based on comparison of the spontaneous firing rates of groups of units before and after exposure. A significant contribution to the understanding of the decreased spontaneous discharge rate after noise exposure is represented by the data of Liberman and Dodds (1984a), who have described a decrease in the spontaneous firing rate after noise exposure in those auditory nerve fibers which are connected with IHCs with the loss of the tallest row of stereocilia. In cochlear regions where virtually all of the tall stereocilia were found missing from the IHCs, the maximum rates of spontaneous discharge are less than 1/3 normal values. Liberman and Dodds (1984a) suggest that the loss of IHC stereocilia results in a decrease of the resting current, which normally flows through stereocilia.

2.9.5 Auditory Brainstem Responses and Noise-Induced Hearing Loss

Auditory brainstem responses (ABR) have been used widely in audiological diagnosis by clinicians, since their description by Jewett (1970). Several studies have reported characteristic changes in the ABR in patients with sensorineural hearing loss (Møller and Blegvad 1976; Coats and Martin 1977; Coats 1978; Galambos and Hecox 1978; Jerger and Mauldin 1978; Rossi et al. 1979; Rosenhamer 1981; Hayes and Jerger 1982). There are several changes of ABR present in patients with cochlear lesions (see review by Rosenhamer 1981): the latency of the most prominent wave V is prolonged, the interval between waves I and V is not changed, and the amplitude of wave V is not reduced relatively to wave I. However, even in the first studies (Møller and Blegvad 1976; Coats and Martin 1977) a correlation was noted between increase in the latency of the wave V and the type of cochlear hearing loss. The latency increase of wave V is less pronounced in patients with flat hearing loss than in patients with gradually sloping high-frequency hearing loss (Møller and Blegvad 1976). The longest latencies are found in patients with a steeply sloping loss. This finding has been confirmed by Jerger and Mauldin (1978), who found a strong correlation between the wave V latency increase and the slope of pure tone audiogram for frequencies between 1 kHz and 4 kHz (the latency increased by 0.2 ms for each 30-dB increase in the threshold difference between 1 kHz and 4 kHz).

However, because of recruitment in hearing loss of cochlear origin, the relationship between the wave V latency and the sound intensity (latency-intensity curve) is not simply a shift to longer latencies, but the curve is also steeper at higher intenties (Coats 1978; Galambos and Hecox 1978; Skinner and Glattke 1977). Whereas the wave V latency at low-intensity levels (relative to hearing threshold) may be considerably longer than in people with normal hearing at equal click intensities, the latency approaches normal values at higher stimulation intensities. Again the latency increases at low levels of stimulus intensity and the steepness of the latency intensity curve increases with the increasing loss at high frequencies (Coats 1978). It must be stressed at this point that the ABR for both filtered and unfiltered clicks and tone pips is relevant for an understanding of the cochlear lesion only for frequencies between 2 and 8 kHz, whereas the prediction of the status of the cochlea at 500 Hz region in man is more precise in the pure-tone audiogram (Hayes and Jerger 1981).

In animal studies the ABR has been used to estimate hearing loss in the guinea pig, particularly because the acquisition of behavioral data is difficult in this species (Prosen et al. 1977; Syka and Popelář 1980). Cook et al. (1981) have measured ABR in infant guinea pigs, which were exposed in utero to high noise levels in an attempt to estimate the damage to the cochlea during development in this precocial species. Pregnant guinea pigs were exposed to 115 dB industrial noise for 7.5 h/day for various periods during the last trimester of pregnancy. The peak IV latency (which seems to correspond to wave V latency in man) of the exposed pups was found to be significantly longer than in normal control pups.

2.10 Influence of Noise on Metabolic Activity in the Auditory System

Since its introduction by Sokoloff et al. (1977), the [^{14}C] 2-deoxy-D-glucose DG autoradiographic method has proven to be a valuable technique for mapping functional activity in neural tissue. Plugging the external auditory canal in the rat decreases metabolic activity in the inferior colliculus (Sokoloff 1977). The effects of acoustic stimulation of different intensities (25 – 105 dB SPL) on the 2 DG uptake in different auditory nuclei of the mongolian gerbil have been studied systematically by Sharp et al. (1981). Animals were injected with [^{14}C] 2-deoxy-glucose and exposed to wide-band noise for about 1 h. Autoradiographs showed regular increases of 2 DG uptake with increasing sound intensities in the ventral and dorsal cochlear nuclei, nuclei of the lateral lemniscus, and in the inferior colliculus. A smaller increase of 2 DG uptake was noted in the medial geniculate body and auditory cortex.

To what extent this method reveals specific changes in auditory nuclei elicited by noise is not clear at present. For example, Woolf et al. (1983) have observed significant increases in 2 DG incorporation in several auditory nuclei in the mongolian gerbil after removal of the tympanic membrane and malleus. This procedure, which results in unilateral conductive hearing loss, unexpectedly increased the 2 DG uptake in the ipsilateral ventral and dorsal cochlear nuclei, ipsilateral lateral superior olivary nucleus, and contralateral dorsal nucleus of the lateral lemniscus. In another study, glucose utilization increased in the mouse cochlea and inferior colliculus at moderate noise intensities (55 – 85 dB SPL), whereas it decreased at high intensities (100 and 115 dB SPL; Canlon and Schacht 1983). These results at high intensities are in disagreement with the autoradiographic data of Sharp et al. (1981), who have found only an increased uptake with increasing intensity.

LeDoux et al. (1983) have investigated the local central blood flow in rats using the ^{14}C-labeled iodoantipyrine technique. Sound stimuli of moderate intensity and lasting only 10 s produced significant increases in the blood flow through all nuclei of the auditory pathway with the exception of the medial geniculate body. An especially large increase was found in the inferior colliculus. This finding must be, however, related to normal physiological conditions and not to the threshold shifts or other effects of the high intensity exposure.

2.11 Cochlear Vascular Changes After Noise Exposure

Since about 1960, many investigators have been interested in the effect of noise on the vascular system of the cochlea. It was assumed that if noise exposure suppresses the circulation through the cochlea vessels the resulting hypoxia and lack of supply of energetic metabolites would result in pathological changes in the receptor elements and surrounding cells of the organ of Corti. There is probably no other area in noise research where so many controversial results have been accumulated in recent years. Several reviews (Bohne 1976 a; Borg 1981; Axelsson and Vertes 1982) support this conclusion.

The pioneering paper in this field was published in 1962 by Perlman and Kimura. These authors, having directly observed cochlear blood vessels, indicated that blood flow in the stria vascularis and spiral ligament of the upper turns is not impeded until the exposure intensity reaches a very high level (153 dB SPL). Cochlear damage occurs at levels of exposure which do not produce any change in the cochlear blood flow. A diminution or total lack of red blood cells in the vessels below the basilar membrane in noise-exposed animals has been observed by others (Lawrence et al. 1967; Lipscomb and Roettger 1973). Similar changes were also noted in vessels of the lateral wall of the cochlea (Hawkins 1971; Lawrence and Yantis 1957), which according to recent studies with microspheres (Angelborg et al. 1985) represents more than half of the blood supply of the cochlea under normal conditions. On the other hand, Kellerhals (1972) has observed the stria vascularis vessels to be tightly packed with red blood cells 24 h after exposure to impulse noise. This abnormality was not found in all other vessels of the cochlea. However, the significance of this finding is problematic in light of the observations of Johnsson and Hawkins (1972) who have found, under normal circumstances, slow blood flow and high hematocrit in the stria. The vessels of the stria were generally tightly packed with red cells.

Other methods employed in studies of the cochlear blood flow during noise exposure were impedance plethysmography and distribution of microspheres. No change in blood flow in the cochlea has been observed in response to tonal and white noise exposures with intensities ranging from 90 dB to 120 dB SPL using impedance plethysmography (Morimitsu et al. 1965). Occasionally an increase in blood flow has occurred. Cochlear blood flow did not show any changes in cats exposed to white noise at 100 dB SPL for 6 min (Angelborg et al. 1979). In this case the distribution of microspheres was measured. Similarly, Hultcrantz (1979) was not able to find changes in cochlear blood flow in anesthetized rabbits exposed to white noise from 100 to 120 dB SPL for 6 min. There was, however, blood flow increase in the inferior colliculus. In contrast to this, Prazma et al.

(1983, 1984) reported blood flow increases in the cochlea of noise-exposed gerbils using the microsphere technique.

In comparison with these conflicting results in acute studies of the cochlear blood flow, the results of histological examinations of chronic changes in the cochlear structures are more consistent. Strong evidence in many studies (especially when high intensity noise exposure was used) confirmed changes in the stria vascularis: swelling of the stria vascularis, marked vasoconstriction in strial capillaries which is later followed by atrophy, and thinning of the stria vascularis (Lawrence and Yantis 1957; Ward and Duvall 1971; Santi and Duvall 1978; Lim and Melnick 1971; Lim 1976; Clark and Bohne 1978).

Extensive studies of the function of the cochlear vasculature under normal and pathological conditions have been performed by Axelsson and Vertes (Axelsson 1968; Axelsson et al. 1980; Vertes et al. 1979, 1981, 1984) and some of them have been reviewed by Axelsson and Vertes (1982). They used the soft-surface specimen preparation, i.e., a histological technique, and observed vessels and such parameters as red blood corpuscles, perivascular cells, vessel lumen, and others. They concluded that noise does not seem to result in any extraordinary pathology which could be discerned by light microscopy. However, if there were any changes in the overall blood supply to the cochlea during and after noise exposure, these tended to decrease the cochlear blood flow. Noise exposure produced degeneration of the strial epithelium together with changes in pigment cells.

2.12 Effects of Noise on the Developing Auditory System

The available experimental data indicate that the immature mammalian auditory system is more susceptible to noise trauma than the adult auditory system. Young guinea pigs exposed to noise showed greater damage to the cochlea than did adult animals (Falk et al. 1974; Douek et al. 1976; Coleman 1976). For example, Falk et al. (1974), who exposed 2-day-, 8-day-, and 8-month-old guinea pigs to 30 h of white noise at 120 dB SPL, observed a 23.7% loss of OHCs for the 2-day-old subjects, 37% in those 8 days old, and only 7.2% for the oldest animals. IHC damage was negligible in all groups. Also in the cat there was enhanced susceptibility to noise (Price 1976). The average loss in CM after 5 kHz 115 dB SPL exposure for 50 min was 15.6 dB in kittens and only 8.7 dB in adult cats.

Whereas the duration of the greater susceptibility to noise seems to be wide in the guinea pig, in rats and mice narrower periods of enhanced susceptibility have been reported. In the mouse this period peaks at 17 days postpartum (Saunders and Hirsch 1976). Similarly, an extremely narrow period of increased sensitivity was found in the rat (Lenoir and Pujol 1979). Young rats were exposed at different ages (from 16 to 40 days) to 120 dB SPL white noise for 30 min. Maximum damage occurred in those animals which were exposed at 22 days postpartum, i.e., 2 days after cochlea maturation was complete. Eight days following noise exposure the changes were minimal, although a few OHCs were already missing. Another group of animals exposed at 22 days of age to white noise showed 80% loss of OHC when killed at 2 months postexposure. Also extensive degeneration of IHCs and unmyelinated parts of the auditory nerve fibers was evident in these animals. A period of greater susceptibility to noise was found also in immature hamsters (Bock and Seifter 1978). The problem of age-dependent susceptibility to noise in experimental animals has been discussed in several recent reviews (Saunders and Bock 1978; Saunders and Chen 1982; Henry 1983).

3 Effects of Ototoxic Drugs

The ototoxicity of some drugs (e.g., salicylates) has been known for many years, but the introduction of aminoglycoside antibiotics just after World War II started an intensive investigation of the mechanisms of ototoxicity. Some aminoglycoside antibiotics in particular (streptomycin, neomycin, kanamycin) appeared to be extremely harmful for the inner ear, and their sometimes unpredictable ototoxicity in patients has stimulated many investigations in animals. New groups of drugs were subsequently recognized as ototoxic, although their effects were generally less harmful than those of the aminoglycosides. Most of these are diuretics, but some anticancer drugs, such as cisplatin, have also been found to be ototoxic.

3.1 Aminoglycoside Antibiotics

The aminoglycoside antibiotics are a class of broad-spectrum antibiotics with activity against gram-positive and gram-negative organisms as well as mycobacteria. They are derivatives either of streptamine (streptomycin, dihydrostreptomycin, neomycin) or of 2-deoxystreptamine (kanamycin, gentamicin, sisomicin, netilmicin, tobramicin, amikacin). Their ototoxic effects were recognized soon after they were introduced into clinical practice. Hinshaw and Feldman (1945) described disturbances of vestibular function and impairment of hearing occurring during treatment of patients with streptomycin, the first effective chemotherapeutic against tuberculosis. Later practically all aminoglycoside antibiotics were found to be potentially ototoxic, although the degree of ototoxicity might differ, as well as the target of the ototoxic action, i.e., the cochlear or the vestibular part of the inner ear. Table 1 shows the potential level of ototoxicity of aminoglycoside antibiotics in current use, together with their main target (cochlear or vestibular part of the inner ear). The list of antibiotics includes also three antibiotics (vancomycin, viomycin, capreomycin) which are not aminoglycosides but which may exert ototoxic effects.

When ranking of the antibiotics is considered, the highest cochleotoxic effects are usually found both in patients and experimental animals with neomycin, although gentamicin, tobramycin, sisomicin, kanamycin, and amikacin have also pronounced cochleotoxic activity. The vestibular part of the inner ear was found to be the chief target of the ototoxic action of streptomycin, gentamicin, and dibekacin. In addition to the ototoxic effects, all these antibiotics are known to produce side effects on the kidney. The nephrotoxicity, which tends to be more reversible than the ototoxicity, decreases the clearance of the aminoglycoside antibiotics and results in further accumulation in body fluids, among them the

Table 1. Ototoxicity of aminoglycoside antibiotics

Antibiotic	Cochleotoxicity	Vestibulotoxicity
Streptomycin	+	+ +
Neomycin	+ + +	+
Gentamicin	+ +	+ +
Tobramycin	+ +	+
Sisomicin	+ +	+
Kanamycin	+ +	+
Amikacin	+ +	+
Dibekacin	+	+ +
Vankomycin	+	+
Viomycin	+	+
Capreomycin	+	+

fluids of the inner ear. The different aspects of aminoglycoside ototoxicity, morphological changes in the cochlea, signs of ototoxicity in electrophysiology, and the possible mechanisms of ototoxicity will be further discussed. Several recent reviews present more detailed information about the problem (Hawkins 1976; Matz et al. 1981; Prazma 1981; Rybak 1985, 1986; Lim 1986).

3.1.1 Morphological Changes in the Cochlea

Histological evidence of the damage of the cochlea caused by treatment with streptomycin was first obtained in 1949 by Berg and later confirmed by many others (as reviewed by Hawkins 1976). The main targets of the ototoxic effects are the hair cells, and pathology is usually more pronounced in the OHCs. Disturbances in the orderly pattern of the stereocilia, nuclear swelling, and vacuolization of cytoplasm after disappearance of its subcuticular organelles have been described by Hawkins and Engström (1964). Similar changes were observed in the vestibular part of the inner ear, particularly after injections of streptomycin or gentamicin (Berg 1949; Hawkins and Lurie 1952, 1953; Wersäll and Hawkins 1962; Farkashidy et al. 1963; Duvall and Wersäll 1964; Wersäll et al. 1969, 1973). Hair cells of type I were usually damaged first, followed by damage to type II hair cells. Darrouzet and Guilhaume (1974) studied changes in hair cells by electron microscopy in guinea pigs treated with 400 mg/kg kanamycin/day for 7 days. At first the Golgi apparatus increased in the subcuticular portion of the OHCs of the basal turn. Later lysosomes accumulated in this area, and stereocilia begun to fuse. The cell content was further extruded through the former cuticular plate and finally a phalangeal scar appeared in place of the hair cell. With severe injury the organ of Corti was replaced by epithelium.

With the advances in scanning electron microscopy it has been possible to observe various changes in the stereocilia after aminoglycoside antibiotic intoxication. Floppy cilia, fusion of stereocilia, and giant stereocilia have been noted in many studies (Lim 1976; Hawkins 1977; Theopold 1977). Lim (1986) has observed complete loss of stereociliary actin filaments in intoxicated cochlear hair cells, together with a softened appearance of the cuticular plate. Using an immunohistochemical method (Hayashida 1985), a heavy accumulation of gentamicin was found in the subcuticular region. The topographical distribution of damaged hair cells in the cochlea after aminoglycoside treatment also deserves attention. After kanamycin, which is predominantly used in animal studies, the majority of authors have found loss of hair cells to occur in the basal high frequency part of the cochlea, independent of the animal species (Kiang et al. 1970, 1976; Dallos and Wang 1974; Hawkins and Johnsson 1976; Dallos and Harris 1978; Stebbins et al. 1969; Ryan et al. 1980; and others). In all of these studies OHCs, were always damaged more than IHCs and in some cases there was practically normal preservation of IHCs in the cochlea after kanamycin treatment (Dallos and Harris 1978; Ryan et al. 1980). IHCs and their innervation were normal in regions of OHC loss when the adjacent supporting cells were undamaged. When IHC supporting cells were damaged or absent, damage to IHCs was commonly observed. Perhaps the only exception to reports of uniform distribution of hair-cell damage in cochlea after aminoglycoside antibiotics treatment has been published by Ylikoski et al. (1974). In guinea pigs treated with 250 mg/kg neomycin per day damage was observed both in basal and apical parts of the cochlea and IHCs were more frequently damaged than OHCs.

The pathological changes in the cochlea after aminoglycoside antibiotic intoxication are, however, not confined only to hair cells. Changes in the stria vascularis have been observed in the guinea pig by Rüedi (1951) and Johnsson and Hawkins (1972). The marginal cells were shrunken with loss of mitochondria, also the intermediate cells were shrunken, and the thickness of the stria was reduced to one-half of the normal value. More pronounced changes were observed after gentamicin and neomycin than after kanamycin and streptomycin. Forge and Fradis (1985) have found lipid bodies in marginal cells and a dilation of the rough reticulum in guinea pigs after gentamicin treatment. The stria was thinner than normal and appeared less structurally complex.

3.1.2 Electrophysiological Changes

First experimental studies (Hawkins 1950; Hawkins and Lurie 1953) already indicated cochlea damage produced by streptomycin and neomycin, expressed as a reduction of the cochlear microphonics (CM) and the auditory nerve action potential (AP). Davis et al. (1958) found that the intratympanic application of streptomycin of guinea pigs results in substantial reduction of the CM and the AP, while the positive endocochlear potential remained unchanged. Characteristic changes were observed in summating potentials (SP). Negative SP was enhanced and positive SP reduced. The authors assumed from the results that negative SP is generated in IHCs, which mostly survive the intoxication whereas positive SP is a product of OHCs, suffering mostly from the ototoxic effects. Recently the data were confirmed by application of toxic doses of kanamycin in guinea pigs (Syka et al., unpublished data). Extensive loss of OHCs was accompanied by disappearance of the positive SP, which was recorded in scala media of four cochlear turns, as already was described by Honrubia and Ward (1969). This finding is not surprising, in light of the available data on distribution of DC receptor potentials in OHCs and IHCs (Russell and Sellick 1978; Cody and Russell 1985; Dallos et al. 1982). OHCs produce both depolarizing and hyperpolarizing DC receptor potentials, whereas IHCs are only depolarized by sound. Receptor potentials of the opposite polarity are recorded in scala media as summating potentials.

Since 1970, when Kiang et al. published their pioneering study on the effect of kanamycin injected in cats on the responsiveness of auditory nerve fibers many other studies have appeared with similar results. Kiang et al. (1970) have produced various degrees of hair-cell destruction, utilizing the ototoxic effect of the kanamycin. The results indicate that nerve fibers that likely originated in cochlear regions with destroyed OHCs do not respond to sound. Fibers that probably arose from the segment where the transition from damaged to normal hair-cell populations occurred had essentially normal low-frequency tail segments and mostly missing tip segments. Later experiments generally confirmed their data and have shown several changes in cochlear function, when the OHCs were missing (Kiang et al. 1976; Dallos et al. 1977; Harrison and Evans 1979; Schmiedt et al. 1980; and others). Dallos and Harris (1978) have found more selective destruction of OHCs in chinchillas treated with kanamycin than in cats, and they have described a partially different picture of tuning curves in auditory nerve fibers which were connected with cochlear region and missing OHCs. These fibers responded to sound, but their tuning curves were abnormal, with elevated threshold at CF and missing tips. Fibers from border regions between normal and damaged parts of the cochlea showed great variability in tuning curves. Other indexes of fiber response such as latency, spontaneous rate, and time pattern of the PST histograms, were not affected by the OHC loss. Changes in frequency tuning of individual fibers of the auditory nerve after aminoglycoside intoxication were generally identical to those observed after noise exposure (see previous chapters). Deterioration in frequency selectivity was confirmed also in man using psychophysical and electrophysiological methods (Harrison et al. 1981b). Elevated thresholds and broadening of the tuning of cochlear fibers were observed also in other pathological states of the cochlea, i.e., in hypoxia (Evans 1976) and after cyanide and furosemide poisoning (Evans and Klinke 1974, 1982).

Dallos et al. (1977) have described some discrepancies between the results obtained with behavioral techniques and single-unit recordings. Frequency tuning curves in chinchillas with significant OHC loss after kanamycin treatment were detuned and elevated, whereas PTCs taken from areas of the organ of Corti devoid of OHCs exhibited increased thresholds, but retained sharp tuning. The question has been examined recently by Smith et al. (1987), who measured PTCs in patas monkeys treated with dihydrostreptomycin-sulfate. At frequencies corresponding to regions of complete OHC loss and complete IHC retention, a lack of selectivity was evident and PTCs closely resembled low-pass filters.

The recent data reported by Liberman (1984) and Liberman and Dodds (1984a, b), which are discussed in sections on the effects of noise, were in fact also obtained in cats treated with kanamycin and therefore provide evidence of chronic cochlear pathology. In kanamycin-treated cats shifts in the CF of auditory nerve fibers with respect to the location of the cochlear damage were also found. The spontaneous discharge rate was low and the decrease was correlated with selective loss of the tallest stereocilia. Tuning curves from fibers

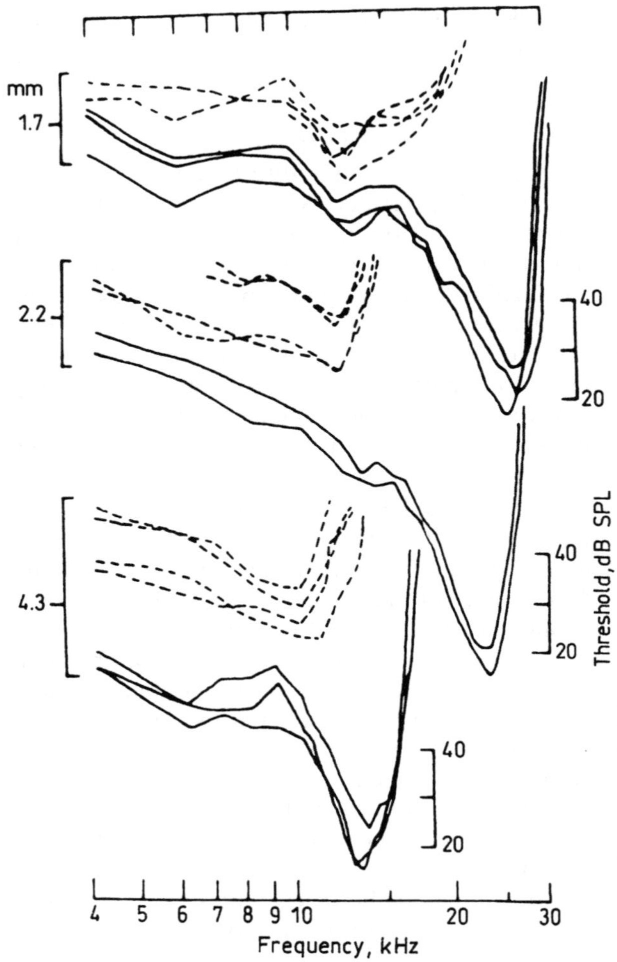

Fig. 14. Neural tuning curves recorded from cochlear ganglion cells of normal cochlea (*solid lines*) and after elimination of OHCs by kanamycin. Numbers at the *left* indicate cochlear location. (Robertson and Johnstone 1979)

connected with cochlear regions virtually devoid of OHCs but with preserved IHCs were characteristically bowl-shaped, lacking the tip. Robertson and Johnstone (1979) described in kanamycin-treated guinea pigs an octave shift toward the low frequencies, in neurons from areas where OHCs were absent (Fig. 14). Also in kanamycin-treated animals it was found that cochlear nonlinearities such as combination tones $2f_1 - f_2$ cannot be elicited if the region denude of OHCs is tuned to primary frequencies f_1 (Dallos et al. 1980).

Practical aspects of the ototoxicity of aminoglycoside antibiotics have been studied with electrophysiological methods by Aran and his colleagues. Aran and Darrouzet (1975) monitored changes in compound action potential in the guinea pig with electrodes chronically implanted on the round window after administration of kanamycin. They described first a period of acute intoxication which was followed by less dramatic changes, supposed to result from metabolic impairment.

Chronic monitoring of the effects of aminoglycoside antibiotics on CM and CAP with the aid of implanted electrodes has also been used by other authors (Simmons et al. 1960; Farkashidy et al. 1963). Electrophysiological recording of the CM, combined with histological control of the sensory hair cells and measurements of concentrations of antibiotics in plasma and perilymph was performed in the guinea pig to determine the relative ototoxic liability of several aminoglycoside antibiotics as tobramycin, gentamicin, amikacin, and sisomicin (Brummet et al. 1978). In this study, tobramycin and amikacin were found to be less ototoxic than gentamicin and sisomicin. Aran et al. (1982) also compared relative cochleo- and vestibulotoxicity of several antibiotics (dibekacin, gentamicin, and tobramycin). They measured evoked responses in the auditory cortex of guinea pigs and monitored eye movements with electronystagmography. The electrophysiological results together with those of morphological examination (including scanning electron microscopy) indicated that gentamicin exerts strong cochleo- and vestibulotoxicity, whereas tobramycin is cochleotoxic and less vestibulotoxic, and dibekacin is vestibulotoxic and less cochleotoxic. Chronic electrophysiological recordings, which can be carried out for a period of up to 1 year, are certainly suitable for dynamic studies of cochlear toxicity and may serve for screening the potential ototoxicity of newly developed agents (Aran 1981).

In acute experiments electrophysiological methods were used for evaluation of the local ototoxicity of aminoglycoside antibiotics. Morizono and Johnstone (1975) instilled a 0.3% solution of gentamicin into the middle ear cavity and found a significant decrease in CM at 1 kHz 24 h after the instillation of gentamicin. Nuttall et al. (1971) have estimated the toxic effects of different concentrations of neomycin administered by a method of acute perilymphatic perfusion of the cochlea in guinea pigs on the CM at 500 Hz and 4 kHz. Perilymphatic and endolymphatic perfusion with solutions containing kanamycin and dihydrostreptomycin led Konishi (1979a) to the conclusion that these antibiotics suppressed CM when applied to the scala media in a very low concentration, whereas it was only gradually suppressed by perilymphatic perfusion with relatively high concentrations.

In contrast to the toxic effects of aminoglycoside antibiotics on the acoustically evoked potentials, produced by hair cells (CM, SP, and consequently also AP) the endocochlear potential (EP) produced by the stria vascularis is changed to a lesser extent. Davis et al. (1958) and Konishi et al. (1967) demonstrated small effects of the systemic injection of streptomycin and kanamycin on the EP. EP decreased during endolymphatic perfusion of aminoglycoside antibiotics (Konishi 1979a), but this effect may be produced by the perfusion. While the resting level of the positive EP was found to be essentially the same in kanamycin-treated and in normal guinea pigs (Konishi 1979b), it had been shown in earlier studies (Konishi et al. 1967) that the rate of the decline of the negative values of EP during anoxia is significantly slower in kanamycin-treated than in normal animals. This effect was later investigated in greater detail by Konishi (1979b) and Konishi and Salt (1980), with the conclusion that in kanamycin-treated guinea pigs (as in noise-exposed guinea pigs) the K^+ conductance and K^+ permeability coefficient of the perilymph-endolymph barrier substantially decreases, when compared to normal guinea pigs. Again these data indicate the changes in hair cells caused by kanamycin and demonstrate the relative robustness of the pumping mechanisms of the stria vascularis, which are essentially unimpaired.

3.1.3 Ototoxic Mechanisms

Slow clearance of aminoglycoside antibiotics from the fluids of the inner ear has been considered to be a major reason for their ototoxicity. Pharmacokinetic studies performed by Voldřich (1965) have shown that streptomycin, kanamycin, and neomycin penetrate slowly into the perilymph in the guinea pig and are also more slowly eliminated from the perilymph than from the plasma. After regularly repeated doses of neomycin, accumulation of this antibiotic occurred in perilymph. Similar results were obtained by Vrabec et al. (1965) and Stupp (1970). The accumulation of the aminoglycosides in the perilymph was postulated to be the primary cause of their ototoxicity. More controversial were data concerning the pharmacokinetics of antibiotics in the endolymph. Due to technical difficulties with endolymph sampling, data are rare. Stupp (1970) found similar concentrations of kanamycin in endolymph and perilymph hours after the subcutaneous injection. Similarly Federspil et al. (1976) found practically identical concentration of gentamicin in endolymph and in perilymph 2 – 5 h after subcutaneous injection. Watanabe et al. (1971) detected kanamycin in endolymph only after prolonged treatment of animals with doses of 400 mg/kg day. Using autoradiography, Balogh et al. (1970) have found lower quantities of tritiated dihydrostreptomycin in endolymph than in perilymph in guinea pigs.

More recent studies cast some doubt on taking the simple explanation of the slow clearance and accumulation of ototoxic antibiotics as the main reason for their toxic effect in cochlea. The data of Tran Ba Huy et al. (1981) in rats suggest that gentamicin does not accumulate in the perilymphatic compartment. The perilymph: plasma ratio remained well correlated regardless of the plasma levels. Gentamicin entered the endolymph very slowly, suggesting that either the surrounding tissues had a great ability to bind aminoglycosides, or saturation of possible mechanisms for transport of the drug into endolymph prevents any further increase in aminoglycoside concentration. In a related study, Tran Ba Huy et al. (1983) observed that ethacrynic acid facilitates penetration of gentamicin into the endolymph, whereas the concentration of gentamicin in the perilymph is not changed by ethacrynic acid. Similarly, the results of the study by Desrochers and Schacht (1982) do not support the explanation of the ototoxic effect of aminoglycoside antibiotics on the basis of their accumulation in the cochlear fluids. Concentration of neomycin was measured in this study in different tissues in the guinea pig after prolonged injection (100 mg/kg body weight). Concentrations in cochlear tissues after 2 weeks of treatment were relatively low (0.7 mg/kg protein) and comparable with concentrations in the heart, liver, lung, and spleen. In contrast to this, the concentration in kidney was 6.8 μg neomycin/mg protein, in accordance with the well-known nephrotoxicity of neomycin.

It seems therefore that specific targets must exist in the cochlea, which are responsible for the high susceptibility of hair cells to toxic effects of aminoglycoside antibiotics. Extensive in vivo and in vitro studies have led Schacht and

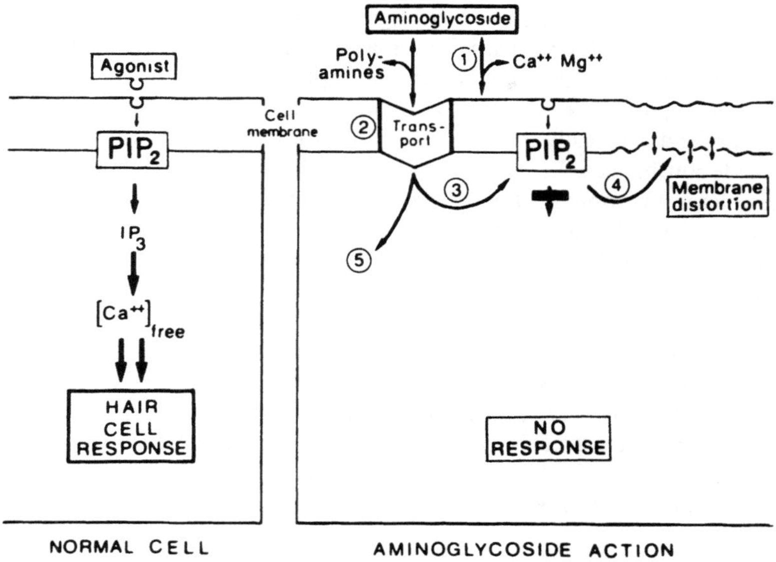

Fig. 15. Molecular mechanisms of aminoglycoside toxicity (*right*). The toxic actions of the drug encompass several sequential steps. *1*, Binding to the plasma membrane and displacement of cations such as calcium. This step should acutely but reversibly interfere with calcium-dependent membrane functions. *2*, Uptake into the cell. This process is suggested to be energy-dependent and competitive with polyamine transport. *3*, Binding to phosphatidylinositol 4,5-biphosphate (PIP2) on the cytoplasmatic side of the membrane. As a consequence, the physiological function of PIP2 is inhibited and the integrity of plasma membrane disrupted. *Left:* the physiological role of PIP_2 in OHCs. (Williams et al. 1987)

his colleagues to the conclusion that phosphatidylinositolbiphosphate may be a specific binding site for aminoglycosides (Schacht et al. 1977; Schacht 1979). Polyphosphoinositides are acid phospholipids of the plasma membrane which provide the structural framework of cell membranes, regulate membrane function, and mediate the communication of the cell with the external environment. Schacht (1986) proposed a model of aminoglycoside toxicity which consists of several steps (Fig. 15). In the first step the aminoglycosides, which are positively charged, react with negatively charged components of the outer plasma membrane. Then the drug is taken up into the cell by an energy-dependent transport process and is bound intracellularly by phosphatidylinositolbiphosphate. The latter process is irreversible and results in pathological changes of hair cells. The first step of binding is reversible and may be antagonised by calcium. When a suppression of CM occurred during perilymphatic perfusion of gentamicin, it could be reversed by perfusion of fluid with Ca^{2+} (Takada and Schacht 1982). Evidence supporting the three-step process of aminoglycoside ototoxic action has recently been presented by Williams et al. (1987).

Schacht's model of the three-step interaction of aminoglycosides with hair cells is discussed in detail by Lim (1986). It is apparent that future research in this field must resolve the problem of the entrance of aminoglycoside antibiotics into hair cells. While the reversible blockage of the transduction channels on the cilia-bearing apex of hair cells, which may occur as a first effect of aminoglycosides, assumes entrance through the endolymph, other routes of access are probable. For example, large basolateral surfaces of the OHCs are exposed to the corti-lymph and are easily accessible. OHCs are usually first targets of the aminoglycoside antibiotics. Last, but not least, it has to be mentioned that there may be problems in applying results from animal experiments in the study of clinical ototoxicity. The doses of antibiotics normally used in clinical practice are in the range of 3 mg/kg/day, whereas significantly higher doses generally have to be used in animal experiments to achieve detectable changes in cochlear structure and function (50 mg/kg/day to 400 mg/kg/day). Although qualitatively the results are usually in good agreement, the mechanisms of the toxicity considering the large differences in doses may not be completely identical. It should be mentioned, however, that in some cases the ototoxic effects in animals occur at doses which are used in clinical practice. For example, in the patas monkey the administration of dihydrostreptomycin in doses lower than those formerly used in patients caused severe hearing loss and hair cell damage (Hawkins et al. 1977).

3.2 Loop Diuretics

In 1965 Maher and Schreiner described transient hearing loss in patients treated with ethacrynic acid for renal dysfunction. Heidland and Wigand (1970) were among the first who observed hearing loss in patients with severe renal failure treated with furosemide infusion. Since that time numerous clinical and experimental studies have been devoted to the problem of the ototoxicity of ethacrynic acid and furosemide. Although they are chemically different (ethacrynic acid is a derivative of aryloxyacetic acid, whereas furosemide is a sulfonamide) they both are involved in the diuretic action in the kidney. Blockage of the transport of Cl^- from the lumen to the interior of tubular epithelial cells of the ascending loop of Henle is recognized as their common effect on kidney function. The ototoxic mechanisms are apparently also similar although they may be different in timing and in degree. Generally the hearing loss after ethacrynic acid is the more serious. Two other sulfonamide diuretics, bumetanide and piretanide, were also found to be ototoxic. In contrast to aminoglycoside antibiotics, which mostly produce permanent hearing loss, treatment with loop diuretics results mostly in temporary hearing loss. However, several patients have been reported to be permanently deaf, especially after ethacrynic acid (Pillay et al. 1969). Histological alterations after administration of loop diuretics in animals include intracellular and intercellular edema of the stria vascularis, and degeneration of OHCs of the basal turn (Quick and Duvall 1970; Nakai 1971; Horn et al. 1977; Brummett et al. 1977; Anniko 1978; Santi and Duvall 1979). Also tight junctions of the strial marginal cells have been found to be modified (Rarey and Ross 1982).

Changes in cochlear potentials with concomitant changes in the ionic concentration in cochlear fluids have been measured in experimental animals since Wilson and Juhn (1970) observed an increased perilymphatic concentration of K^+ after repeated administration of ethacrynic acid for several days. A decrease in the positive endocochlear potential has uniformly been found to be a sign of patho-

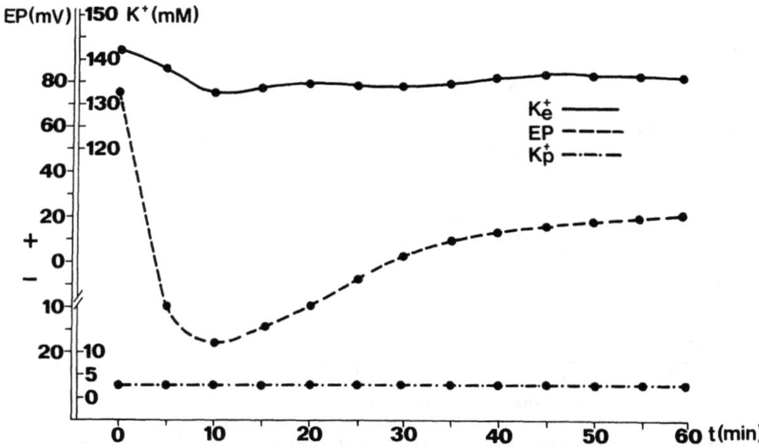

Fig. 16. Changes in the endocochlear potential (*EP*), K^+ concentration in endolymph (Ke^+) and perilymph (Kp^+) after injection of 50 mg/kg ethacrynic acid in guinea pig. (Melichar and Syka 1978)

logical changes in the stria vascularis (Prazma et al. 1972; Bosher et al. 1973; Sellick and Johnstone 1974; Melichar and Syka 1978; Bosher 1980a). The decrease, which was dose-dependent, was observed also after furosemide administration (Brusilow and Gordes 1973; Rybak et al. 1979; Kusakari et al. 1978; Pike and Bosher 1980). Juhn et al. (1981) have shown that the EP is depressed only after systemic furosemide doses sufficient to produce a perilymphatic concentration of furosemide greater than 1.3 µg/ml.

Figure 16 shows that the EP after 50 mg/kg ethacrynic acid decreases to approximately −20 mV and later returns to the original value. The decrease in the K^+ level in endolymph measured with an ion-selective microelectrode is less pronounced, and the K^+ level in perilymph does not change (Melichar and Syka 1978). When anoxia was introduced after EA administration, the decrease in K^+ level in endolymph, or the increase in perilymph, was slower than in normal control animals. It may be assumed from the results that ethacrynic acid reduces the passive permeability of the walls of the cochlear duct to the potassium ions.

Similar results have been reported by Bosher (1980a, b), who also measured Na^+ concentration in cochlear fluids using ion-selective electrodes, and in addition they correlated the electrophysiological findings with a morphological investigation of the stria after 60 mg/kg ethacrynic acid injection in the rat. Initially the drug completely inhibited the strial potential-producing and cation-transporting processes. The recovery process, which began quickly, later declined and coincidentally developed a decrease in overall cation permeability of the endolymph-perilymph barrier. K^+ concentration in endolymph declined slowly (as shown also by Melichar and Syka 1978), whereas the Na^+ level in endolymph slowly increased (in agreement with the results of Sellick and Johnstone 1974). The development of changes in the cation-transporting system of the stria was correlated with the occurrence of the swelling of the marginal cells and later reduction of the swelling; however, no unequivocal structural correlates were found

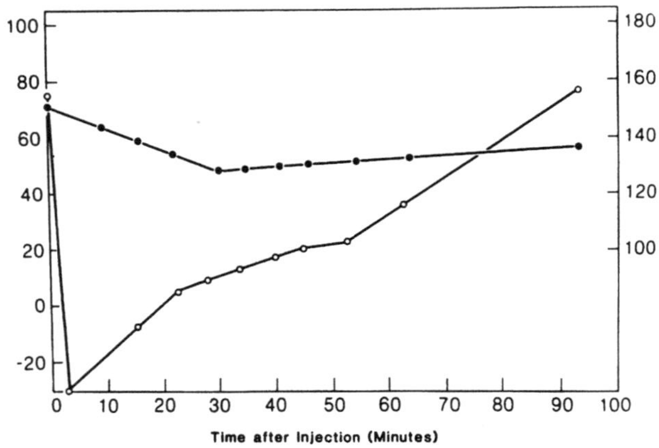

Fig. 17. Endocochlear potential (EP; *open circles*) and K^+ concentration in endolymph (Ke^+; *filled circles*) after injection of 100 mg/kg furosemide in chinchilla. (Rybak and Morizono 1982)

for the diminution in membrane permeability and the probable late decrease in energy production. Similar development of ultrastructural abnormalities was described by Pike and Bosher (1980) in the stria vascularis of the guinea pig after intravenous injection of furosemide (80 mg/kg). Alterations consisted of marginal cell swelling, shrinkage of the intermediate cells, and enlargement of the intracellular spaces. Again, the time course of the development of morphological changes was correlated with the decline and later return of the EP. The decline in the EP after furosemide is similar to that after ethacrynic acid, and is combined with a small decrease in the K^+ level in the endolymph (Syka and Melichar 1981; Rybak and Morizono 1982; Fig. 17). Also, gradual decrease and later return of the Cl^- concentration after furosemide administration was described by Rybak and Whitworth (1986).

The original hypothesis, that the mode of action of ethacrynic acid on the stria vascularis is similar to that imposed by ouabain, namely inhibition of the Na^+ K^+/ATPase system, has not been confirmed by subsequent findings: the inhibition of strial Na^+K^+/ATPase by ethacrynic acid occurs at much higher concentration than the inhibition of the endolymphatic potential (Paloheimo and Thalmann 1977; Kusakari et al. 1978). Also the participation of adenylatecyclase as a possible mediator in the ototoxic action of loop diuretics was later disproved (Thalmann et al. 1982).

Monitoring of the CM, SP, and AP was routinely used in studies of the effects of furosemide and ethacrynic acid on hearing. The CM, SP, and AP values depend on the positive polarization on the apical membrane of the hair cells due to presence of the positive EP. Consequently, any changes in the generation of EP result in such alterations. Loop diuretics may also influence hair cells directly, an effect which is at least partially demonstrated in the changed permeability for K^+ after their systemic administration (Melichar and Syka 1978; Bosher 1980a). Aran and Charlet de Sauvage (1977) described the sequence of changes in the CM, SP, and AP after ethacrynic acid administration. While CM at first

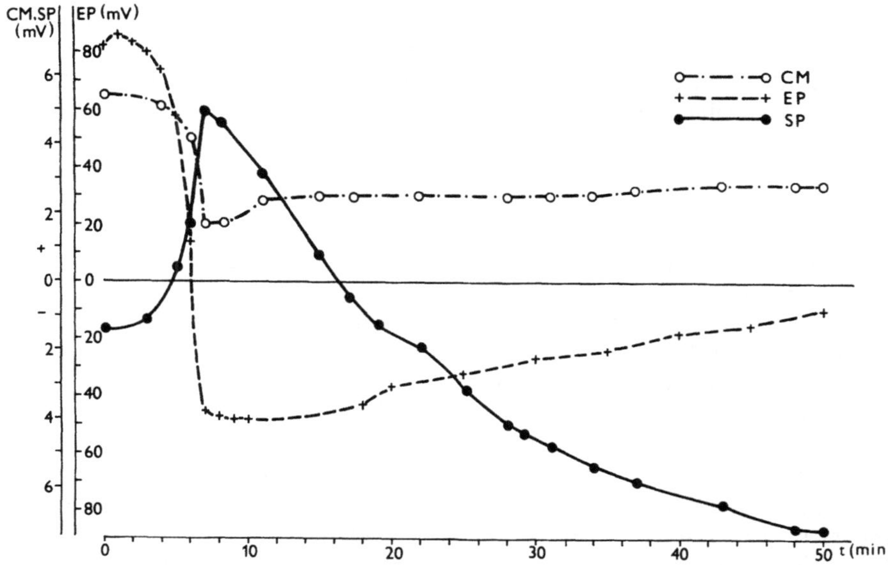

Fig. 18. Changes in the EP, CM, and SP (summating potential) values after injection of 50 mg/kg ethacrynic acid. The *abscissa* indicates the time after the injection in minutes. Stimulus: tone 1 kHz, 105 dB SPL. (Syka and Melichar 1985)

decreased and later slowly recovered, never reaching the original values, the SP changes were more complex, with two reversals of polarity and a supernormal amplitude. The SP changes resembled those found in other cochlear pathologies in patients, e.g., in Menière's disease (Eggermont 1976). Syka and Melichar (1981, 1985) recorded CM, SP, and EP in guinea pigs after ethacrynic acid and furosemide administration. In contrast to recording from implanted electrodes by Aran and Charles de Sauvage (1977), potentials were recorded from scala media in acute experiments. EP and CM changes were different. In the first phase all SPs irrespective of the original polarity attained high positive values; 12 – 18 min after injection SPs reversed polarity, and in the late phase all sounds evoked high negative SPs (Fig. 18). No explanation of the SP changes is available at the present time.

The effect of furosemide on the tuning properties of single cochlear nerve fibers in the cat has been reported in two recent studies (Evans and Klinke 1982; Sewell 1984). In both cases significant reversible elevation of the tips of tuning curves was observed. There was, however, disagreement as to the changes in the tails of tuning curves: Evans and Klinke (1982) reported no significant changes in the tail segments, whereas Sewell (1984) found threshold elevations also in the tail segment. According to Sewell (1984), the threshold shifts were correlated with the amount of decrease in the EP. Approximately 1 dB of threshold elevation corresponded with every millivolt of decrease in the EP. Also the shifts in the CF were observed: for units with CF higher than 1 kHz the CF shifted downwards. In contrast to changes in tuning curves after ototoxic antibiotics and noise (e.g., Liberman and Mulroy 1982), neither study reported hypersensitivity in the tail region after furosemide. In addition to the changes in tuning properties of the cochlear nerve fibers the micromechanics of the cochlea were also found to be changed after intravenous furosemide administration. Kemp (1979) has described a reversible reduction in the evoked cochlear emission after furosemide.

3.3 Salicylates

The effects of salicylates on the hearing of human subjects are well known. A hearing loss of 20 – 40 dB which is mostly flat across the audible frequency range occurs in normal subjects (McCabe and Day 1965; Myers and Bernstein 1965; Mongan et al. 1973; Bonding 1979b). As a rule, the threshold shift is accompanied by tinnitus when blood concentrations of salicylate exceed 200 – 400 mg (Mongan et al. 1973). Hearing returns to the pretreatment levels within 2 or 3 days after the drug is withdrawn.

Several parameters of auditory functions were disturbed in patients after several days of aspirin use with doses exceeding 2 g per day. McCabe and Day (1965) and Pedersen (1974) found hearing losses accompanied by changes in temporal integration that were similar to those seen in sensorineural deafness of other origin. These findings were recently confirmed by McFadden et al. (1984) who found subjects treated with aspirin to have undisturbed values of forward masking, flatter than normal temporal integration functions, and poorer detection of a temporal gap at low sound pressure levels. Deterioration in frequency resolution was also observed: Bonding (1979b) has found, by means of a loudness summation method, a wider critical band at 1 kHz, in subjects with salicylate-induced hearing loss.

In animals, disturbances in auditory function induced by salicylates have frequently been demonstrated: Meyers and Bernstein (1965) measured reversible hearing losses after acute injections in behaviorally trained monkeys, correlated with plasma-salicylate levels. They were not able, however, to demonstrate cochlear changes using light or electron microscopy. More recently changes have been reported in the stiffness of the stereocilia and vacuolation of the smooth endoplasmatic reticulum of OHCs and IHCs after multiple doses of sodium salicylate (Douek et al. 1983). Silverstein et al. (1967) has found in cats a decrease in the auditory nerve potential after aspirin treatment. The ototoxic effects of salicylates on the responses of single cochlear nerve fibers have been studied by Evans and Borerwe (1982). Intravenous injection of 400 mg/kg sodium salicylate produced a threshold shift in fibers of 13 to 21 dB. Several parameters of the fiber responses were significantly changed for a period of about 10 h: Q_{10} dB values of hearing curves decreased and the dynamic range was reduced. In contrast to findings in other cochlear pathologies with aminoglycoside antibiotics, the spontaneous activity of some fibers significantly increased. This finding is interesting with respect to explanation of the mechanisms of tinnitus. It has been proposed by Evans et al. (1981) that the increase in spontaneous firing occurring with higher serum salicylate levels may be the neural correlate of tinnitus. In addition to the increased spontaneous firing rate, more than half of the fibers displayed anomalous temporal patterns of spontaneous activity.

The mechanism of the ototoxic effect of salicylates remains, however, unknown. A significant contribution to the understanding of the ototoxicity of salicylates has been made by Hawkins (1967, 1973) who described vasoconstriction in capillaries of the spiral ligament, the stria vascularis, the tympanic lip, and the basilar membrane after intoxication with sodium salicylate or quinine. The changes in capillaries involved local narrowings of the capillary lumen by endothelial swelling with participation of pericytes. The role of ischemia in the ototoxic effects of salicylates and quinine had been proposed already by Gradenigo (1892), Witmaack (1903), and Blau (1904). Hawkins (1976) also speculated about possible involvement of the prostaglandins in the effect of salicylates.

3.4 Cisplatin

Cisplatin is an anticancer drug used for treatment of testicular and ovarian tumors, head and neck carcinoma, and hemoblastoma. Ototoxicity has been reported as a toxic side effect of the drug, together with bone marrow inhibition, nephrotoxicity and gastrointestinal disorders. Since the report of Helson et al. (1978), several reports have appeared about the defects in patient's hearing after treatment with this drug. The hearing loss was present mostly at higher frequencies and was usually permanent, although in some cases it was found to be reversible.

After cisplatin treatment of experimental animals the primary histopathologic changes in the inner ear were destruction of OHCs, mainly in the basal part of the cochlea. The first row of OHCs was more damaged, and moderate changes were also found in the stria vascularis. Pathological alterations were similar in the guinea pig and monkey (Standnicki et al. 1975; Fleischman et al. 1975; Estrem et al. 1981; Tange et al. 1983; Wright and Schaefer 1982). Laurell and Borg (1986) studied the effect of cisplatin in guinea pigs previously exposed to noise. Frequency-specific brainstem audiometry was used for evaluation of the hearing changes. The interaction was characterized as an addition and not a potentiation of the ototoxic effects.

3.5 Ototoxicity and Development of Auditory Function

Aminoglycoside antibiotics and loop diuretics penetrate the placental barrier, and therefore present a potential danger for the hearing of the newborn. For example, Jones (1973) has reported a striking case of maternal and fetal ototoxicity attributed to combined treatment with kanamycin and ethacrynic acid.

The problem of early susceptibility to ototoxic effects of drugs, particularly aminoglycoside antibiotics, was investigated in animal experiments. The guinea pig may serve as a model of transplacental ototoxicity, because the maturation of cochlear receptors occurs mainly during intrauterine life, as in humans. Uziel et al. (1979a, b) injected pregnant guinea pigs with 200 to 400 mg/kg/day kanamycin for 8 days at different stages of gestation. Recording of cochlear microphonics and compound action potential from the round window revealed ab-

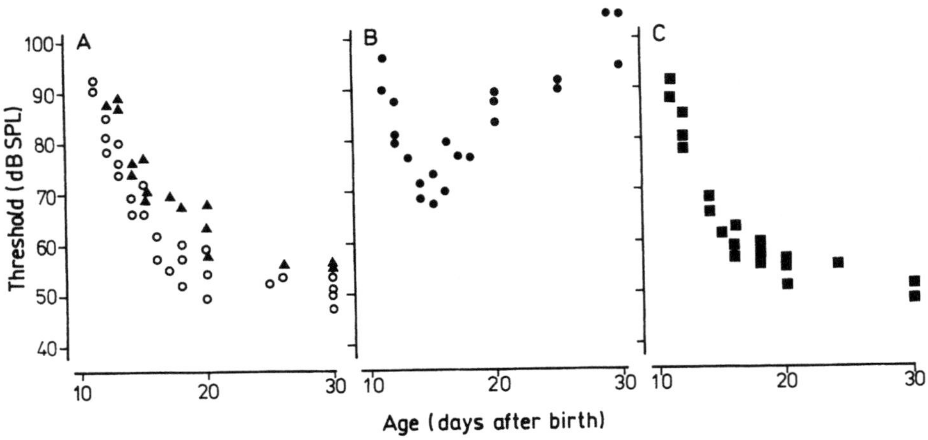

Fig. 19 A – C. Brainstem response thresholds in rats treated with kanamycin (400 mg/kg daily for 10 days) with three different time schedules. Kanamycin was applied from 1st to 10th day (**A**), from 11th to 20th day (**B**), and from 15th to 23rd day (**C**). *Open circles* in (**A**) denote control obtained from normal rats. (Osako et al. 1979)

normalities of hearing function similar to those observed in adult animals treated with aminoglycoside antibiotics: increased thresholds of CM and AP and recruiting amplitude-intensity functions, predominantly for high frequencies. Also, morphological abnormalities were analogous to those in adult animals: ultrastructural changes in the apical OHCs, degenerated or bent stereocilia, vacuolation of the cytoplasm, ruptures of the cell plasma membranes, and finally degeneration of the organ of Corti. In cases of severe injury, the OHC damage extended to the second and third turns and the IHCs in the basal turn were also injured. The damage was present only in the case where pregnant guinea pigs were treated with kanamycin during the last 15 days of gestation. This corresponds with the onset of hearing in guinea pigs (Romand 1971). The results have been confirmed by Dumas and Charachon (1982).

A related study in rats (Osako et al. 1979) has shown somewhat similar data (Fig. 19). Also in rat pups, kanamycin (400 mg/kg) did not produce any immediate alterations of hearing function when administered earlier than the 11th day after birth. The 11th day is considered the day of onset of auditory function in the rat. When kanamycin was injected between days 11 and 20 after birth, the auditory potentials were almost completely abolished 10 days after the beginning of the kanamycin treatment. Correlated with this was the occurrence of morphological changes in the organ of Corti. In contrast, the administration of kanamycin beginning 20 days after birth or later resulted in delayed alterations in hearing function. The decrease in evoked potentials and increase in hearing thresholds started to appear at least 3 weeks after the beginning of kanamycin treatment. The results suggest that: (a) the rat may be less susceptible to kanamycin intoxication than the guinea pig; (b) the period between day 11 and 20, i.e., the period of the greatest development of auditory function in the rat, may be considered as a period of great sensitivity to aminoglycoside ototoxicity. Carlier and Pujol (1980) examined the ototoxic effects of 50 – 200 mg/kg/day injections of amikacin in young and adult Sprague-Dawley rats. Again the sensitive period for ototoxicity, characterized by elevation of AP thresholds, was between 10 and 25 days after birth. Similar results were obtained in the C57 mouse strain (Chen and Saunders 1983). Daily injections of kanamycin during the period between 10 and 13 days caused great OHC loss over much of the cochlea, whereas earlier or later treatment had no effect or had smaller effects. The loss in OHCs was matched by a loss in auditory sensitivity (Chen and Saunders 1984).

3.6 Interaction of Ototoxic Drugs

Clinically, aminoglycoside antibiotics are often used in combination with loop diuretics. It has been shown that very soon after the introduction of loop diuretics in patients receiving simultaneous therapy with an aminoglycoside antibiotic there is severe permanent hearing losses of rapid onset (Mathog and Klein 1969; Johnson and Hamilton 1970; Meriwether et al. 1971). The interaction of aminoglycoside antibiotics with loop diuretics has been subsequently examined in numerous animals experiments (West et al. 1973; Prazma et al. 1974; Nakai 1977; Brummett et al. 1979; Russell et al. 1979). Generally, the results have shown unequivocally that there is a potentiation of the ototoxic effects of both groups of drugs, i.e., the damage was more severe than when the drugs were given individually. The pattern of structural damage and the changes in cochlear potentials were essentially similar to those produced by individual ototoxic agents. For example, severe swelling was seen in the stria vascularis and destruction of outer hair cells occurred in the basal turn of the cochlea.

Mechanisms of the interaction have been studied in vitro and in vivo by Orsulakova and Schacht (1981). In vitro, the addition of neomycin to incubations of synaptosomes and inner ear tissues of guinea pigs raised the concentration of bound ethacrynate approximately by twofold. Treatment of animals with neomycin in vivo (100 mg/kg) 60 min prior to ethacrynate led to a three- to fivefold higher accumulation of ethacrynate in cochlear tissues. The authors suggested that neomycin breaks down hemolabyrinthine or tissue permeability barriers and allows increased penetration of the other drug into the inner ear.

The effect of timing was noted even in the initial studies: the interaction is maximal when an injection of the aminoglycoside antibiotics precedes the ethacrynic acid but not if the order is reversed (West et al. 1973; Prazma et al. 1974; Hawkins et al. 1975). Yet an interesting interaction has been observed by Takada et al. (1985). When gentamicin was administered to guinea pigs by perilymphatic perfusion at the time when the EP was maximally reduced by previous intravenous administration of furosemide (or during anoxia lasting 3 min), the reduction in CM was prevented. This finding suggests that entry of gentamicin into hair cells requires specific conditions of the plasma membrane and its polarization.

The interaction demonstrated for loop diuretics and aminoglycoside antibiotics does not exist for kanamycin and nonloop diuretics, such as hydrodiuril, mercuhydrin, and mannitol (Brummett et al. 1974). Other ototoxic antibiotics such as viomycin, capreomycin, and polymyxin B, interact with ethacrynic acid

(Davis et al. 1982). Potentiation of the ototoxic effects was observed when cisplatin and ethacrynic acid were given simultaneously (Komune and Snow 1981).

A problem related to the interaction between ototoxic drugs is the problem of interaction between noise and ototoxic drug. In general, concurrent administration of aminoglycoside antibiotics and noise exposure results in a cochlear damage level exceeding that predicted from the effects of either agent given alone. For example, Brown et al. (1980) have reported augmentation of the effect of 300 mg/kg kanamycin sulfate injection for 7 days when combined with 115 dB SPL broad-band noise. The interaction was masked at 400 mg/kg by ceiling effects. Similar results were reported in many other studies (Jauhiainen et al. 1972; Dayal et al. 1971; Dodson et al. 1982). When 200 mg/kg neomycin was combined with 115 dB broad-band noise, the impairment of the cochlea was four times that found in animals given either agent alone. Less consistent were data when the noise exposure was combined with administration of aspirin. In a psychophysical study McFadden and Plattsmier (1983) have found larger temporary threshold shifts when subjects exposed to intense sound had taken 3.9 g aspirin for 2 days or more. Similar results were obtained by McCabe and Day (1965). In animal experiments potentiation of noise-induced threshold shifts by aspirin was observed in chinchillas by Eddy et al. (1976). However, Woodford et al. (1978) have not been able to show any potentiation of impulse noise-induced hearing loss by aspirin in the same animal species. Similarly, negative results have been reported for the interaction of noise and aspirin in the basilar papilla of the chicken (Lambert et al. 1986).

4 Concluding Remarks

Remarkable progress has been achieved in understanding the structural and functional basis of sensorineural hearing disorders during the past 20 years. Two main achievements have resulted from these studies: (a) The pathological mechanisms of noise-induced and drug-induced hearing impairment are better understood and their prevention, treatment and rehabilitation can be more efficient. (b) The lesions produced by noise and ototoxic drugs help to elucidate the function of individual cochlear structures and their role in auditory perception.

First studies, mostly performed with high level noise exposures, have demonstrated the basic pathological features of impaired cochlear function, particularly the high correlation between the spectral content of the noise and localization of the lesion in the cochlea. Also, the susceptibilities of individual types of receptors to noise and ototoxic drugs have been assessed, mostly in exposures which resulted in permanent hearing losses. Application of different micromethods in cochlear research (scanning and transmission electron microscopy, intracellular recording from hair cells and auditory nerve fibers with injection of horseradish peroxidase, measurement of ionic concentration with ion-selective microelectrodes, measurement of stereocilia stiffness change etc.) have opened new perspectives in the field. Subtle changes in the stereocilia of hair cells were found to be associated with low intensity noise exposure. Broken tip and lateral links of stereocilia, changes in their rootlets, disarray and final fusion, and reorganization into giant stereocilia are characteristic pathologies of stereocilia. They are combined with changes in stereocilia stiffness and, probably, also in the permeability of ion channels, localized in the stereocilia. Similar changes can be found after administration of many ototoxic drugs. It is apparently within the possibilities of contemporary research to explain the subtle changes causing the temporary threshold shifts elicited by minimum noise exposures.

However, noise exposure induces changes not only in the organ of Corti, but also in the stria vascularis and in the central part of the auditory pathway. Stria vascularis activity, which maintains the unique potentials and ionic concentration of the endolymph, is less endangered by the noise and ototoxic drug effects, unless they specifically interfere with the stria metabolism as loop diuretics. Stria vascularis function also recovers more easily than the function of the organ of Corti. Full recovery of the stria function occurs after exposures to noise or ototoxic drugs which result in permanent destruction of the hair cells.

Limited information is available on pathologies induced by noise exposures in the central part of the auditory system. The results of psychoacoustical tests in hearing-impaired persons unequivocally demonstrate that some changes also develop in the central part of the auditory system as a consequence of partial loss of receptor function in the periphery. For example, disturbances in the temporal analysis, as detected in gap tests, point to pathologies in the central auditory pathway. The complexity of the central auditory structure and function, which is not fully understood at the present time, prevents elucidation of pathological changes, which can be subtle and compensated by powerful central mechanisms. However, recent progress in auditory neurochemistry, particularly in regard to neurotransmitter distribution, promises new achievements even in this difficult area.

5 References

Anderson SD, Kemp DT (1979) The evoked cochlear mechanical response in laboratory primates. Arch Otorhinolaryngol 224:47–54

Angelborg C, Engström H (1973) The normal organ of Corti. In: Møller AR (ed) Basic mechanisms in hearing. Academic, New York, pp 125–182

Angelborg C, Hultcrantz E, Beausang-Linder M (1979) The cochlear blood flow in relation to noise and cervical sympathectomy. Adv Otorhinolaryngol 25:41–48

Angelborg C, Larsen HC, Slepecky N (1985) Regional cochlear blood flow studied by observation of microspheres in serial section. Ann Otol Rhinol Laryngol 94: 181–185

Anniko M (1978) Reversible and irreversible changes of the stria vascularis. Acta Otolaryngol (Stockh) 85:349

Aran JM (1981) Electrophysiology of cochlear toxicity. In: Matz CJ, Lerner S, Hawkins JE (eds) Aminoglycoside ototoxicity. Little Brown, Chicago, pp 31–50

Aran JM, Darrouzet J (1975) Observation of click-evoked compound VIII nerve responses before, during and over seven months after kanamycin treatment in the guinea pig. Acta Otolaryngol (Stockh) 79:24–32

Aran JM, De Sauvage RC (1977) Evolution of CM, SP and AP during ethacrynic acid intoxication in the guinea pig. Acta Otolaryngol (Stockh) 83:153–159

Aran JM, Erre JP, Guilhaume A, Aurousseau C (1982) The comparative ototoxicities of gentamicin, tobramycin and dibekacin in the guinea pig. Acta Otolaryngol (Stockh) 390:1 – 30

Arehole A, Salvi RS, Saunders SS, Henderson D (1987) Evoked response "forward masking" in chinchillas with temporary hearing loss. Hear Res 2:193 – 205

Ashmore JF, Meech RW (1986) Ionic basis of membrane potential in outer hair cells of guinea pig cochlea. Nature 322:368 – 371

Axelsson A (1968) The vascular anatomy of the cochlea in the guinea pig and in man. Acta Otolaryngol (Stockh) 243:1 – 134

Axelsson A, Vertes D (1982) Histological findings in cochlear vessels after noise. In: Hamernik RP, Henderson D, Salvi R (eds) New perspectives on noise-induced hearing loss. Raven, New York, pp 49 – 67

Axelsson A, Vertes D, Miller J (1980) Immediate noise effects on cochlear vasculature in the guinea pig. Acta Otolaryngol (Stockh) 91:237

Babighian G, Moushegian G, Rupert AL (1975) Central auditory fatigue. Audiology 14:72 – 83

Balogh K Jr, Hiraide F, Ishii D (1970) Distribution of radioactive dihydrostreptomycin in the cochlea: an autoradiographic study. Ann Otol Rhinol Laryngol 79:641 – 652

Beagley HA (1965a) Acoustic trauma in the guinea pig: I Electrophysiology and histology. Acta Otolaryngol (Stockh) 60:437 – 451

Beagley HA (1965b) Acoustic trauma in the guinea pig: II Electron microscopy including the morphology of cell junctions in the organ of Corti. Acta Otolaryngol (Stockh) 60:479 – 495

Békésy G von (1963) Three experiments concerned with pitch perception. J Acoust Soc Am 35:602 – 606

Benitez LD, Eldredge DH, Templer JW (1972) Temporary threshold shifts in chinchilla: electrophysiological correlates. J Acoust Soc Am 52:1115 – 1123

Berg K (1949) The toxic effect of streptomycin on the eighth cranial nerve. Ann Otol Rhinol Laryngol 58:448 – 456

Blau A (1904) Experimentelle Studien über die Veränderungen im Gehörorgan nach Vergiftung mit Salizylsäuren Natrium. Arch Ohrenheilkd 61:220 – 233

Bock GR, Seifter EJ (1978) Developmental changes of susceptibility to auditory fatigue in young hamsters. Audiology 17:193 – 203

Boer E de, Bouwmeester J (1974) Critical bands and sensorineural hearing loss. Audiology 13:236 – 259

Bohne BA (1976a) Mechanisms of noise damage in the inner ear. In: Henderson D, Hamerník RP, Dosanjh DS (eds) Effects of noise in hearing. Raven, New York, pp 41 – 68

Bohne BA (1976b) Hearing of the noise damaged inner ear. In: Hirsh SK, Eldredge DH, Hirsh IJ, Silverman SR (eds) Hearing and Davis: essays honoring Halowell Davis. Washington University Press, St. Louis, pp 85 – 96

Bohne BA, Clark WW (1982) Growth of hearing loss and cochlear lesions with increasing duration of noise exposure. In: Hamernik RP, Henderson D, Salvi R (eds) New perspectives of noise-induced hearing loss. Raven, New York, pp 183 – 302

Bohne BA, Rabbitt KD (1983) Holes in the reticular lamina after noise exposure: implication for continuing damage in the organ of Corti. Hear Res 11:41 – 53

Bohne BA, Yohman L, Gruner M (1987) Cochlear damage following exposure to high-frequency noise. Hear Res 29:251 – 264

Bonding P (1979a) Critical bandwidth in loudness summation in sensorineural hearing loss. Br J Audiol 13:23 – 30

Bonding P (1979b) Critical bandwidth in patients with a hearing loss induced by salicylates. Audiology 18:133 – 144

Borg E (1981) Physiological and pathogenic effect of sound. Acta Otolaryngol (Stockh) 381:1 – 68
Bosher SK (1979) The nature of the negative endocochlear potential produced by anoxia and ethacrynic acid in the rat and guinea pig. J Physiol (Lond) 293:329 – 345
Bosher SK (1980a) The nature of the ototoxic actions of ethacrynic acid upon the mammalian endolymph system I. Functional aspects. Acta Otolaryngol (Stockh) 89: 407 – 418
Bosher SK (1980b) The nature of the ototoxic actions of ethacrynic acid upon the mammalian endolymph system II. Structural-functional correlates in the stria vascularis. Acta Otolaryngol (Stockh) 90:40 – 54
Bosher SK, Smith C, Waren RL (1973) The effects of ethacrynic acid upon the cochlear endolymph and stria vascularis. A preliminary report. Acta Otolaryngol (Stockh) 75:184
Brandt JF (1967) Frequency discrimination following exposure to noise. J Acoust Soc Am 41:448 – 457
Brown JJ, Brummett RE, Fox KE, Bendrick TW (1980) Combined effects of noise and kanamycin. Arch Otolaryngol 106:744 – 750
Brummett RE, West B, Traynor J, Manor N (1974) Ototoxic interaction between aminoglycoside antibiotics and diuretics. Toxicol Appl Pharmacol 29:45
Brummett RE, Smith CA, Uemo Y, Cameron S, Richter R (1977) The delayed effects of ethacrynic acid on the stria vascularis of the guinea pig. Acta Otolaryngol (Stockh) 83:98
Brummett RE, Fox KE, Bendrick TW, Himes DL (1978) Ototoxicity of tobramycin, gentamicin, amikacin and sisomicin in the guinea pig. J Antimicrob Chemother 4:73 – 83
Brummett E, Brown T, Himes L (1979) Quantitative relationships of the ototoxic interaction of kanamycin and ethacrynic acid. Arch Otolaryngol 105:240 – 246
Brusilow SW, Gordes E (1973) The mutual independence of the endolymphatic potential and the concentrations of sodium and potassium in endolymph. J Clin Invest 52:2517 – 2521
Canlon B, Schacht J (1983) Acoustic stimulation alters deoxyglucose uptake in the mouse cochlea and inferior colliculus. Hear Res 10:217 – 226
Canlon B, Miller J, Flock A, Borg E (1987) Pure tone overstimulation changes the micromechanical properties of the inner hair cell stereocilia. Hear Res 30:65 – 72
Carder HM, Miller JD (1971) Temporary threshold shifts produced by noise-exposure of long duration. Trans Am Acad Ophthalmol Otolaryngol 75:1346 – 1354
Carder HM, Miller JD (1972) Temporary threshold shifts from prolonged exposure to noise. J Speech Hear Res 15:603 – 623
Carlier E, Pujol R (1980) Supra-normal sensitivity to ototoxic antibiotic of the developing rat cochlea. Arch Otorhinolaryngol 226: 129 – 133
Chen C-S, Saunders JC (1983) The sensitive period for ototoxicity of kanamycin mice. Arch Otorhinolarygol 238:217 – 223
Chen C-S, Saunders JC (1984) Effects of kanamycin on cochlear nuclear evoked responses and behavioral responses in C57 BL/68 mice. Exp Neurol 11:461 – 467
Clark WW, Bohne BA (1978) Animal model for the 4-kHz tonal dip. Ann Otol Rhinol Laryngol 51:1 – 16
Clark WW, Bohne BA, Boettcher FA (1987) Effects of periodic rest on hearing loss cochlear damage following exposure to noise. J Acoust Soc Am 82:1253 – 1264
Coats AC (1978) Human auditory nerve action potentials and brainstem evoked responses. Latency-intensity functions in detection of cochlear and retrocochlear abnormality. Arch Otolaryngol 104:709
Coats AC, Martin JL (1977) Human auditory nerve action potentials and brainstem evoked responses. Effects of audiogram shape and lesion location. Arch Otolaryngol 103:605

Cody AR, Johnstone BM (1980) Single auditory neuron response during acute acoustic trauma. Hear Res 3:3 – 16
Cody AR, Johnstone BM (1982a) Temporary threshold shifts modified by binaural acoustic stimulation. Hear Res 6:199 – 205
Cody AR, Johnstone BM (1982b) Reduced temporary and permanent hearing losses with multiple tone exposures. Hear Res 6:291 – 301
Cody AR, Russel IJ (1985) Outer hair cells in the cochlea and noise hearing loss. Nature 315:662 – 665
Coleman JW (1976) Age dependent changes and acoustic trauma in the spiral organ of the guinea pig. Scand Audiol 5:63 – 68
Comis SD, Pickles JD, Osborne MP (1985) Osmium tetroxide postfixation in relation to the linkage and spatial organization of stereocilia in the guinea pig cochlea. J Neurocytol 14:113 – 130
Cook RO, Konishi T, Salt AN, Hamm CW, Labetkin EH, Koo J (1981) Brainstem-evoked responses of guinea pigs exposed to high noise levels in utero. Dev Psychol 15:95 – 104
Dallos P, Cheatham MA (1976) Compound action potential (AP) tuning curves. J Acoust Soc Am 59:591 – 597
Dallos P, Cheatham MA (1977) Analog of two-tone suppression in whole nerve responses. J Acoust Soc Am 62:1048 – 1051
Dallos P, Harris D (1978) Properties of auditory nerve responses in absence of outer hair cells. J Neurophysiol 41:365 – 383
Dallos P, Wang CY (1974) Bioelectric correlates of kanamycin intoxication. Audiology 13:277 – 289
Dallos R, Ryan A, Harris D, McGee T, Özdamar O (1977) Cochlear frequency selectivity in the presence of hair cell damage. In: Evans EF, Wilson JP (eds) Psychophysics and physiology of hearing. Academic, London, pp 249 – 258
Dallos P, Harris D, Relkin E, Cheatham MA (1980) Two-tone suppression and intermodulation distortions in the cochlea: effect of outer hair cell lesion. In: Van den Brink G, Bilsen FA (eds) Psychophysical, physiological and behavioral studies in hearing. Delft, the Netherlands, pp 242 – 249
Dallos P, Santos-Sacchi J, Flock A (1982) Intracellular recordings from cochlear outer hair cells. Science 218:582 – 584
Darrouzet J, Guilhaume A (1974) Ototoxicity de la kanamycine au jour. Étude expérimentale en microscopie électronique. Rev Laryngolotol Rhinol (Bord) 95:601 – 621
Davis H, Morgan CT, Hawkins JE, Galambos R, Smith FW (1950) Temporary deafness following exposure to loud tones and noise. Acta Otolaryngol (Stockh) 88:1 – 57
Davis RR, Brummett RE, Bendrick TW, Himes DL (1982) The ototoxic interaction of viomycin, capreomycin and polymyxin B with ethacrynic acid. Acta Otolaryngol (Stockh) 93:211 – 217
Davis H, Deatherage BH, Rosenblut B, Fernandez C, Kimura R, Smith CA (1958) Modification of cochlear potentials produced by streptomycin poisoning and by extensive venous obstruction. Laryngoscope 68:596
Dayal VS, Koksanian A, Mitchell DP (1971) Combined effects of noise and kanamycin. Ann Otol Rhinol Laryngol 80:897 – 902
Desrochers CS, Schacht J (1982) Neomycin concentrations in inner ear tissues and other organs of the guinea pig after chronic drug administration. Acta Otolaryngol (Stockh) 93:233 – 236
Dodson HC, Bannister LH, Douek EE (1982) The effects of combined gentamicin and white noise on the spiral organ of young guinea pigs. Acta Otolaryngol (Stockh) 94:193 – 202
Douek E, Bannister LH, Dodson HC, Ashcroft P, Humphries KN (1976) Effects of incubator noise on the cochlea of the newborn. Lancet 20:1110 – 1113

Douek ES, Dodson HC, Bannister LH (1983) The effects of sodium salicylate on the cochlea of guinea pigs. J Laryngol Otol 93:793 – 799
Drescher MJ, Drescher DG, Medina JE (1983) Effects of sound stimulation at several levels on concentration of primary amines, including neurotransmitter candidates, in perilymph of the guinea pig inner ear. J Neurochem 41:309 – 320
Dumas G, Charachon R (1982) Ototoxicity of kanamycin in developing guinea pigs. Acta Otolaryngol (Stockh) 94:203 – 212
Duvall AJ, Wersäll J (1964) Site of action of streptomycin upon inner ear sensory cell. Acta Otolaryngol (Stockh) 57:581 – 598
Duvall AJ, Ward WD, Lauhala KE (1974) Stria ultrastructure and vessels transport in acoustic trauma. Ann Otol Rhinol Laryngol 83:498 – 514
Eddy LB, Morgan RJ, Carney HC (1976) Hearing loss due to combined effects of noise and sodium salicylate. ISA Trans 15:103 – 108
Eggermont JJ (1976) Electrocochleography. In: Keidel WD, Neff WD (eds) Anatomy, physiology. Springer, Berlin Heidelberg New York, pp 626 – 705 (Handbook of sensory physiology. Auditory system, vol V/3)
Eggermont JJ (1977) Compound action potential tuning curves in normal and pathological human ears. J Acoust Soc Am 62:1247 – 1251
Elliott LL (1975) Temporal and masking phenomena in persons with sensorineural hearing loss. Audiology 14:336 – 353
Engström B (1983) Stereocilia of sensory cells in normal and hearing impaired ears. Scand Audiol [Suppl] 19, pp 1 – 34
Engström H, Ades HW (1960) Effects of high-intensity noise on inner ear sensory epithelia. Acta Otolaryngol (Stockh) 158:219 – 229
Engström H, Ades HW, Bredberg G (1970) Normal structure of the organ of Corti and the effect of noise-induced cochlear damage. In: Wolstenholm GEW, Knight T (eds) Sensorineural hearing loss. Churchill, London, pp 127 – 152
Erlandsson B, Hakånson H, Ivarson A, Nilson P, Wersäll J (1980) Hair cell damage in the guinea pig due to different kinds of noise. Acta Otolaryngol (Stockh) 367:1 – 43
Estrem SA, Babin RW, Ryu JH (1981) Cisdiammminedichloroplatinum (II) ototoxicity in the guinea pig. Otolaryngol Head Neck Surg 89:638 – 645
Evans EF (1974) Effects of hypoxia on the tuning of single cochlear nerve fibers. J Physiol (Lond) 238:65 – 67 p
Evans EF (1976) Temporal sensorineural hearing losses and VIIIth nerve changes. In: Henderson D, Hamernik R, Disanjh D, Mills J (eds) Effects of noise on hearing. Raven, New York, pp 199 – 224
Evans EF, Borerwe TA (1982) Ototoxic effects of salicylates on the responses of single cochlear nerve fibers and on cochlear potentials. Br J Audiol 16:101 – 108
Evans EF, Klinke R (1974) Reversible effects of cyanide and furosemide on the tuning of single cochlear fibers. J Physiol (Lond) 242:129 – 131
Evans EF, Klinke R (1982) The effects of intracochlear and systemic furosemide on the properties of single cochlear nerve fibers in the cat. J Physiol (Lond) 331:409 – 427
Evans EF, Wilson JP, Borerwe TA (1981) Animal models of tinnitus. In: Evered D, Lawrence C (eds) Tinnitus. Pitman, London, pp 217 – 224
Eybalin M, Rebillard G, Jarry T, Cupo A (1987) Effect of noise level on the met-enkephalin content of the guinea pig cochlea. Brain Res 418:189 – 192
Falk SA, Cook RO, Haseman JK, Sanders GM (1974) Noise induced inner ear damage in newborn and adult guinea pigs. Laryngoscope 84:444 – 453
Farkashidy J, Black RG, Briant TDR (1963) Effect of kanamycin on the internal ear: electrophysiological and electron microscopic study. Laryngoscope 73:713 – 727
Federspil P, Schätzle W, Tiesler E (1976) Pharmacokinetics and ototoxicity of gentamicin, tobramycin and amikacin. J Infect Dis 134:200

Fitzgibbons P, Wightman FL (1982) Temporal resolution in listeners with sensorineural hearing loss. J Acoust Soc Am 72:761 – 765
Fitzgibbons PJ, Gordon-Salant S (1987a) Temporal gap resolution in listeners with high-frequency sensorineural hearing loss. J Acoust Soc Am 81:133 – 137
Fitzgibbons PJ, Gordon-Salant S (1987b) Minimum stimulus levels for temporal gap resolution in listeners with sensorineural hearing loss. J Acoust Soc Am 81:1542 – 1545
Fleischman RW, Standnicki SW, Ethier MF (1975) Ototoxicity of cisdichlorodiammine-platinum (II) in the guinea pig. Toxicol Appl Pharmacol 33:320 – 332
Forge A, Fradis M (1985) Structural abnormalities in the stria vascularis following chronic gentamicin treatment. Hear Res 20:233 – 244
Galambos R, Hecox KE (1978) Clinical applications of the auditory brain stem response. Otolaryngol Clin North Am 11:709
Glasberg BR, Moore BCJ (1986) Auditory filter shapes in subjects with unilateral and bilateral cochlear impairments. J Acoust Soc Am 79:1020 – 1033
Glasberg BR, Moore CJ, Bacon SP (1987) Gap detection and masking in hearing-impaired and normal-hearing subjects. J Acoust Soc Am 81:1546 – 1556
Goulios H, Robertson D (1983) Noise induced cochlear damage assessed using electro-physiological and morphological criteria: an examination of the equal energy principle. Hear Res 11:327 – 341
Gradenigo G (1892) Krankheiten des Labyrinths und des Nervus Acusticus. In: Vogel FCW (ed) Handbuch der Ohrenheilkunde II. Schwartze, Leipzig, pp 474 – 479
Hamernik RP, Henderson D, Salvi R (1982) New perspectives on noise-induced hearing loss. Raven, New York
Harris DM (1979) Action potential suppression, tuning curves, and threshold: comparison with single fiber data. Hear Res 1:133 – 154
Harrison RV (1981) Rate-versus-intensity functions and related AP responses in normal and pathological guinea pig and human cochleas. J Acoust Soc Am 70:1036 – 1044
Harrison RV, Evans EF (1979) Cochlear fibre responses in guinea pigs with well-determined cochlear lesions. In: Hoke M, De Boer E (eds) Models of the auditory system and related signal processing techniques. Scand Audiol 9:83 – 92
Harrison RV, Aran JM, Erre J-P (1981a) AP tuning curves from normal and pathological human and guinea pig. J Acoust Soc Am 69 (5):1374 – 1385
Harrison RV, Aran JM, Negrevergne M (1981b) The frequency selectivity of the normal and pathological human cochlea. Arch Otorhinolaryngol 230:221 – 227
Hawkins JE Jr (1950) Cochlear signs of streptomycin intoxication. J Pharmacol Exp Ther 100:38 – 44
Hawkins JE Jr (1967) Vascular patterns of membranous labyrinth. In: Graybiel A (ed) Third symposium on the role of vestibular organs in space exploration. National Aeronautics and Space Administration, Washington DC p 241
Hawkins JE Jr (1971) The role of vasoconstriction in noise-induced hearing loss. Ann Otol Rhinol Laryngol 80:903 – 913
Hawkins JE Jr (1973) Ototoxic mechanisms: a working hypothesis. Audiology 12: 383 – 393
Hawkins JE Jr (1976) Drug ototoxicity. In: Keidel WD, Neff WD (eds) Anatomy, physiology. Springer, Berlin Heidelberg New York, pp 707 – 748 (Handbook of Sensory Physiology. Auditory system, vol V/3)
Hawkins JE Jr (1977) Condition of the inner hair cells after aminoglycoside intoxication. In: Portmann M, Aran JM (eds) Inner ear biology. Paris, Colloques INSERM 68, pp 237 – 334
Hawkins JE Jr, Engström H (1964) Effect of kanamycin on cochlear cytoarchitecture. Acta Otolaryngol (Stockh) 188:100 – 106

Hawkins JE Jr, Johnsson LG (1976) Patterns of sensorineural degeneration in human ears exposed to noise. In: Henderson D, Hamernik RP, Dosanjh DS (eds) Effects of noise on hearing. Raven, New York, pp 91–110
Hawkins JE Jr, Lurie MH (1952) The ototoxicity of streptomycin. Ann Otol Rhinol Laryngol 61:789–806
Hawkins JE Jr, Lurie MH (1953) The ototoxicity of dihydrostreptomycin and neomycin in the cat. Ann Otol Rhinol Laryngol 62:1128–1148
Hawkins JE Jr, Marques DM, Clark CS, Preston RE (1975) Ototoxic potentiation between ethacrynic acid and aminoglycoside antibiotics in guinea pig. J Acoust Soc Am 57:60
Hawkins JE Jr, Stebbins WC, Johnsson LG, Moody DB, Muraski A (1977) The patas monkey as a model for dihydrostreptomycin ototoxicity. Acta Otolaryngol (Stockh) 83:123–129
Hayashida T (1985) Distribution of gentamicin in guinea pig's inner ear revealed by immunohistochemical method. Ear Res Jpn 16:4–8
Hayes D, Jerger J (1982) Auditory brainstem response (ABR) to tone-pips: results in normal and hearing-impaired subjects. Scand Audiol 11:133–142
Heidland A, Wigand ME (1970) Einfluss hoher Furosemiddosen auf die Gehörfunktion bei Urämie. Klin Wochenschr 48:1052–1056
Helson L, Okonkwo E, Anton L (1978) *Cis*-platinum ototoxicity. Clin Toxicol 13:469–478
Henderson D (1969) Temporal summation of acoustic signals by the chinchilla. J Acoust Soc Am 46:474–475
Henderson D, Hamernik RP, Dosanjh DS, Mills JH (1976) Effects of noise on kearing. Raven, New York
Henry KR (1980) Effects of noise, hypothermia and barbiturate on cochlea electrical activity. Audiology 19:44–56
Henry KR (1983) Abnormal auditory development resulting from exposure to ototoxic chemicals, noise and auditory restrictions. In: Romand R (ed) Development of the auditory system and vestibular system. Academic, New York, pp 273–308
Hinshaw HC, Feldman WH (1945) Streptomycin in the treatment of clinical tuberculosis: a preliminary report. Proc Mayo Clin 20:313–318
Honrubia V, Ward PH (1969) Properties of the summating potential of the guinea pig cochlea. J Acoust Soc Am 45:1443–1450
Horn KL, Langley LR, Gates GA (1977) Effect of ethacrynic acid on the stria vascularis. Arch Otorhinolaryngol 103:539
Hultcrantz E (1979) The effect of noise on cochlear blood flow in the conscious rabbit. Acta Physiol Scand 106:29–37
Hunter-Duvar IM (1977) Morphology of the normal and the acoustically damaged cochlea. Scan Electron Microsc 2421:428
Hunter-Duvar IM, Suzuki M, Mount RJ (1982) Anatomical changes in the organ of Corti after acoustic stimulation. In: Hamernik RP, Henderson D, Salvi R (eds) New perspectives on noise induced hearing loss. Raven, New York, pp 3–22
Irwin RJ, Hinchcliff LK, Kemp S (1981) Temporal activity in normal and hearing-impaired listeners. Audiology 20:234–243
Iurato S (1961) Submicroscopic structures of the membranous labyrinths. Z Zellforsch 53:259–298
Jauhiainen T, Kohonen J, Jauhiainen M (1972) Combined effect of noise and neomycin on the cochlea. Acta Otolaryngol (Stockh) 73:387–390
Jerger JF (1955) Influence of stimulus duration on the pure-tone threshold during recovery from auditory fatigue. J Acoust Soc Am 27:121–124
Jerger J, Mauldin L (1978) Prediction of sensorineural hearing level from the brain stem evoked response. Arch Otorhinolaryngol 104:456

Jesteadt W, Bilger RC, Green DM, Patterson JH (1976) Temporal acuity in listeners with sensorineural hearing loss. J Speech Hear Res 19:357–370
Jewett OL (1970) Volume conducted potentials in response to auditory stimuli as detected by averaging in the cat. Electroencephalogr Clin Neurophysiol 28:609
Johnson AH, Hamilton CH (1970) Kanamycin ototoxicity-possible potentiation by other drugs. South Med J 63:511–513
Jones HC (1973) Intrauterine ototoxicity: A case report and review of literature. J Nat Med Ass 65:201–203
Johnsson LG, Hawkins JE Jr (1972) Strial atrophy in clinical and experimental deafness. Laryngoscope 82:1105–1125
Juhn SK, Rybak LP, Morizono T, Green LP (1981) Pharmacokinetics of furosemide in relation to the alteration of endocochlear potential. Scand Audiol [Suppl] 14:39–50
Kellerhals B (1972) Acoustic trauma and cochlear microcirculation. Adv Otorhinolaryngol 18:91–168
Kemp DT (1978) Stimulated acoustic emissions from within the human auditory system. J Acoust Soc Am 64:1386–1391
Kemp DT (1979) Evidence of mechanical nonlinearity and frequency selective amplification in the cochlea. Arch Otorhinolaryngol 224:37–45
Kemp DT (1981) Physiologically active cochlear micromechanics: one source of tinnitus. In: Evered D, Lawrenson G (eds) Tinnitus. Pitman, London, pp 54–81
Kemp DT (1982) Cochlear echoes: implications for noise induced hearing loss. In: Hamernik RP, Henderson R, Salvi R (eds) New perspectives on noise-induced hearing loss, pp 189–206
Kiang NYS, Moxon EC, Levine RA (1970) Auditory nerve activity in cats with normal and abnormal cochlea. In: Wolstenholme GEW, Knight T (eds) Sensorineural hearing loss. Churchill, London, pp 241–268
Kiang NYS, Liberman MC, Levine RA (1976) Auditory nerve activity in cats exposed to ototoxic drugs and high-intensity sounds. Ann Otol Rhinol Laryngol 75:1–17
Komune S, Snow JB (1981) Potentiating effects of cisplatin and ethacrynic acid in ototoxicity. Arch Otorhinolaryngol 107:594–597
Konishi T (1979a) Effects of local application of ototoxic antibiotics on cochlear potentials in guinea pigs. Acta Otolaryngol (Stockh) 88:41–46
Konishi T (1979b) Some observations on negative endocochlear potential during anoxia. Acta Otolaryngol (Stockh) 87:506
Konishi T, Kelsey E, Singleton GT (1967) Negative potential in scala media during early stage of anoxia. Acta Otolaryngol (Stockh) 64:107
Konishi T, Salt AN (1980) Permeability to potassium of the endolymph-perilymph barrier and its possible relation to hair cell function. Exp Brain Res 40:457–463
Konishi T, Salt AN, Hamrick PE (1979) Effects of exposure to noise on ion movement in guinea pig cochlea. Hear Res 1:325–342
Konishi T, Salt AN, Hamrick PE (1982) Effects of exposure to noise on permeability to potassium of the endolymph-perilymph barrier in guinea pigs. Acta Otolaryngol (Stockh) 94:395–401
Kryter KD (1963) Exposure to steady-state noise and impairment of hearing. J Acoust Soc Am 35:1515–1525
Kryter KD, Ward WD, Miller JD, Eldredge DH (1966) Hazardous exposure to intermittent and steady-state noise. J Acoust Soc Am 39:451–464
Kusakari J, Ise I, Comegys TH, Thalmann I, Thalmann R (1978) Effect of ethacrynic acid, furosemide and ouabain upon the endolymphatic potential and upon high frequency phosphates of the stria vascularis. Laryngoscope 88:12–37
Lambert PR, Palmer PE, Rubel EW (1986) The interaction of noise and aspirin in the chick basilar papilla. Arch Otorhinolaryngol 112:1043–1049

Laurell G, Borg E (1986) *Cis*-platin ototoxicity in previously noise exposed guinea-pigs. Acta Otorhinolaryngol 101:66–74

Lawrence M, Yantis PS (1957) Individual differences in functional recovery and structural repair following overstimulation of the guinea pig. Ann Otol Rhinol Laryngol 66:595–621

Lawrence M, Gonzales G, Hawkins JE Jr (1967) Some physiological factors in noise-induced hearing loss. Am Ind Hyg Assoc J 28:425–430

LeDoux E, Thompson M, Iadecola C, Tucker W, Reis J (1983) Local cerebral blood flow increases during auditory and emotional processing in the conscious rat. Science 221:576–578

Lenoir M, Pujol R (1980) Sensitive period to acoustic trauma in the rat pup. Histological findings. Acta Otolaryngol (Stockh) 89:317–322

Lerner SA, Matz GS, Hawkins JE Jr (1981) Aminoglycoside ototoxicity. Little Brown, Boston

Liberman MC (1984) Single neuron labeling and chronic cochlear pathology I. Threshold shift and characteristic-frequency shift. Hear Res 16:33–41

Liberman MC (1987) Chronic ultrastructural changes in acoustic trauma: serial-section reconstruction of stereocilia and cuticular plates. Hear Res 26:65–88

Liberman MC, Dodds LW (1984a) Single neuron labeling and chronic cochlear pathology II. Stereocilia damage and alterations of spontaneous discharge rates. Hear Res 16:43–53

Liberman MC, Dodds LW (1984b) Single-neuron labeling and chronic cochlea pathology III. Stereocilia damage and alterations of threshold tuning curves. Hear Res 16: 55–74

Liberman MC, Dodds LW (1987) Acute ultrastructural changes in acoustic trauma: serial-section reconstruction of stereocilia and cuticular plates. Hear Res 26:45–64

Liberman MC, Kiang NYS (1978) Trauma in cats: cochlear pathology and auditory nerve activity. Acta Otolaryngol 358:1–63

Liberman MC, Kiang NYS (1984) Single-neuron and chronic cochlear pathology IV. Stereocilia damage and alterations in rate- and phase-level functions. Hear Res: 16: 75–90

Liberman MC, Mulroy MJ (1982) Acute and chronic effects of acoustic trauma: cochlear pathology and auditory nerve pathophysiology. In: Hamernik RP, Henderson D, Salvi R (eds) New perspectives on noise-induced hearing loss. Raven, New York, pp 105–134

Lim DJ (1976) Ultrastructural cochlear changes following acoustic hyperstimulation and ototoxicity. Ann Otol Rhinol Laryngol 85:740–751

Lim DJ (1980) Cochlear anatomy related to cochlear micromechanics. A review. J Acoust Soc Am 67:1686–1695

Lim DJ (1986) Effects of noise and ototoxic drugs at the cellular level in the cochlea. A review. Am J Otolaryngol 7:73–99

Lim DJ, Dunn DE (1979) Anatomic correlates of noise induced hearing loss. Otolaryngol Clin North Am 12:493–513

Lim DJ, Lane W (1969) Three-dimensional observations of the inner ear with the scanning electron microscope. Trans Am Acad Ophthal Otolaryngol 73:842–872

Lim DJ, Melnick W (1971) Acoustic damage to the cochlea: a scanning and transmission electron microscopic observation. Arch Otorhinolaryngol 94:294–305

Lindeman HH, Bredberg G (1972) Scanning electron microscopy of the organ of Corti after intense stimulation effects on stereocilia and cuticular surface of hair cells. Arch Klin Exp Ohren Nasen Kehlkopf Heilkd 203:1–15

Lipscomb DM, Roettger R (1973) Capillary constriction in cochlear and vestibular tissues during intense noise stimulation. Laryngoscope 83:259–263

Lonsbury-Martin BL, Martin GK (1981) Effects of moderately intense sound on auditory sensitivity in rhesus monkeys: behavioral and neural observations. J Neurophysiol 46:563 – 586

Lonsbury-Martin BL, Martin GK, Bohne BA (1987) Repeated TTS exposures in monkeys: alterations in hearing, cochlear structure, and single unit thresholds. J Acoust Soc Am 81:1507 – 1518

Maher JF, Schreiner GE (1965) Studies on ethacrynic acid in patients with refractory edema. Med Internol 62:15 – 29

Mathog RH, Klein WJ Jr (1969) Ototoxicity of ethacrynic acid and aminoglycoside antibiotics in uremia. N Engl J Med 280:1223 – 1224

Matz CJ, Lerner S, Hawkins JE (eds) (1981) Aminoglycoside ototoxicity. Little Brown, Chicago

McCabe PA, Day FL (1965) The effect of aspirin upon auditory sensitivity. Ann Otol Rhinol Laryngol 74:312 – 325

McFadden D, Plattsmier HS (1982) Suprathreshold aftereffects of exposure to intense sounds. In: Hamernik RP, Henderson D, Salvi R, New perspectives on noise induced hearing loss. Raven, New York, pp 347 – 362

McFadden D, Plattsmier HS (1983) Aspirin can potentiate temporary hearing loss induced by intense sounds. Hear Res 9:295 – 316

McFadden D, Wightman FL (1983) Audition: some relations between normal and pathological hearing. Annu Rev Psychol 34:95 – 128

McFadden, Plattsmier HS, Pasanen EG (1984) Aspirin-induced hearing loss as a model of sensorineural hearing loss. Hear Res 16:251 – 260

Melichar I, Syka J (1977) Time course of anoxia-induced K^+ concentration changes in the cochlea measured with K^+ specific microelectrodes. Pflugers Arch 372:207 – 213

Melichar I, Syka J (1978) The effects of ethacrynic acid upon the potassium concentration in guinea pig cochlear fluids. Hear Res 1:35 – 41

Melichar I, Syka J (1987 a) DC potentials in different cells of the stria vascularis measured in vitro. Hear Res 25:23 – 33

Melichar I, Syka J (1987 b) Electrophysiological measurements of the stria vascularis potentials in vivo. Hear Res 25:35 – 43

Melichar I, Syka J, Úlehlová L (1980) Recovery of the endocochlear potential and the K^+ concentrations in the cochlear fluids after acoustic trauma. Hear Res 2:55 – 63

Melnick W (1977) Temporary threshold shifts following 24-hour noise exposure. Ann Otol Rhinol Laryngol 86:821 – 826

Melnick W, Maves SM (1974) Asymptotic threshold shift (ATS) in man from 24 hour exposure to continuous noise. Ann Otol Rhinol Laryngol 83:820 – 829

Meriwether WD, Maugi RJ, Serpick AA (1971) Deafness following standard intravenous doses of ethacrynic acid. JAMA 216:795

Miller JD (1974) Effects of noise on people. J Acoust Soc Am 56:729 – 764

Miller J, Canlon B, Flock A (1985) High intensity noise effects on stereocilia mechanics. Assoc Res Otolaryngol 8:50

Mills JH (1973) Threshold shifts produced by exposure to noise in chinchillas with noise-induced hearing losses. J Speech Hear Res 16:700 – 708

Mills JH, Gengel RW, Watson CH, Miller J (1970) Temporary changes of the auditory system due to exposure to noise for one or two days. J Acoust Soc Am 2:524 – 530

Mills JH (1976) Threshold shifts produced by 90 day exposure to noise. In: Henderson D, Hamernik RP, Disanjh OS, Mills JM, Effects of noise on hearing. Raven, New York, pp 265 – 276

Mills JH, Talo SA (1972) Temporary threshold shifts produced by exposure high-frequency noise. J Speech Hear Res 15:624 – 631

Möller K, Blegvad B (1976) Brain stem responses in patients with sensorineural hearing loss. Monaural versus binaural stimulation. The significance of the audiogram configuration. Scand Audiol 5:115

Mongan E, Kelly P, Nies K, Porter WW, Paulus HF (1973) Tinnitus as an indication of therapeutic serum salicylate levels. J Am Med Wom Assoc 226:142–145

Moody DB, Stebbins WC, Hawkins JE Jr (1976) Noise-induced hearing loss in the monkey. In: Henderson O, Hamernik RP, Dosanjh DS, Mills JH (eds) Effects of noise on hearing. Raven, New York, pp 309–326

Moore BCJ, Glasberg BR (1986) Comparison of frequency selectivity in spontaneous and forward masking for subjects with unilateral cochlear impairments. J Acoust Soc Am 80:93–107

Morest DK (1982) Degeneration in the brain following exposure to noise. In: Hamernik RP, Henderson D, Salvi R (eds) New perspectives on noise-induced hearing loss. Raven, New York, pp 87–93

Morest D, Bohne BA (1983) Noise-induced degeneration in the brain and representation of inner and outer hair cells. Hear Res 6:61–82

Morimitsu T, Matsuo K, Suga F (1965) Behavior of the cochlear blood flow. Ann Otol Rhinol Laryngol 74:22–52

Morizono T, Johnstone BM (1975) Ototoxicity of topically applied gentamicin using a statistical analysis of electrophysiological measurements. Acta Otolaryngol (Stockh) 80:389–393

Myers EN, Bernstein JM (1965) Salicylate ototoxicity. Arch Otorhinolaryngol 82: 483–493

Nakai Y (1971) Electron microscopic study of the inner ear after ethacrynic acid intoxication. Pract Otol Rhinol Laryngol 33:366–376

Nakai Y (1977) Combined effect 3′,4′-dideoxykanamycin B and potent diuretics on the cochlea. Laryngoscope 87:1548

Nakai Y, Konishi K, Chang KC, Ohashi K, Morisaki N, Minowa Y, Morimoto A (1982) Ototoxicity of the anticancer drug cisplatin. Acta Otolaryngol 93:227–232

Nelson DA, Turner CW (1980) Decay of masking and frequency resolution in sensorineural and hearing-impaired listeners. In: Van den Brink G, Bilsen FA (eds) Psychoacoustical, physiological and behavioral studies in hearing. Delft University Press, The Netherlands

Nelson DA, Freyman RL (1986) Psychometric functions for frequency discrimination from listeners with sensorineural hearing loss. J Acoust Soc Am 79:799–805

Nelson DA, Freyman RL (1987) Temporal resolution in sensorineural hearing-impaired listeners. J Acoust Soc Am 81:709–720

Nielson DW (1982) Asymptotic threshold shifts in the squirrel monkey. In: Hamernik RP, Henderson D, Salvi R (eds) New perspective on noise-induced hearing loss. Raven, New York, pp 303–319

Nuttall AL, Marques DM, Lawrence M (1971) Effects of perilymphatic perfusion with neomycin on the cochlear potential in the guinea pig. Acta Otolaryngol (Stockh) 83:393–400

Orsulakova A, Schacht J (1981) A biochemical mechanism of the ototoxic interaction between neomycin and ethacrynic acid. Acta Otolaryngol (Stockh) 93:43–48

Osako S, Tokimoto T, Matsuura S (1979) Effects of kanamycin on the auditory evoked responses during postnatal development of the hearing of the rat. Acta Otolaryngol (Stockh) 88:359–368

Özdamar O, Dallos P (1976) Input-output functions of cochlear nerve action potential: interpretation in terms of the one population of neurons. J Acoust Soc Am 59:143–147

Paloheimo S, Thalmann R (1977) Influence of "loop" diuretics upon Na^+K^+-ATPase and adenylate cyclase of the stria vascularis. Arch Otol Rhinol Laryngol 217:347–359

Patuzzi R, Sellick PM (1983) A comparison between basilar membrane and inner hair cell receptor potential input-output functions in the guinea pig cochlea. J Acoust Soc Am 74:1734 – 1741
Pederson CB (1974) Brief-tone audiometry in persons treated with salicylate. Audiology 13:311 – 319
Perlman HB, Kimura RS (1962) Cochlear blood flow in acoustic trauma. Acta Otolaryngol (Stockh) 54:99 – 110
Petersen MR, Prosen CA, Moody DB, Stebbins WC (1977) Operant conditioning in the guinea pig. J Exp Anal Behav 27:529 – 532
Pickles JC, Comis SD, Osborne MP (1984) Cross links between stereocilia in the guinea pig organ of Corti and their possible relation to sensory transduction. Hear Res 15:103 – 112
Pickles JO, Osborne MP, Comis SD (1987) Vulnerability of tip links between stereocilia to acoustic trauma in the guinea pig. Hear Res 25:177 – 183
Pike A, Bosher SK (1980) The time course of the strial changes produced by intravenous furosemide. Hear Res 8:79 – 89
Pillay VKG, Schwartz FD, Aimi K, Kark RM (1969) Transient and permanent deafness following treatment with ethacrynic acid in renal failure. Lancet 1:77
Popelář J, Syka J (1982) Noise impairment in the guinea pig II. Changes of single unit response in the inferior colliculus. Hear Res 8:273 – 283
Popelář J, Syka J, Berndt H (1987) Effect of noise on auditory evoked responses in awake guinea pig. Hear Res 26:239 – 247
Popelář J, Syka J, Úlehlová L (1978) Effect of high-frequency sound on cochlear microphonics and activity of inferior colliculus neurons in the guinea pig. Arch Otorhino laryngol 221:115 – 122
Prazma J (1981) Ototoxicity of aminoglycoside antibiotics. In: Brown RD, Daigneault EA (eds) Pharmacology of hearing: experimental and clinical bases. Wiley, New York, pp 153 – 196
Prazma J, Browder JP, Fischer ND (1974) Ethacrynic acid ototoxicity potentiation by kanamycin. Ann Otol Rhinol Laryngol 83:1
Prazma J, Thomas WG, Fischer ND, Preslar MJ (1972) Ototoxicity of ethacrynic acid. Arch Otorhinolaryngol 95:448 – 456
Prazma J, Rodgers GK, Pillsbury HC (1983) Cochlear blood flow. Arch Otorhinolaryngol 109:611 – 615
Prazma J, Vance SG, Rodgers G (1984) Measurements of the cochlear blood flow – new technique. Hear Res 14:21 – 28
Price GR (1976) Age as a factor in susceptibility to hearing loss: young versus adult ears. J Acoust Soc Am 60:886 – 892
Prosen CA, Petersen MR, Moody DB, Stebbins WC (1977) Auditory thresholds and kanamycin-induced hearing loss in the guinea pig assessed by a positive reinforcement procedure. J Acoust Soc Am 63:559 – 566
Quick CA, Duvall AJ (1970) III. Early changes in the cochlear duct from ethacrynic acid: an electromicroscopic evaluation. Laryngoscope 80:954 – 965
Rajan R, Johnstone BM (1983) Residual effects in monaural temporary threshold shift to pure tones. Hear Res 12:185 – 197
Rarey KE, Ross MD (1982) A survey of the effects of loop diuretics on the zonulae occludentes of the perilymph-endolymph barrier by freeze fracture. Acta Otolaryngol (Stockh) 94:307 – 316
Robertson D (1976) Correspondence between sharp tuning and two-tone inhibition in primary auditory neurons. Nature 259:477 – 478
Robertson D (1982) Effects of acoustic trauma on stereocilia structure and spiral ganglion cell tuning properties in the guinea pig cochlea. Hear Res 7:55 – 74

Robertson D (1983) Functional significance of dendritic swelling after loud sounds on the guinea pig cochlea. Hear Res 9:263 – 278

Robertson D, Johnstone BM (1979) Aberrant tonotopic organisation in the inner ear damaged by kanamycin. J Acoust Soc Am 66:466

Robertson D, Johnstone BM (1980) Acoustic trauma in the guinea pig cochlea, early changes in ultrastructure and neuronal threshold. Hear Res 3:167 – 179

Robertson D, Johnstone BM (1981) Primary auditory neurons: nonlinear responses altered without changes in sharp tuning. J Acoust Soc Am 69:1096 – 1098

Robertson D, Johnstone BM, McGill TJ (1980) Effects of loud tones on the inner ear: a combined electrophysiological and ultrastructural study. Hear Res 2:39 – 53

Romand R (1971) Maturation des potentiels cochlearies dans la periode perinatale chez le chat et chez le cobaye. J Physiol (Paris) 63:763 – 782

Rosenhamer H (1981) The auditory evoked brainstem electric response (ABR) in cochlear hearing loss. Scand Audiology 13:83 – 93

Rossi G, Robechi MG, Penna M (1976) Effects of acoustic trauma on Corti's ganglion. Acta Otolaryngol (Stockh) 81:270 – 277

Rossi G, Solero P, Cortesina MF (1979) Brainstem electric response audiometry. Value and significance of latency and amplitude in absolute sense and in relation to the auditory threshold. Acta Otolaryngol (Stockh) [Suppl] 364:1 – 13

Rüedi L (1951) Some animal experimental findings on the functions of the inner ear. Ann Otol Rhinol Laryngol 60:993 – 1023

Ruggero MA, Rich NC, Freyman R (1983) Spontaneous and impulsively evoked otoacoustic emissions: indicators of cochlear pathology. Hear Res 10:283 – 300

Russell JJ (1983) The origin of receptor potential in inner hair cell of the mammalian cochlea – evidence for Davis theory. Nature 301:334 – 336

Russell JJ, Sellick PM (1978) Intracellular studies in hair cells in the mammalian cochlea. J Physiol (Lond) 284:261 – 290

Russell NJ, Fox KE, Brummett RE (1979) Ototoxic effects of the interaction between kanamycin and ethacrynic acid. Acta Otolaryngol (Stockh) 88:369 – 381

Rutten WLC (1980) Evoked acoustic emissions from within normal and abnormal human ears: comparison with audiometric and electrocochleographic findings. Hear Res 2:263 – 271

Ryan AF, Woolf NK, Bone RC (1980) Ultrastructural correlates of selective outer hair cell destruction following kanamycin intoxication in the chinchilla. Hear Res 3:335 – 351

Rybak LP (1985) Furosemide ototoxicity: clinical and experimental aspects. Laryngoscope 195:38

Rybak LP (1986) Ototoxic mechanisms. Neurobiology of hearing. In: Altschuler RA, Hoffman DW, Bobbin RP (eds) The cochlea, Raven, New York, pp 441 – 453

Rybak LP, Morizono T (1982) Effect of furosemide upon endolymph potassium concentration. Hear Res 7:223 – 231

Rybak LP, Whitworth C (1986) Changes in endolymph chloride concentrations following furosemide injection. Hear Res 24:133 – 136

Rybak LP, Green TP, Juhn SK, Morizono T, Mirkin BL (1979) Elimination kinetics of furosemide in perilymph and serum of the chinchilla. Acta Otolaryngol (Stockh) 88:382 – 387

Sachs MB, Abbas PJ (1974) Rate versus level functions for auditory nerve fibers in cats: tone burst stimuli. J Acoust Soc Am 65:1835 – 1847

Salt AN, Konishi T (1979) Effects of noise on cochlear potential and endolymph potassium concentration recorded with potassium-selective electrodes. Hear Res 1:343 – 363

Salvi RJ (1976) Central components of the temporary threshold shift. In: Henderson D, Hamernik RP, Dosanjh DS, Mills JH (eds) Effects of noise on hearing. Raven, New York, pp 247 – 260

Salvi RJ, Henderson D, Hamernik R (1975) Auditory fatigue: retrocochlear components. Science 190:486 – 487
Salvi RJ, Hamernik RP, Henderson D (1978) Discharge patterns in the cochlear nucleus of the chinchilla following noise induced asymptotic threshold shift. Exp Brain Res 32:301 – 320
Salvi RJ, Henderson D, Hamernik RP (1979) Single auditory nerve fiber and action potential latencies in normal and noise-treated chinchillas. Hear Res 1:237 – 251
Salvi RJ, Perry J, Hamernik RP, Henderson D (1982) Relationship between cochlear pathologies and auditory nerve behavioral responses following acoustic trauma. In: Hamernik RP, Henderson D, Salvi R (eds) New perspectives on noise-induced hearing loss. Raven, New York, pp 165 – 188
Salvi RJ, Hamernik RP, Henderson D (1983) Response patterns of auditory nerve fibers during temporary threshold shift. Hear Res 10:37 – 67
Santi PA, Duvall AJ (1978) Stria vascularis pathology and recovery following noise exposure. Trans Am Acad Ophthalmol Otolaryngol 86:354 – 361
Santi PA, Duvall AJ (1979) Morphological alteration of the stria vascularis after administration of the diuretic bumetanide. Acta Otolaryngol (Stockh) 88:1
Saunders JC, Bock GR (1978) Influence of early auditory trauma on auditory development. In: Gottlieb G (ed) Study on the development of behavior and the nervous system, vol 4. Early influences. Academic, New York, pp 249 – 287
Saunders JC, Chen CS (1982) Sensitive periods of susceptibility of auditory trauma in mammals. Environ Health Perspect 44:63 – 66
Saunders JC, Flock Å (1986) Recovery of threshold shift in hair-cell stereocilia following exposure to intense stimulation. Hear Res 23:233 – 243
Saunders JC, Hirsch KA (1976) Changes in cochlear microphonics sensitivity after priming C57BL/6J mice at various ages for audiogenic seizures. J Comp Physiol Psychol 90:212 – 220
Saunders JC, Dear SP, Schneider ME (1985) The anatomical consequences of acoustic injury. A review and tutorial. J Acoust Soc Am 78:833 – 860
Saunders JC, Canlon B, Flock Å (1986a) Growth of threshold shift in hair-cell stereocilia following overstimulation. Hear Res 23:245 – 255
Saunders JC, Canlon B, Flock Å (1986b) Changes in stereocilia micromechanics following overstimulation in metabolically blocked hair cells. Hear Res 24:217 – 225
Schacht J (1979) Isolation of an aminoglycoside receptor from guinea pig inner ear tissues and kidney. Arch Otol Rhinol Laryngol 224:129 – 134
Schacht J (1986) Molecular mechanics of drug-induced hearing loss. Hear Res 22:297 – 304
Schacht J, Lodhi S, Weiner ND (1977) Effects of neomycin on polyphosphoinosides in inner ear tissues and monomolecular films. In: Miller MW, Shamoo EA (eds) Membrane Toxicity. Plenum New York, pp 191 – 208
Schmiedt RA (1984) Acoustic injury and the physiology of hearing. J Acoust Soc Am 76:1293 – 1317
Schmiedt RA, Zwislocki JJ (1980) Effects of hair cell lesions on responses of cochlear nerve fibers. II. Single and two-tone intensity functions in relation to tuning curves. J Neurophysiol 43:1390 – 1405
Schmiedt RA, Zwislocki JJ, Hamernik RP (1980) Effect of hair cell lesions on response of cochlear nerve fibers. I. Lesions, tuning curves, two tone inhibition and responses to trapeziodal wave patterns. J Neurophysiol 43:1367 – 1389
Sellick PM, Johnstone BM (1974) Differential effects of ouabain and ethacrynic acid on the labyrinthine potentials. Pflugers Arch 352:339 – 350
Sewell WF (1984) The effects of furosemide on the endocochlear potential and auditory-nerve tuning curves in cats. Hear Res 14:305 – 314

Sharp FR, Ryan AF, Goodwin P, Woolf NK (1981) Increasing intensities of wide band noise increase ^{14}C/2-deoxyglucose uptake in gerbil central auditory structures. Brain Res 230:87–96

Siegel JH, Kim DO, Molnar CE (1982) Effects of altering organ of Corti on cochlear distortion products $f_2 - f_1$ and $2f_1 - f_2$. J Neurophysiol 47:303–328

Sikora MA, Morizono T (1983) N_1 latency following acute pure-tone trauma. Hear Res 11:93–101

Silverstein H, Bernstein JM, Davies DG (1967) Salicylate ototoxicity: a biochemical and electrophysiological study. Ann Otol Rhinol Laryngol 76:118–128

Simmons FB, Galambos B, Albriet JP (1960) Serial studies of the onset and progression of drug-induced cochlear damage in cats. Arch Otorhinolaryngol 72:233–239

Sinex DG, Clark WW, Bohne BA (1987) Effects of periodic rest on physiological measures of auditory sensitivity following exposure to noise. J Acoust Soc Am 82:1265–1273

Skinner B, Glattke TJ (1977) Electrophysiologic response audiometry: state of the art. J Speech Hear Disord 41:179

Slepecky N (1986) Overview of mechanical damage to the inner ear: noise as a tool probe of cochlear function. In: Flock Å, Wersäll J (eds) Cellular mechanisms in hearing. Elsevier, Amsterdam, pp 307–321

Slepecky N, Hamernik RP, Henderson D, Coling D (1981) Ultrastructural changes to the cochlea resulting from impulse noise. Arch Otorhinolaryngol 230:273–278

Slepecky N, Hamernik RP, Henderson D, Coling D (1982) Correlation of audiometric data with changes in cochlear hair cell stereocilia resulting from impulse noise trauma. Acta Otolaryngol (Stockh) 93:329–340

Smith CA, Lowry OH, Wu Meiling (1954) The electrolytes of the labyrinthine fluids. Laryngoscope 64:141–153

Smith DW, Moody DB, Stebbins WC, Norat MA (1987) Effects of outer hair cell loss on the frequency selectivity of the patas monkey auditory system. Hear Res 29:125–138

Smoorenburg CF (1972) Combination tones and their origin. J Acoust Soc Am 52:615–632

Smoorenburg CF (1980) Effects of temporary threshold shift on combination-tone generation and two-tone suppression. Hear Res 2:347–355

Sokoloff L (1977) Relation between physiological function and energy metabolism in the central nervous system. J Neurochem 29:13–26

Sokoloff L, Reivich M, Kennedy C, Des Rosiers MH, Patlak CS, Pettigrew KD, Sakurada O, Shinohava M (1977) The ^{14}C deoxyglucose method for the measurements of local cerebral glucose utilisation: theory, procedure and normal values in the conscious and anaesthetized albino rat. J Neurochem 28:197–216

Spoendlin H (1971) Primary structural changes in the organ of Corti after acoustic overstimulation. Acta Otolaryngol (Stockh) 71:166–176

Spoendlin H (1975) Retrograde denervation of the cochlear nerve. Acta Otolaryngol (Stockh) 79:266–275

Standnicki SW, Fleischman RW, Schaeppi U (1975) *Cis*-dichlorodiamine platinum (II) (NSC-119875): hearing loss and other toxic effects in rhesus monkeys. Cancer Treat Rep 59:467–480

Starr A (1965) Suppression of single neuron activity in the cochlear nucleus of the cat following sound stimulation. J Neurophysiol 28:850–862

Starr A, Livingston RB (1963) Long-lasting nervous system responses to prolonged sound stimulation in waking cats. J Neurophysiol 26:416–431

Stebbins WC, Miller JM, Johnsson L-G, Hawkins JE Jr (1969) Ototoxic hearing loss and cochlear pathology in the monkey. Ann Otol Rhinol Laryngol 78:598–602

Strauss M, Towfighi J, Lord S, Brown B, Lipton A, Harvey HA (1983) *Cis*-platinum ototoxicity: clinical experience and temporal bone histopathology. Laryngoscope 93:1554–1559

Stupp HF (1970) Untersuchung der Antibiotikaspiegel in den Innenohrflüssigkeiten und ihre Bedeutung für die Spezifische Ototoxizität der Aminoglykosidantibiotika. Acta Otolaryngol (Stockh) 262:85
Strelioff D, Flock Å (1984) Stiffness of sensory-cell hair bundles in the isolated guinea pig cochlea. Hear Res 15:19 – 28
Syka J, Melichar I (1981) Comparison of the effects of furosemide and ethacrynic acid upon the cochlear function in the guinea pig. Scand Audiol 14:63 – 70
Syka J, Melichar I (1985) The effect of loop diuretics upon summating potentials in the guinea pig. Hear Res 20:267 – 273
Syka J, Popelář J (1980) Hearing threshold shifts from prolonged exposure to noise in guinea pig. Hear Res 3:205 – 213
Syka J, Popelář J (1982) Noise impairment in the guinea pig. I. Changes in electrical evoked activity along the auditory pathway. Hear Res 8:263 – 272
Syka J, Melichar I, Úlehlová L (1981) Longitudinal distribution of cochlear potentials and the K^+ concentration in the endolymph after acoustic trauma. Hear Res 4:287 – 298
Syka J, Syková E, Patuzzi R, Johnstone BM (1987) Acoustic stimulation evokes potassium concentration changes in the organ of Corti in guinea pig. Neuroscience [Suppl] 22:129
Takada A, Schacht J (1982) Calcium antagonism and reversibility of gentamicin-induced loss of cochlear microphonics in the guinea pig. Hear Res 8:179 – 186
Takada A, Bledsoe S, Schacht J (1985) An energy-dependent step in aminoglycoside ototoxicity: prevention of gentamicin ototoxicity during reduced endolymphatic potential. Hear Res 19:245 – 251
Tange RA, Conijn EAJG, Van Zeyl LEPM (1983) Differences in the cochlear degeneration pattern in the guinea pig as a result of gentamicin and *cis*-platin intoxication. Clin Otolaryngol 8:138
Thalmann I, Kobayashi T, Thalmann R (1982) Arguments against a mediating role of the adenylate cyclase-cyclic AMP system in the ototoxic action of loop diuretics. Laryngoscope 92:589 – 593
Theopold HM (1977) Comparative surface studies of ototoxic effects of various aminoglycoside antibiotics on the organ of Corti in the guinea pig. Acta Otolaryngol (Stockh) 84:57 – 64
Tilney LG, Saunders JC, Engelman E, De Roiser DJ (1982) Changes in the organization of actin filaments in the stereocilia of noise-damaged lizzard cochlea. Hear Res 7: 181 – 197
Tran Ba Huy P, Manuel C, Meulemans A, Sterkers O, Amiel C (1981) Pharmacokinetics of gentamicin in perilymph and endolymph of the rat as determined by radioimmunoassay. J Infect Dis 143:476 – 486
Tran Ba Huy P, Manuel C, Meulemans A, Sterkers O, Wassef M, Amiel C (1983) Ethacrynic acid facilitates gentamicin entry into endolymph of the rat. Hear Res 11:191 – 202
Trittipoe WJ (1959) Residual effects at longer preexposure durations. J Acoust Soc Am 31:244
Turner CW (1987) Effects of noise and hearing loss upon frequency discrimination. Audiology 26:133 – 140
Úlehlová L (1983) Stria vascularis in acoustic trauma. Arch Otol Rhinol Laryngol 237:133 – 138
Uziel A, Gabrion J, Romand R (1979a) Hair cell degeneration in guinea pigs intoxicated with kanamycin during intrauterine life. Arch Otorhinolaryngol 224:187 – 191
Uziel A, Romand R, Marot M (1979b) Electrophysiological study of the ototoxicity of kanamycin during development in guinea pig. Hear Res 1:203 – 211
Van Heusden E, Smoorenburg CF (1981a) Eight-nerve action potentials evoked by tone burst in cats before and after inducement of an acute noise trauma. Hear Res 5:1 – 23

Van Heusden E, Smoorenburg CF (1981b) Eight-nerve action-potential tuning curves in cats before and after inducement of an acute noise trauma. Hear Res 5:25–48
Vertes D, Axelsson A, Lipscomb DM (1979) Some vascular effects of noise exposure in the chinchilla cochlea. Acta Otolaryngol (Stockh) 88:47–55
Vertes D, Axelsson A, Miller J, Lidén G (1981) Cochlear vascular and electrophysiological effects in the guinea pig to 4 kHz pure tones of different durations and intensities. Acta Otolaryngol (Stockh) 92:15–24
Vertes D, Axelsson A, Hornstrand C, Nilsson P (1984) The effect of impulse noise on cochlear vessels. Arch Otorhinolaryngol 110:111–115
Voldřich L (1965) The kinetics of streptomycin, kanamycin and neomycin in the inner ear. Acta Otolaryngol (Stockh) 60:243–248
Vrabec DP, Cody DTR, Ulrich JA (1965) A study of the relative concentrations of antibiotics in the blood, spinal fluid and perilymph in animals. Ann Otol Rhinol Laryngol 74:688–705
Wang CY, Dallos P (1972) Latency of whole nerve action potentials: influence of hair cell normally. J Acoust Soc Am 52:1678–1686
Ward WD (1960) Latent and residual effects in temporary threshold shift. J Acoust Soc Am 32:135–137
Ward WD (1963) Diplacusis and auditory theory. J Acoust Soc Am 35:1746–1747
Ward WD, Duvall AJ (1971) Behavioral and ultrastructural correlates of acoustic trauma. Ann Otol Rhinol Laryngol 80:881–896
Watanabe Y, Nakajima R, Oda R, Uno M, Saito T (1971) Experimental study on the transfer of kanamycin to the inner ear fluids. Med J Osaka Univ 21:257–263
Wernick JS, Tobias JV (1963) Central factor in pure-tone auditory fatigue. J Acoust Soc Am 35:1967–1971
Wersäll J, Hawkins JE Jr (1962) The vestibular sensory epithelia in the cat labyrinth and their reaction in chronic streptomycin intoxication. Acta Otolaryngol (Stockh) 54:1–22
Wersäll J, Lundquist PG, Björkroth B (1969) Ototoxicity of gentamicin. J Infect Dis 119:410–416
Wersäll J, Björkroth B, Flock Å, Lundquist PG (1973) Experiments on ototoxic effects of antibiotics. Adv Otorhinolaryngol 20:14–41
West BA, Brummett RE, Himes DL (1973) Interaction of kanamycin and ethacrynic acid severe cochlear damage in guinea pigs. Arch Otorhinolaryngol 98:32
Wightman FL (1982) Psychoacoustic correlates of hearing loss. In: Hamernik RP, Henderson D, Salvi R (eds) New perspectives on noise induced hearing loss. Raven, New York, pp 375–394
Williams JA, Zenner HP, Schacht J (1987) Three molecular steps of aminoglycoside ototoxicity demonstrated in outer hair cells. Hear Res 30:11–18
Wilson JP (1986) Otoacoustic emissions and tinnitus. Scand Audiol 25:109–119
Wilson KS, Juhn SK (1970) The effect of ethacrynic acid on perilymph Na and K. Pract Otol Rhinol Laryngol 32:279–282
Wilson JP, Sutton GJ (1981) Acoustic correlates of tonal tinnitus. In: Evered D, Lawrenson G (eds) Tinnitus. Pitman, London pp 82–101
Witmaack K (1903) Beiträge zur Kenntnis der Wirkung des Chinins auf das Gehörorgan. Pflugers Arch 95:203–233
Witmaak K (1907) Über Schädigung des Gehörs durch Schalleinwirkung. Z Ohrenheilkd 54:37–80
Woodford CM, Henderson D, Hamernik RP (1978) Effects of combination of sodium salycilate and noise on the auditory threshold. Ann Otol Rhinol Laryngol 87:117–122
Woolf NK, Sharp FR, Davidson TM, Ryan AF (1983) Cochlear and middle ear effects on metabolism in the central auditory pathway during silence: A 2-deoxyglucose study. Brain Res 274:119–127

Wright H (1968) The effect of sensorineural hearing loss on threshold-duration functions. J Speech Hear Res 11:842–852
Wright CG (1976) Neural damage in the guinea pig cochlea after noise exposure. Acta Otolaryngol (Stockh) 82:82–94
Wright A (1981) Scanning electron microscopy of the human cochlea – the organ of Corti. Arch Otorhinolaryngol 230:11–19
Wright A (1984) Dimensions of the cochlear stereocilia in man and the guinea pig. Hear Res 13:89–98
Wright CG, Schaefer SD (1982) Inner ear histopathology in patients treated with *cis*-platinum. Laryngoscope 92:1408–1413
Ylikoski J, Wersäll J, Björkroth B (1974) Correlative studies on the cochlear pathology and hearing loss in guinea pigs after intoxication with ototoxic antibiotics. Acta Otolaryngol (Stockh) 326:5–59
Yoshii U (1909) Experimentelle Untersuchungen über die Schädigung des Gehörorganes durch Schalleinwirkung. Z Ohrenheilkd 58:201–251
Zurek PM (1981) Spontaneous narrow-band acoustic signals emitted by human ears. J Acoust Soc Am 69:514–523
Zurek PM, Clark WW (1981) Narrow-band acoustic signals emitted by chinchilla ear after noise exposure. J Acoust Soc Am 70:446–450

Sensing of Endogenous Chemicals in Control of Feeding

Y. Oomura

Institute of Bio-Active Science, Nippon Zoki Pharmaceutical Co., Ltd., Yashiro-cho, Kato-gun, Hyogo 673 – 14, Japan

Progress in Sensory Physiology, Vol 9

1 Rationale for Considering the Subject of This Review to be One of Sensory Physiology

In addition to the receptors for the commonly considered senses – sight, taste, smell, ect. – there are receptors associated with our nervous system that respond qualitatively and quantitatively to endogenous compounds and thereby initiate sequences of neuronal events that influence feeding behavior. Although they are found in diverse locations in the nervous system and visceral organs, these chemical sensors are confined to distinct neuronal centers or organ sites.

2 Nature and Extent of the Internal Chemical Sense in the Control of Feeding Behavior

In 1955, J. Mayer suggested that neurons in the ventromedial hypothalamus (VMH) might respond to blood glucose level. In 1969, we defined glucoreceptor neurons (GRN) in the VMH and glucose-sensitive neurons (GSN) in the lateral hypothalamic area (LHA). A GRN was defined as one whose activity increased dose-dependently when it was subjected to direct electrophoretic application of glucose through a micropipette. The activity increase by glucose is due to membrane depolarization caused by a decrease in K^+ permeability (Minami et al. 1986). A neuron was defined as a GSN if its spontaneous activity was dose-dependently depressed by similar application of glucose; inhibition by glucose is due to membrane hyperpolarization caused by Na-K pump activation (Oomura et al. 1974).

Although GRN and GSN quite accurately describe these neurons, the names were incidental to the hypothesis of glucostatic control of feeding, which was in vogue at the time when the neurons were identified. If GSNs and GRNs were to be renamed, what we now know of their other properties might suggest that a GSN be called a "feeding control" or "hunger-related" neuron, and a GRN migth be called a "satiation control" or "satiety-related" neuron. These neurons respond with excitation or suppression of their activity to many metabolites and hormones which behavior experiments have shown to change in level as a result of feeding and to affect feeding behavior. Insulin, glucagon, calcitonin, and other hormones affect GRNs and GSNs in ways that reflect the degree of satiation and the presence of other hormones or metabolites (Oomura 1983). GRNs and GSNs also respond to metabolites other than glucose, such as free fatty acids, in ways that are consistent with the physiological levels of these metabolites in the diurnal cycle and during feeding behavior.

About 25% of the neurons in the VMH are GRNs, and about the same proportion of neurons in the LHA are GSNs. Most of the neuronal signals that are generated by ongoing food intake originate in or pass through the VMH, where at least one step in their integration takes place (Oomura 1983). These signals then go on to other regions where the ongoing feeding is controlled or halted. Similarly, most neuronal events and chemical responses that appear to inititate feeding behavior pass through or originate in the LHA. Other regions in the upper brainstem that contain GRNs include the dorsomedial nucleus of the hypothalamus (DMH), the dorsal motor nucleus of the vagus (DMV), the area postrema, and the nucleus of the solitary tract (NTS). Regions that contain GSNs include the paraventricular nucleus (PVN), the DMH, the area postrema, and the NTS. The NTS, and area postrema located in the medulla oblongata between the visceral organs and the hypothalamus, acts as both a relay and a lower integrating center, and complements the function of the hypothalamus. In the periphery, glucose-responsive units can be found in the hepatic portal vein, the pancreas, the duodenum, and the jejunum and ileum.

3 Effects of Endogenous Substances on Feeding and Internal Sensors

The number of excessively overweight or underweight people in the world must be in the millions. The number of drugs, compounds, folk cures, herbs, medicines, and pharmaceuticals that are recommended, sold, prescribed, or advocated must run to thousands. Because of space and other practical restrictions, we will limit this review to some endogenous substances the mechanisms of whose functions are known; a few that are not yet well characterized are included for the sake of comparison.

4 Hormones

4.1 Bombesin

Bombesin, a potent gastric acid secretagogue, is found in the viscera and brain of mammals. In relatively high doses, it is a powerful depressant of feeding (Gibbs et al. 1979) and induces hyperglycemia and hyperglucagonemia (Brown et al. 1979). Alone, it has no effect on LHA neurons when applied electrophoretically, but it modulates the effects that other substances have on GSNs in the LHA (Shiraishi et al. 1980).

4.2 Calcitonin

Calcitonin is released into the blood from the thyroid gland. Its principal function is maintenance of the concentration of Ca^{2+} in the blood within very narrow limits, and it accomplishes this by controlling the uptake or release of Ca^{2+} by bones. The calcitonin concentration in the blood doubles during feeding. A small amount of calcitonin injected into the third ventricle (0.32 ng/rat) reduced 24-h food intake by 50%. This effect is limited to the third ventricle and is not seen after administration in the fourth ventricle (Plata-Salamán and Oomura 1987). These results indicate that calcitonin is a satiety-inducing substance and its effect is apparently mediated through direct inhibition of GSNs in the LHA. The blood concentration of calcitonin is of the order of 0.1 U/ml (0.16 ng/ml). Assumption of 1% diffusion from the thrid ventricle to the LHA (de Wildt et al. 1982) suggests that 0.3 ng calcitonin was available to act on GSNs in the LHA. The effective dose was thus physiological, since it was comparable to the approximately 0.21 ng/hypothalamus reported by Flynn et al. (1981).

Electrophoretically applied calcitonin suppresses GSN but not glucose-insensitive neuron (GIS) activity in the rat LHA. Application of isobutylmethylxanthine or papaverine (phosphodiesterase inhibitors) augments inhibition by calcitonin, but nicotinic acid (an adenylate cyclase inhibitor) blocks this, so GSN inhibition by calcitonin is probably mediated by increased intracellular cyclic adenosine monophosphate (cAMP). Dibutyryl cAMP inhibits GSN activity. Calcitonin hyperpolarizes GSNs about -15 mV and reduces membrane conductivity about by 75%, which may be due to passive decrease of Na^+ ions, possibly because of decrease in Na^+ or Ca^{2+} permeability or both (Shimizu and Oomura 1986). However, there is no indication of how cAMP affects such changes in ionic conductance.

There is immunoreactive calcitonin in the LHA but not in the VMH. The VMH does, however, contains a calcitonin gene-related peptide which actually suppresses feeding when applied in a cerebral ventricle (Krahn et al. 1984).

4.3 Cholecystokinin

After saying that cholecystokinin (CCK) suppresses feeding, very little more can be said with certainty. Reports have attributed the CCK effect to satiety (Gibbs et al. 1973; Smith et al. 1985; Antin et al. 1975; Della-Fera and Baile 1981), and the evidence for this is not without merit. However, evidence is also reported that indicates some effect of CCK other than satiety. Moore and Deutsch (1985) pointed out that an antiemetic blocked the effects of all doses of exogenous CCK, which ranged from physiological to pharmacological (250 ng – 2 μg/ml · kg), but apparently did not block endogenous CCK. Ettinger et al. (1986) concluded that CCK might suppress feeding by degrading palatability, but not by satiety, discomfort, or illness. Verbalis et al. (1986) reported that the effect of CCK appeared to be mediated by the same neuronal paths as oxytocin release. The plasma level of oxytocin is considered to be an indicator of nausea. After vagotomy, peripheral CCK had no effect on food intake (Smith et al. 1981) and oxytocin did not increase. The effects of CCK in the PVN are unclear in rats, but it clearly suppresses feeding in sheep (Della-Fera et al. 1981). Proglumide, a CCK antagonist, induced feeding when injected into the PVN, and CCK injected at the same time reduced the effect of the proglumide (Hoebel et al. 1984). Electrophoretically applied CCK was reported to have no effect on LHA and VMH neurons, but excited 30% of the cortical neurons tested (Ishibashi et al. 1979b). CCK suppressed the activity of hepatic vagal afferents (Niijima 1981), and this would suppress the activity of GSNs in the NTS and the LHA (Adachi et al. 1984; Shimizu et al. 1983).

4.4 Glucagon

Food intake was suppressed to 70% of control for 3 h after injection of glucagon into the third ventricle at 1900 hours to yield a concentration of 15 ng/ml in the CSF (Inokuchi et al. 1984). The concentration reaching the LHA 20 min after the injection was assumed to be 150 pg/ml, which would be comparable to the physiological concentration (100 pg/ml) of blood glucagon. Since the concentration of glucagon during feeding reaches double the inter-meal concentration, hematogenic pancreatic glucagon may suppress GSNs in the LHA to terminate feeding (Geary and Smith 1982; Langhans et al. 1982), and glucagon-like immunoreactivity (GLI) in the brian could act as an inhibitory neurotransmitter or neuromodulator in the hypothalamus.

Measurement of GLI by glucagon N-terminal specific antiserum revealed GLI in the periventricular region, the PVN, the supraoptic nucleus, the acurate nucleus, the median eminence, and the VMH of the rat, and very little in the pons or olfactory bulb. No immunoreactive pancreatic glucagon has been found in the brain. Concentration of GLI increased about 1.3 times in the VMH and LHA after 48 h of food deprivation (Inokuchi et al. 1986b). No GLI immunoreactive cell bodies have been seen in the hypothalamus, but they have been found in the NTS (Inokuchi et al. 1986b). After 24 h of food deprivation, GLI immunoreactive cell bodies were found at the border between the arcuate nucleus and the ventromedial part of the VMH, and immunostained fibers and terminals increased in the PVN, the dorsolateral part of the VMH, and the ventrolateral part of the LHA.

Electrophoretic application of pancreatic glucagon suppressed the activity of GSNs, but not that of GISs, in the LHA, and higher doses suppressed the activity of GRNs in the VMH (Inokuchi et al. 1986a). Inhibition of GSNs by glucagon was blocked by ouabain, so glucagon may regulate cAMP and the resultant Na^{+}-K^{+} pump activity. Electrophoretic glucagon mimicked glucose; it hyperpolarized GSNs without affecting membrane conductivity. The effectiveness of blood glucagon was demonstrated when 100 µg of glucagon injected into the carotid artery suppressed the activity of glucagon-sensitive neurons.

4.5 Insulin

Acute intracerebroventricular applications of insulin suppressed food intake either at night only or throughout a circadian cycle, depending on the dose and time of application. Chronic intracerebroventricular infusion of insulin over 7 – 8 days suppressed food intake throughout the infusion period (Plata-Salamán et al. 1986a). Small doses of insulin did not depress food intake of food-deprived rats, whereas large doses did. The overall results led to the conclusion that small intra-third ventricular administration of insulin decreases food intake only in non-food-deprived rats (Plata-Salamán and Oomura 1986a). Peripheral administration of insulin leads to hyperphagia and obesity (Booth and Pitt 1968). This is apparently due to hypoglycemia caused by hyperinsulinemia, since hyperinsulinemia without hypoglycemia depresses food intake (Nicolaidis and Rowland 1976; VanderWeele et al. 1980). An analogous effect occurs upon injection of 2-deoxy-D-glucose (2-DG) into the third cerebroventricle which elicits transient feeding and simultaneous hyperglycemia. The hyperglycemia is a secondary effect due to release of adrenaline and norepinephrine (NE) which are produced by LHA excitation of the efferent splanchnic nerve to the adrenal gland (Yoshimatsu et al. 1987) a result of 2-DG excitation of GSN neurons (Oomura 1976). This is consistent with the conclusions of Smith and Epstein (1969).

Central administration of insulin in an amount which was not enough to elicit general hypoglycemia apparently deceived the brain into thinking the subjects were obese, so feeding decreased (Woods et al. 1979; Brief and Davis 1984; Nicolaidis 1981). This reasoning was verified when insulin antibodies in the hypothalamus increased food intake (Strubbe and Mein 1977). Intracerebroventricular infusion of insulin from osmotic minipumps (2.0 mIU/rat/day) reduced food intake and body weight of lean but not obese Zucker rats. This suggests that genetically obese Zucker rats have reduced sensitivity to insulin in the central nervous system, possibly because of a defect in some post-receptor mechanism (Ikeda et al. 1986).

The activity of GRNs in the VMH is slightly suppressed by insulin alone, but greatly facilitated by simultaneous electrophoretic application of insulin and glucose. Activity of GSNs in the LHA is dose-dependently facilitated by insulin (Oomura 1976) and insulin-binding sites in the brain have been identified (Havrankova et al. 1979; Szabo and Szabo 1975; Oomura and Kita 1981; Baskin et al. 1986). Measurements of insulin in the blood plasma and brain of rats vary widely (Havrankova et al. 1979; Oomura and Kita 1981; Stein et al. 1982), but two points are consistent; (1) plasma insulin levels vary widely from lean to obese animals and from one strain to another, and, (2) brain insulin content seems to depend on the strain or the investigator, but it appears to be independent of plasma level and remains stable within each individual – although recent results suggest that insulin level in the CSF may reflect the time-integral of the plasma level and as such, might signal over- or underweight (Woods and Porte 1977; Porte and Woods 1981). Specific insulin-binding sites, which are most abundant in the hypothalamus and olfactory bulb, are independent of peripheral insulin concentration (Havrankova et al. 1979; Sakamoto et al. 1980).

4.6 Neurotensin

Neurotensin, found in various parts of the brain and gut, suppresses feeding when injected centrally (Stanley et al. 1982; Levine et al. 1983), without eliciting malaise (Luttinger et al. 1982). Peripheral application is ineffective except in very large doses, which produce general lethargy (Stanley et al. 1983). The central effects of neurotensin seem to be through the hypothalamus, PVN, VMH, and LHA; NE is also closely involved (Stanley et al. 1985b; Levine et al. 1983). Neurotensin increases NE release from the PVN and VMH in sated rats and suppresses NE release from these centers in fasted rats; its effects in the LHA are these opposite (Myers et al. 1986; see NE effect on food intake, Sect. 5.1). The effects of centrally applied neurotensin on food intake appear to be generally similar to those of CCK (Leibowitz and Stanley 1986) – it suppresses gastric acid secretion (Shiraishi et al. 1980) and insulin secretion, and induces hyperglycemia and hyperglucagonemia (Nagai and Frohman 1976).

When applied alone, electrophoretically, it has no effect on GSN in the LHA, but it modulates the effects of insulin or 2-DG (Shiraishi et al. 1980).

4.7 Neuropetide Y

Suggested to be one of the most potent stimulators of food intake, neuropeptide Y (NPY) is found throughout the central and peripheral nervous systems. Centrally administered in cerebroventricles (Levine and Morley 1984), the PVN and the VMH (Stanley et al. 1985a), it significantly increased feeding in rats. Although it coexists in terminals with NE, and its mode of action is similar to that of NE, phentolamine blocks NE but not NPY effects (Lundberg et al. 1984). The action of NPY is blocked by naloxone, but this is a nonspecific effect (Levine and Morely 1984).

4.8 Opioids

Addiction and withdrawal effects tend to cast some doubt on most opioid results. Morphine and met-enkephalin have been reported to depress food intake under acute conditions (King et al. 1979). When applied electrophoretically, they had the same effects as glucose in the LHA and VMH of naive rats (Ono et al. 1980). The effects of morphine and met-enkephalin, but not those of glucose, were blocked by naloxone. Endogenous opioids selectively suppressed GSNs in the LHA during noxious stimulation (Sikdar and Oomura 1985). Opioid receptors on GRNs in the VMH were facilitated by interferon-α, which suppresses feeding, and this facilitation was blocked by naloxone (Nakashima et al. 1987).

On the other hand, opioid injection into the same brain regions, especially the PVN, greatly increased food intake (Grandison and Guidotti 1977; Lanthier et al. 1985; McLean and Hoebel 1983), and naloxone blocks these effects (Gosnell et al. 1984). Opioid terminals and receptor sites have been reported in the PVN (Di Figlia and Aronin 1984), and morphine suppresses the activity of PVN neurons (Pittman et al. 1980). Opioids and NE appeared to have similar effects and to interact (Leibowitz and Hor 1982; see NE effect on food intake, Sect. 5.1). Feeding enhancement by either NA or some opioids via the PVN or a peripheral route is reduced by adrenalectomy and restored by corticosterone (Bhakthavatsalam and Leibowitz 1984; see NE effect on food intake). However, some opioids can act independently of NE since κ and δ agonists are not blocked by adrenalectomy (Levine and Morley 1983; McLean and Hoebel 1982).

4.9 Thyrotropin-Releasing Hormone and an Analog

Thyrotropin-releasing hormone (TRH, pyroglu-his-pro) inhibits feeding and electrophoretically administered it stimulates GRNs in the VMH (Ishibashi et al. 1979a). Its concentration is highest in the hypothalamus. The effects of TRH in the LHA are not clear. TRHA (pyroglu-his-gly), a TRH analog found in the urine of anorexia nervosa patients, depressed mouse body weight when injected intraperitoneally. It also suppresses GSNs and facilitates GRNs when it is administered electrophoretically (Oomura 1981).

4.10 Estradiol

Medial amygdala implants of estradiol modulated feeding behavior as well as gonadotropin release and ovulation (Donohoe and Stevens 1981; Kawakami et al. 1975; Krieger et al. 1976; Pfaff and Keiver 1973; Terasawa and Kawakami 1974). There is evidence that amygdalar neurons respond to affective stimuli (Nakano et al. 1987). Neurons in the medial amygdala differ morphologically between the sexes, and those in the medial but not the basolateral amygdala provide estrogenic feedback for gonadotropin and prolactin release in the female, but not in the male. The medial amygdala facilitates LHA feeding activity. The activity of neurons in the medial amygdala was rapidly inhibited by 17β-estradiol but not by 17α-estradiol. This effect was dose-dependent and did not densensitize (Nabekura et al. 1986). Intracellular investigation revealed that the effect was due to hyperpolarization accompanied by an increase in membrane conductance for K^+ ions, so it was a direct effect of 17β-estradiol on the postsynaptic membrane. This was further verified by results in the presence of actinomycin D and cyclohexamide. The occurrence of responses to estradiol in females was more than three times that in males. It was concluded that neurons in the medial amygdala might monitor estradiol levels in the blood or CSF and transmit information for its use in functions such as gonadotropin release and feeding behavior (Nabekura et al. 1986).

In the VMH, neurons were either depolarized, with a decrease in membrane conductance (mostly GRNs), or hyperpolarized (non-GRNs) by 17β-estradiol (Oomura et al. 1986). Depolarization of the GRNs with membrance conductance decrease was augmented and prolonged by isobutyl-methylxanthine, a phosphodiesterase inhibitor, and by forskolin, an adenylate cyclase activator. Reversal potential for the GRN's depolarization was about -90 mV, so the K^+ channel was closed by estradiol. These effects might have been mediated by increased cAMP. Bilateral and unilateral implants of estradiol benzoate into the VMH decreased food intake in female and in castrated male rats (Wade and Zucker 1970; Nunez et al. 1980). These results suggest that the VMH is also an estradiol target.

4.11 Prolactin

Increased food intake is one characteristic of lactation in the rat, and this may be a result of increased prolactin level. Treatment of female rats with physiological amounts of prolactin significantly increased food intake, dose dependently, above that of untreated controls. The prolactin-treated rats also gained more body weight than the controls during the 11 days of the experiment (Gerardo-Gettens et al. 1986). The neuronal mechanism has not been studied.

4.12 Testosterone

Bilateral implants of testosterone propionate in the VMH of castrated male rats depressed food intake. Implants of testosterone propionate in the anterior hypothalamus-preoptic area or 5α-dihydrotestosterone in the VMH did not affect food intake (Nunez et al. 1980).

5 Transmitters

5.1 Norepinephrine

Injection of NE into the hypothalamus enhances feeding (Grossman, 1960). The site of this effect may well be the PVN (Leibowitz and Stanley 1986): more specifically, enhancement of feeding by α_2-noradrenergic agonists but not by β-noradrenergic agonists, and opposite effects by the respective antagonists, indicates NE inhibition of PVN neurons through α_2-noradrenergic receptors to accomplish this (Leibowitz and Stanley 1986). Parvocellular neurons in the rat PVN were inhibited through α_2-receptors by intracerebroventricular administration of NE, with elicitation of food intake (Shibata et al. 1988). Subdiaphragmatic vagotomy attenuated feeding elicited by NE injection into the PVN, possibly due to increase in insulin secretion, since lesion of the PVN facilitates efferent vagal activity and suppresses efferent splanchnic activity to the pancreas (Yoshimatsu et al. 1984). The NE effect was also blocked by adrenalectomy and restored by injection of corticosterone. Adrenalectomy reduced the number of α_2-receptors in the PVN, and these receptors were restored by injection of corticosterone. The effect thus appears to be related to both corticosterone and blood glucose levels, since food deprivation, which depressed the blood glucose level, also reduced the number of α_2-receptors, specifically in the PVN. (Jhanwar-Uniyal and Leibowitz 1986). The mutual effects betwen the PVN and these other factors might also suggest a feedback mechanism.

5.2 Serotonin

Direct central or peripheral administration of serotonin (5-hydroxytriptamine, 5-HT) depresses food intake (Blundell 1984), and depletion of 5-HT enhances food intake (Breish et al. 1976; Saller and Stricker 1976). Manipulations such as stress, that increase endogenous 5-hydroxyindole ascetic acid into the LHA (Oomura et al. 1988) also suppress food intake. Electrophoretic administration of 5-HT suppresses GSNs specifically through the 5-HT-1 receptors and produces membrane hyperpolarization caused by K^+ permeability increase (Kai et al. 1988).

5.3 Histamine

The VMH and PVN contain histamine receptors (Palacios et al. 1981). Feeding was induced in the early light, but not in the early dark, by intraventricular injection of H_1- but not H_2-antagonists (Orr and Quay 1975). These effects were shown in the VMH but not in the PVN or the LHA, and were abolished by fluoromethylhistidine (Kollonitsch et al. 1978). Sakata et al. (1988) concluded that histamine suppresses feeding by its action in the VMH. They also found that H_1-antagonists elicited feeding by inhibition of VMH neurons, but did not effect LHA neurons, so the histamine effect is specific to H_1-receptors in the VMH and depends on the phase of the circadian feeding cycle.

6 Sugar Acid Metabolites

Three recently identified metabolites [1], the sugar acids 2,4,5-trihydroxypentanoic acid γ-lactone (2,4,5-TP), 3,4-dihydroxybutanoic acid γ-lactone (3,4-DB), and 2-buten-4-olide (2-B4O), effectively modulate feeding behavior. The first, 2,4,5-TP, induces and the last two potently suppress feeding. The actions of all three are mediated by GSNs and GRNs in hypothalamic and medullary sites. Figure 1 shows the chemical structures of these three sugar acids.

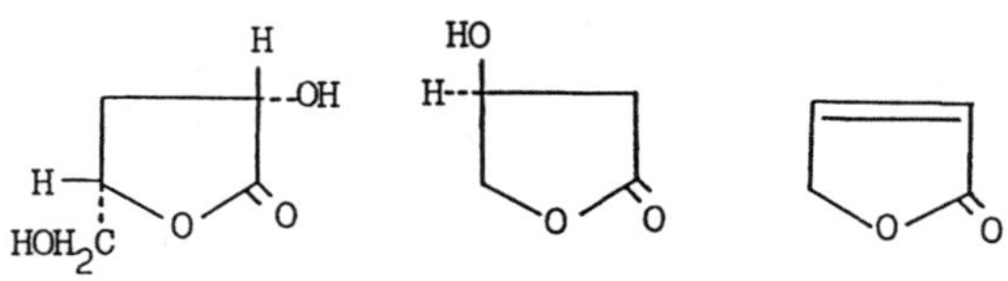

2,4,5-trihydroxy-pentanoic acid γ-lactone   3,4-dihydroxy-butanoic acid γ-lactone   2-buten-4-olide

Fig. 1. Endogenous modulators of feeding behavior. The "hunger substance" 2,4,5-TP (*left*) induces feeding; the two "satiety substances" 3,4-DB and 2-BO (*right*) suppress its

6.1 2,4,5-Trihydroxypentanoic Acid γ-Lactone

During food deprivation, the concentration of 2,4,5-TP in the blood increased from an initial predeprivation value of 200 μ*M* to a first peak of about 360 μ*M* at 12 h of fasting, decreased to slightly above the starting level after 24 h, and then continued to rise for at least the next 40 h. The concentration at 60 h was about 350 μ*M*. This (Oomura 1987) course of change over time suggested that 2,4,5-TP might increase feeding, a supposition ultimately found to be correct, although the effect depended on the time of application and the phase in the circadian cycle. When injected into the third cerebral ventricle at 1030 hours, 2.5 μmole 2,4,5-TP transiently induced feeding, but the increase over 24 h (13%) was not significant (Shimizu et al. 1984). Injection of 2,4,5-TP at 1900 hours increased food intake significantly until, but not after, 2300 hours (46%) (Shimizu et al. 1984). When 2,4,5-TP was injected intravenoulsy or intraperitoneally at 1900 hours, food consumption increased dose-dependently (20 – 40%).
In anesthetized rats, GSNs in the LHA and GRNs in the VMH were tested for their responses to electrophoretic administration of 2,4,5-TP (Shimizu et al.

[1] In prior publications 2,4,5-TP was abbreviated as 3DPA; and 3,4-DB was abreviated as 2DTA. The new abbreviations are used to conform with biochemical usage.

1984; Puthuraya et al. 1985). Activity of GSNs was enhanced by 2,4,5-TP, and its effects on GISs were negligible. Activity of GRNs was dose-dependently inhibited by 2,4,5-TP, and no non-GRNs were affected. No cortical neuron tested responded to 2,4,5-TP administration. GSN activity was facilitated by 2,4,5-TP due to membrane depolarization with a membrane conductance increase. The reversal potential for depolarization was approximately +40 mV, which is close to the Na^+ and Ca^{2+} equilibrium potentials. Inhibition of GRN activity by 2,4,5-TP was due to hyperpolarization with membrane conductance increase. The reversal potential for hyperpolarization was approximately −90 mV, so 2,4,5-TP increased membrane K^+ permeability (Oomura 1986).

Injection of 2.5 μmole 2,4,5-TP into the third cerebroventricle transiently depressed plasma glucose 20% for a few hours and increased insulin to 2.6 times the control level. Increase in insulin level induced by 2,4,5-TP agrees with increased vagal and decreased splanchnic efferent activity to the pancreas that appeared after 2,4,5-TP induced an increase of LHA activity and a decrease of VMH activity (Oomura and Kita 1981). Changes induced in plasma levels of glucose and insulin by 2,4,5-TP were consistent with its effect on feeding behavior. Gastric acid secretion was elicited by electrophoretic administration or microinjection of 2,4,5-TP in the LHA (Shiraishi et al. 1985). It appears that 2,4,5-TP modulates feeding behavior by modifying LHA and VMH neuronal activity.

6.2 3,4-Dihydroxybutanoic Acid γ-Lactone and 2-Buten-4-Olide

The concentration of 3,4-DB increased from an initial level of about 120 μ*M* to a peak of about 160 μ*M* after 48 h of food deprivation (Oomura 1987). Food consumption after injection of 2.5 μmole 3,4-DB into the third cerebroventricle at 1630 hours decreased significantly for 12 h, and recovered after 24 h (Shimizu et al. 1984). Motor activity was not affected and there was no other overt evidence of toxicity. This effect was also observed in rats deprived for 72 h.

Why 2,4,5-TP, 3,4-DB and 2-B4O should increase after 48−60 h of food deprivation, although their effects on feeding are different, cannot yet be answered for certain. We know that food deprivation up to about 60 h tends to elicit feeding, but thereafter causes anorexia in rats (Hamilton 1969) and humans (Bruch 1973). We can only speculate that the effects might be due to relative levels of several metabolites, including the three discussed in this section, rather than to the absolute level of any one of the metabolites alone. Intravenous or intraperitoneal injection of 100−500 μmole 3,4-DB at 1900 hours did not affect food consumption, although massive 2.5 mmole doses of 3,4-DB injected into the carotid artery slightly suppressed feeding. In view of the effectiveness of injections into the third cerebroventricle, the absence of results of peripheral injection of 3,4-DB was attributed to either its failure to cross the blood-brain barrier or its adhesion to blood vessels or other organ tissue.

Two methods were used to correct the failure of 3,4-DB administered intravenously or intraperitoneally to affect feeding. One method was encapsulation in liposome vesicles. After encapsulation, 1.25 and 2.5 mmole 3,4-DB significantly depressed feeding dose-dependently. It was estimated that 8% of the 3,4-DB used was actually encapsulated (0.1 and 0.2 mmole), and the 1% of this that reached the brian (1 and 2 μmole) was comparable to the dose injected into the third cerebroventricle (Terada et al. 1986).

The second method used 2-B4O, a physicochemical derivative of 3,4-DB with comparable biological activity but increased lipophilicity. Injection of 2-B4O into the third cerebroventricle of 1.2 – 5.0 μmole per rat produced effects similar in nature and comparable in magnitude to those of 3,4-DB in both normal and 72-h deprived rats.

When administered intraperitoneally, 2-B4O significantly depressed feeding dose-dependently up to 0.4 mmole/kg. This amounts to about 0.13 mmole/rat, compared to the 0.1 – 0.2 mmole/rat of encapsulated 3,4-DB that was effective (Terada et al. 1986). Intragastric administration of 0.6 – 3.8 mmole had similar, dose-dependent effects (Plata-Salamán et al. 1986). Thus, 2-B4O has biological activity equal to that of 3,4-DB and has the advantage of being effective by either a central or a peripheral route. Interestingly, although 2-B4O was initially synthesized for testing, it was later isolated from blood serum. It has been found to increase from a prefasting level of 3.5 μ*M* in nondeprived rats up to 13.5 μ*M* after 48 – 60 h of food deprivation (Oomura 1987).

In anesthetized rats, GSNs in the LHA and GISs in the VMH were tested for responses to direct application of 3,4-DB and 2-B4O (Shimizu et al. 1984; Puthuraya et al. 1985). Electrophoretic application of either of these sugar acids suppressed GSN activity dose-dependently. Effects of 3,4-DB or 2-B4O on GISs were negligible. The suppressive effects of 3,4-DB and glucose on the GSNs were reversibly blocked by ouabain. Activity of GRNs was dose-dependently facilitated by 3,4-DB. No non-GRNs were affected by either of the compounds, and no cortical neuron tested responded to 3,4-DB or 2-B4O.

The effect of 3,4-DB on membrane potentials and conductance was investigated in guinea pig and rat hypothalamic slice preparations (Oomura 1986). GSNs in the LHA were hyperpolarized by 3,4-DB while membrane conductance remained constant. Attenuation of the 3,4-DB hyperpolarization by ouabain indicates that its effect, like that of glucose, was due to Na^+-K^+ pump activation through ATP formation. In the VMH, GRNs were depolarized by 3,4-DB or 2-B4O the same as by glucose, while membrane conductance decreased. Since the reversal potential for both glucose, 3,4-DB, and 2-B4O depolarization was approximately -90 mV, close to the K^+ equilibrium potential, depolarization by 3,4-DB or 2-B4O, like that by glucose, is caused by K^+ permeability decrease (Minami et al. 1986; Fukuda et al. 1988). This again suggests that 3,4-DB is metabolized in the brain, as are ketone bodies, if there is a shortage of glucose. Levels of plasma glucose and insulin were changed by injections of 3,4-DB or 2-B4O into the third cerebroventricle at 2.5 μmole concentration. Blood glucose was increased about 20% by 3,4-DB, first transiently and then persistently for 8 h after injection. Insulin concentration was not significantly affected by 3,4-DB. Blood glucose was increased for approximately 5 h to 1.5 times control by intaperitoneal injection of 5 nmole 2-B4O. During that time, vagal efferent activity decreased and splanchnic efferent activity to the pancreas increased.

Changes induced in plasma levels of glucose and insulin by 3,4-DB and 2-B4O were consistent with the effect that each drug has on feeding behavior. Gastric acid secretion induced by 2-DG was suppressed by administration of 3,4-DB or

2-B4O in the LHA (Shiraishi et al. 1985); this effect is the same as that of glucose. The concentrations of these sugar acids in CSF were estimated to be 4.2 – 8.2 m*M* after 1.2- to 2.5-μmole injection into the third cerebroventricle if they were not metabolized. If CSF is produced at a rate of 0.5% of the total volume per min (Davson 1967), the concentration of sugar acids would be diluted to 22% of the initial concentration, or 0.9 to 1.8 m*M* approximately 5 h after injection. The concentration of ketone bodies in CSF is about 50% of that in the blood, and increases from 0.12 to 1.6 m*M* after 96 h of food deprivation. This 1.6 m*M* value is comparable to the 5-h concentration estimated for the sugar acids. Because diffusion is 1% from the cerebral ventricle to the brain parenchyma after 20 min (De Wildt 1982), concentrations of 3,4-DB, 2-B4O, or 2,4,5-TP available to act on LHA and VMH neurons should be no more than 50 – 100 μ*M* (10 – 20 μ*M*, 5 h after injection).

It appears that these sugar acids modulate feeding behavior by modifying LHA and VMH neuronal activity. Substances such as calcitonin and glucagon mimic glucose in the rat LHA; sugar acids modify plasma glucose, insulin levels, and feeding behavior. It is suggested that 3,4-DB and 2-B4O mediate satiety through their effects on the activity of GSNs and GRNs.

6.3 Sugar Derivative

D-glucosamine (2-amino-2-deoxy-D-glucose), an endogenous glucose analogue, and 1-deoxy-D-glucosamine induced transient feeding episodes and then suppressed food intake for a long time when applied into the third cerebroventricle (6 – 24 μmole/rat). The suppressive effect of 1-deoxy-D-glucosamine was stronger than that of D-glucosamine. GSNs were inhibited and GRNs were facilitated by electrophoretically administered 1-deoxy-D-glucosamine (Fujimoto et al. 1986).

6.4 Other Factors

Two other factors have recently been observed that appear to be related to feeding. An increase in acidic fibroblast growth factor (aFGF) activity was detected in rat CSF after eating, or after intraperitoneal injection of glucose. Heparin affinity chromatographic analysis of rat CSF after feeding revealed a chemical similar in nature to aFGF. This aFGF-like activity was investigated by bioassay – aFGF specific depression of the feeding response of hydra. Intracerebroventricular microinfusion of aFGF (100 – 200 ng/rat) suppressed food intake in rats. Central infusion of inactivated aFGF, or peripheral administration of aFGF in doses equal to or higher than those administered centrally, had no effect. Electrophoretically applied aFGF specifically suppressed the activity of GSNs in the LHA, but only slightly affected GINs. The results suggest that aFGF might be involved in the regulation of feeding through its action in the central nervous system (Hanai et al. 1989). It is interesting that aFGF is a potent mitogen for smooth muscle cells in vessels and endothelial cells of capillaries and thus might participate in the development of cerebral atherosclerosis.

Quite similar to the amino acids sequence of aFGF is interleukin-1β (IL-1β). IL-1β coexists with tumor necrosis factor (TNF) in monocytes. Intracerebroventricular microinfusion of recombinant human TNF (rhTNF, 50 – 500 ng/rat) and of recombinant human IL-1β (rhIL-1β, 1 – 10 ng/rat) suppressed food intake in rats. Central infusion of heat-inactivated rhTNF or rhIL-1β had no such effect, and rhIL-1β did not affect the dipsogenic response to centrally administered angiotension II. Peripheral administration of rhTNF and rhIL-1β in doses equivalent to or higher than those administered centrally had no effect. Electrophoretically administered rhTNF and rhIL-1β specifically suppressed the activity of GSNs in the LHA but did not significantly affect GINs. The results suggest that TNF and IL-1β act directly in the central nervous system to suppress feeding, and this effect might be evident during acute and chronic disease (Plata-Salamán et al. 1988).

7 Organization of the Feeding-Related Chemical Sense

Figure 2 charts the hierarchy of chemoresponsive components that contribute to the control of feeding through chemical sensation. Each component in the organization, except the association cortex, contains chemoresponsive elements. Elements in the DMH (Fukuda et al. 1984), and PVN (Shiraishi 1987), however, despite strong evidence of their existence, have not yet been thoroughly characterized. Glucose on the tongue and on peripheral glucose-sensitive and glucoreceptor vagal units in the hepatic portal vein (Shimizu et al. 1983), the duodenum, the lower intestine, and the pancreas affects neural activity. The glucose-sensitive hepatic vagal units respond to insulin, glucagon, 3,4-DB, 2-B4O, and 2,4,5-TP in the same way as GSNs in the LHA (A. Niijima, unpublished data). This information is relayed in the NTS (Mizuno and Oomura 1984) and area postrema (Adachi and Kobashi 1988), which also contains GSNs and GRNs, and the DMV, which contains GRNs. Initial integration of this information with information on vascular or ventricular (fourth ventricle) chemicals probably occurs in the NTS. This integrated information can move in two directions: rostrally for higher integration and caudally, by way of the DMV vagal nucleus, as a reflex from the NTS to produce responses in the appropriate viscera, or to control activity of hepatic enzymes or pancreatic hormones.

Information in the rostral direction goes, either directly or through the parabrachial nucleus, to the hypothalamus (Norgren and Pfaffman 1975; Nishimura and Oomura 1987). Glucose-responsive cells in these higher regions, the LHA and the ventral hypothalamus, sense systemic and third ventricular chemical information. The association areas, the rostral part of the dorsolateral prefrontal area and the orbitofrontal area, actively communicate with the LHA and VMH, and respond in synchronism with the hypothalamus during feeding behavior (Inoue et al. 1985; Ohta and Oomura 1979; Kita and Oomura 1981). This communication between the above-mentioned association areas and the hypothalamus, plus further integration of information from other modalities, leads to feeding-related decision that go along paths parallel to those over which the afferent signals originally ascended in order to control visceral organs, and along other paths to the motor system by way of the motor cortex.

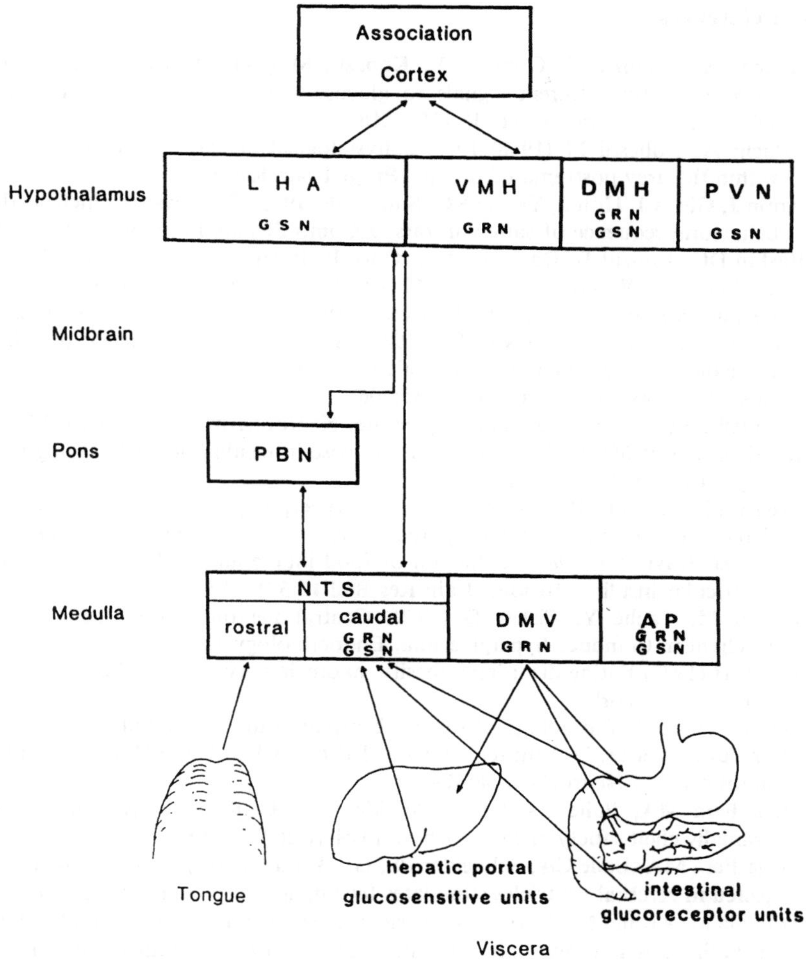

Fig. 2. Hierarchial organization of chemical information processing by central and peripheral nervous system: schema of centers in which chemosensitive neurons and sensors have been identified, and paths over which they are known to be connected. Viscera communicate with lower brain stem centers which relay information in both directions, to higher centers and to viscera. Signals to viscera can be both reflexes from lower brain stem centers and more integrated signals relayed from higher centers. *AP*, areas postrema; *DMH*, dorsomedial hypothalamic nucleus; *DMV*, dorsal motor nucleus of the vagus; *GRN*, glucoreceptor neurons; *GSN*, glucose-sensitive neurons; NTS, nucleus of the tractus solitarius; *PBN*, parabrachial nucleus; *PVN*, paraventricular nucleus

Feeding-related regions of the central nervous system include the prefrontal cortex, the amygdala, and other parts of the reticular formation which mediate signals from various sources and contribute, along with the hypothalamus, to integration of hunger and satiation signals.

8 References

Adachi A, Shimizu N, Oomura Y, Kobashi M (1984) Convergence of hepatoportal glucose-sensitive afferent signals to glucose-sensitive units with the nucleus of the solitary tract. Neurosci Lett 46:215 – 218

Adachi A, Kobashi M (1988) Electrophysiological analysis of chemosensitive neurons within the arey postrema of the rat. Progr Brain Res 74:77 – 83

Antin J, Gibbs J, Holt J, Young RC, Smith GP (1975) Cholecystokininelicits the complete behavioral sequence of satiety in rats. J Comp Physiol Psychol 89:784 – 790

Baskin DG, Brewitt B, Davidson DA, Corp E, Paquette T, Figlewicz DP, Lewellen TK, Graham MK, Woods SC, Dorsa DM (1986) Quantitative autoradiographic evidence for insulin receptors in the choroid plexus of the rat brain. Diabetes 35:246-249

Bhakthavatsalam P, Leibowitz SF (1984) Studies on the influence of adrenal glucocorticoids on the alpha-noradrenergic and opiate feeding effects of different drugs in rats. Proc Eastern Pyschol Assoc 55:106

Blundell JE (1984) Serotonin and appeatite. Neuropharmacology 23:1537 – 1552

Booth DA, Pitt ME (1968) The role of glucose in insulin-induced feeding and drinking. Physiol Behav 3:447 – 453

Breish ST, Zemlan FP, Hoebel BG (1976) Hyperphagia and obesity following serotonin depletion with intraventricular parachlorophenylalanine. Science 192:283 – 285

Brief DJ, Davis JD (1984) Reduction of food intake and body weight by chronic intraventricular insulin infusion. Brain Res Bull 12:571 – 575

Brown M, Taché Y, Fisher D (1979) Central nervous system action of bombesin: mechanism to induce hyperglycemia. Endocrinology 105:660 – 665

Bruch H (1973) Eating disorders: obesity, anorexia nervosa, and the person within. Basic Books, New York

Davson H (1967) Physiology of the cerebrospinal fluid. Churchill, London

De Wildt D, Verhoef J, Whitter A (1982) H-Pro-[^{3}H]Leu-Gly-NH_2 uptake and metabolism in rat brain. J Neurochem 38:67 – 74

Della-Fera MA, Baile CA (1981) Peptides with CCK-like activity administered intracranially elicit satiety in sheep. Physiol Behav 26:979 – 983

Della-Fera MA, Baile CA, Schneider BS, Grinker JA (1981) Cholecystokinin antibody injected in cerebral ventricles stimulates feeding in sheep. Science 212:687 – 689

DiFiglia M, Aronin N (1984) Immunoreactive leu-enkephalin in the monkey hypothalamus including observations on its ultra-structural localization in the paraventricular nucleus. J Comp Neurol 225:313 – 326

Donohoe TP, Stevens R (1981) Modulation of food intake by amygdaloid estradiol benzoate implants in female rats. Physiol Behav 27:105 – 114

Ettinger RH, Thomspon S, Staddon JER (1986) Cholecystokinin, diet palatability, and feeding regulation in rats. Physiol Behav 36:801 – 809

Flynn JJ, Margules DL, Cooper CW (1981) Presence of immunoreactive calcitonin in the hypothalamus and pituitary lobes of rats. Brain Res Bull 6:547 – 549

Fujimoto K, Sakata T, Shiraishi T, Kurata K, Terada K, Etou H (1986) Anorexia induced in rat by D-glucosami deoxidized at C-1. Am J Physiol 251:R481 – R491

Fukuda M, Ono T, Nishino H, Sasaki K (1984) Independent glucose effects on rat hypothalamic neurons: an in-vitro study. J Auton Nerv Syst 10:373 – 381

Fukuda A, Oomura Y, Minami T, Ito C (1988) A novel endogenous sugar acid depolarizes ventromedial hypothalamic neurons *in vitro*. Am J Physiol 255:R134 – R140

Geary N, Smith GP (1982) Pancreatic glucagon and postprandial satiety in the rat. Physiol Behav 28:313 – 322

Gerardo-Gettens T, Moore BJ, Stern JS, Horwitz BA (1986) Exogenous prolactin administration stimulates food intake in a dose-dependent manner. IXth Int Conf Physiol Food and Fluid Intake, Abstr 28

Gibbs J, Fauser DJ, Rowe EA, Rolls BJ, Rolls ET, Maddison SP (1979) Bombesin suppresses feeding in rats. Nature (Lond) 282:208 – 210

Gibbs J, Young RC, Smith GP (1973) Cholecystokinin elicits satiety in rats with open gastric fistulas. Nature (Lond) 245:323 – 325

Gosnell BG, Morley JE, Levine AS (1984) Localization of naloxone-sensitive brain areas in relation to food intake. Soc Neurosci Abstr 10:306

Grandison L, Guidotti A (1977) Stimulation of food intake by muscimol and beta endorphin. Neuropharmacology 16:533 – 536

Grossman SP (1960) Eating or drinking by direct adrenergic or cholinergic stimulation of the hypothalamus. Science 132:301 – 302

Hamilton CL (1969) Problems of refeeding after starvation in the rat. Ann NY Acad Sci 157:1004 – 1017

Hanai K, Oomura Y, Kai Y, Shimizu N, Nishikawa K, Morita H, Plata-Salamán Cr (1989) Central action of acidic fibroblast growth factor in feeding regulation. Am J Physiol 256: in press

Havrankova J, Roth J, Brownstein MJ (1979) Concentrations of insulin and insulin receptors in the brain are independent of peripheral insulin levels. Studies of obese and streptozotocin-treated rodents. J Clin Invest 64:636 – 642

Hoebel BG, Dorfman D, Scott P (1984) Effects of CCK in the brain. In: The neural and metabolic bases of feeding. Satellite Symposium of Neuroscience Society Meeting Napa Valley, CA, Oct. 7 – 9, 1984. Abstr. p8

Ikeda H, West DB, Pustek JJ, Figlewicz DP, Greenwood MRC, Porte D Jr, Woods SC (1986) Intraventricular insulin reduces food intake and body weight of lean but not obese Zucker rats. Appetite 7:381 – 386

Inokuchi A, Oomura Y, Nishimura H (1984) Effect of intracerebroventricular infused glucagon on feeding behavior. Physiol Behav 33:397 – 400

Inokuchi A, Oomura Y, Shimizu N, Yamamoto T (1986a) Central action of glucagon in the rat hypothalamus. Am J Physiol 250:R120 – R126

Inokuchi A, Tomita Y, Yanaihara C, Yui R, Oomura Y, Kimura H, Hase T, Matsumoto T, Yanaihara N (1986b) Glucagon-related peptides in the rat hypothalamus. Cell Tiss Res 246:71 – 75

Inoue M, Oomura Y, Aou S, Nishino H, Sikdar SK (1985) Reward related neuronal activity in monkey dorsolateral prefrontal cortex during feeding behavior. Brain Res 326:307 – 312

Ishibashi S, Oomura Y, Okajima T (1979a) Facilitatory and inhibitory effects of TRH on lateral hypothalamic and ventromedial neurons. Physiol Behav 22:785 – 787

Ishibashi S, Oomura Y, Okajima T, Shibata S (1979b) Cholecystokinin, motilin and secretin effects on the central nervous system. Physiol Behav 23:401 – 403

Jhanwar-Uniyal M, Leibowitz SF (1986) Impact of circulating corticosterone an α_1- and α_2-noradrenergic receptors in discrete brain areas. Brain Res 368:404 – 408

Kai Y, Shimizu N, Oomura Y (1988) Pain transmission to rat lateral hypothalamus. J Neurophysiol 60:524 – 536

Kawakami M, Kimura F, Higuchi T (1975) Localization and mechanism of stimulatory feedback action of estrogen: effect of limbic forebrain implantation of estradiol benzoate on advancement of ovulation. Endocrinol Jpn 22:327 – 328

King MG, Kastin AJ, Olson RD, Coy DH (1979) Systemic administration of met-enkephalin, (D-Ala^2)-met-enkephalin, beta-endorphin, and (D-Ala^2)-beta-endorphin: effect on eating, drinking and activity measures in rats. Pharm Biochem Behav 11:407 – 411

Kita H, Oomura Y (1981) Reciprocal connections between the lateral hypothalamus and the frontal cortex in the rat: electrophysiological and anatomical observations. Brain Res 213:1 – 16

Kollonitsch J, Perkins LM, Patchett AA, Doldouras GA, Marburg S, Duggan DE, Maycock AL, Aster SD (1978) Selective inhibitors of biosynthesis of aminergic neurotransmitters. Nature (Lond) 274:906–908
Krahn DD, Gosnell BA, Levine AS, Morley JE (1984) Effects of calcitonin gene-related peptide on food intake. Peptides 5:861–864
Krieger MS, Morrell JI, Pfaff DW (1976) Autoradiographic localization of estradiol-concentrating cells in the female hamster brain. Neuroendocrinology 22:193–205
Langhans W, Zieger V, Scharrer E, Geary N (1982) Stimulation of feeding in rats by intraperitoneal injection of antibodies to glucagon. Science 218:894–896
Lanthier D, Stanley BG, Leibowitz SF (1985) Feeding elicited by central morphine injection: sites of action in the brain. Proc Eastern Psychol Assoc 56:30
Leibowitz SF, Hor L (1982) Endorphinergic and α-adrenergic systems in the paraventricular nucleus: effects on eating behavior. Peptides 3:421–428
Leibowitz SF, Stanley G (1986) Neurochemical controls of appetite. In: Ritter RC, Ritter S, Barner CD (eds) Feeding behavior: neuronal and humoral controls. Academic, New York, pp 191–234
Levine AS, Morley JE (1983) Adrenal modulation of opiate-induced feeding. Pharmacol Biochem Behav 19:403–406
Levine AS, Morley JE (1984) Neuropeptide Y: a potent inducer of consummatory behavior in rats. Peptides 5:1025–1029
Levine AS, Kneip J, Grace M, Morley JE (1983) Effect of centrally administered neurotensin on multiple feeding paradigms. Pharmacol Biochem Behav 18:19–23
Lundberg JM, Terenius L, Hokfelt T, Tatemoto K (1984) Catecholamines, neuropeptide Y (NPY), and the pancreatic polypeptide family: coexistence and interaction in the sympathetic response. In: Usdin R (ed) Catecholamines: neuropharmacology and central nervous system – Theoretical Aspects. Liss, New York, pp 179–189
Luttinger D, King RA, Sheppard D, Struff J, Nemeroff CB, Prange AJ Jr (1982) The effect of neurotensin on food consumption in the rat. Eur J Pharmacol 81: 499–503
McLean S, Hoebel BG (1982) Opiate and norepinephrine-induced feeding from the paraventricular nucleus of the hypothalamus are dissociable. Life Sci 31:2379–2382
McLean S, Hoebel BG (1983) Feeding induced by opiates into the paraventricular hypothalamus. Peptides 4:287–292
Minami T, Oomura Y, Sugimori M (1986) Electrophysiological properties and glucose responsiveness of guinea-pig ventromedial hypothalamic neurones *in vitro*. J Physiol 380:127–143
Mizuno Y, Oomura Y (1984) Glucose-responding neurons in the nucleus tractus solitarius of the rat: in-vitro study. Brain Res 301:109–116
Moore BO, Deutsch JA (1985) An antiemetic is antidotal to the satiety effects of cholecystokinin. Nature (Lond) 315:321–322
Myers RD, Swartzwelder HS, Peinodo JM, Lee TF, Hepler JR, Denbow DM, Ferrer JMR (1986) CCK and other peptides modulate hypothalamic norepinephrine release in the rat: dependence on hunger or satiety. Brain Res Bull 17:583–597
Nabekura J, Oomura Y, Minami T, Mizuno Y, Fukuda A (1986) Mechanism of the rapid effect of 17β-estradiol on medial amygdala neurons. Science 233:226–228
Nagai K, Frohman LA (1976) Hyperglycemia and hyperglucagonemia following neurotensin administration. Life Sci 19:273–279
Nakano Y, Lénárd L, Oomura Y, Nishino H, Aou S, Yamamoto T (1987) Functional involvement of catecholamines in reward related neuronal activity of the monkey amygdala. J Neurophysiol 57:72–91
Nakashima T, Hori T, Kuriyama K, Kiyohara T (1987) Naloxone blocks the interferon-α induced changes in hypothalamic neuronal activity. Neurosci Lett 82:332–336

Nicolaidis S (1981) Lateral hypothalamic control of metabolic factors related to feeding. Diabetologia 20:426 – 434

Nicolaidis S, Rowland N (1976) Metering of intravenous versus oral nutrients and regulation of energy balance. Am J Physiol 231:661 – 668

Niijima A (1981) Visceral afferents and metabolic function. Diabetologia [Suppl] 20:325 – 330

Nishimura H, Oomura Y (1987) Effects of hypothalamic stimulation on activity of dorsomedial medulla neurons that respond to subdiaphragmatic vagal stimulation. J Neurophysiol 58:655 – 675

Norgren R, Pfaffmann C (1975) The pontine taste area in the rat. Brain Res 91:99 – 117

Nunez AA, Siegel LI, Wade GN (1980) Central effects of testosterone on food intake in male rats. Physiol Behav 24:469 – 472

Ohta M, Oomura Y (1979) Monosynaptic facilitatory pathway from the hypothalamic ventromedial nucleus to the frontal cortex in the rat. Brain Res Bull 4:223 – 229

Ono T, Oomura Y, Nishino H, Sasaki K, Muramoto K, Yano I (1980) Morphine and enkephalin effects on hypothalamic glucoresponsive neurons. Brain Res 185:208 – 212

Oomura Y (1976) Significance of glucose, insulin, and free fatty acid on the hypothalamic feeding and satiety neurons. In: Novin D, Wyricka W, Bray G (eds) Hunger: basic mechanisms and clinical implications. Raven, New York, pp 145 – 157

Oomura Y (1981) Chemosensitive neuron in the hypothalamus related to food intake behavior. Jpn J Pharmacol 31, Suppl 1:12

Oomura Y (1983) Glucose as a regulator of neuronal activity. In: Advances in metabolic disorders, vol 10. Academic, New York, pp 31 – 65

Ommura Y (1986) Feeding regulation by endogenous sugar acids through hypothalamic chemosensitive neurons. Brain Res Bull 17:551 – 562

Oomura Y (1987) Feeding regulatin by neural reaction to endogenous factors. News in Physiol Sci 2:199 – 203

Oomura Y, Kita H (1981) Insulin acting as a modulator of feeding through the hypothalamus. Diabetologia [Suppl] 20:290 – 298

Oomura Y, Minami T, Nabekura J (1986) Effect of estradiol on the amygdala and ventromedial hypothalamic neurons. Soc Neurosci Abstr 12:1330

Oomura Y, Shimizu N, Kai Y (1988) Neuronal activity in feeding center by biowarning process In: Takagi H, Oomura Y, Ito M (eds) Biowarning system in the brain. Tokyo University Press, Tokyo, pp 243 – 259

Oomura Y, Ooyama H, Sugimori M, Nakamura T, Yamada Y (1974) Glucose inhibition of the glucose-sensitive neurone in the rat lateral hypothalamus. Nature (Lond) 247: 284 – 286

Orr E, Quay WB (1975) Hypothalamic 24-hour rhythms in histamine, histidine, decarboxylase and histamine-*N*-methyltransferase. Endocrinology 96:941 – 945

Palacios J, Wamsley JK, Kuhar MJ (1981) The distribution of histamine H_1-receptors in the rat brain: an autoradiographic study. Neuroscience 6:15 – 37

Pfaff DW, Keiver M (1973) Atlas of estradiol-concentrating cells in the central nervous system of the female rat. J Comp Neurol 151:121 – 158

Pittman QJ, Hatton JD, Bloom FE (1980) Morphine and opioid peptides reduce paraventricular neuronal activity: studies on the rat hypothalamic slice preparation. Proc Natl Acad Sci USA 77:5527 – 5531

Plata-Salamán CR, Oomura Y (1986) Effect of intra-third ventricular administration of insulin on food intake after food deprivation. Physiol Behav 37:735 – 739

Plata-Salamán CR, Oomura Y (1987) Calcitonin as a feeding suppressant: localization of control action of the cerebral III ventricle. Physiol Behav 40:501 – 513

Plata-Salamán CR, Oomura Y, Shimizu N (1986a) Dependence of food intake on acute and chronic ventricular administration of insulin. Physiol Behav 37:717 – 734

Plata-Salamán CR, Oomura Y, Shimizu N (1986b) Endogenous sugar acid derivative acting as a feeding suppressant. Physiol Behav 38:359 – 373
Plata-Salamán CR, Oomura Y, Kai Y (1988) Tumor necrosis factor and interleukin-1β: suppression of food intake by direct action in the central nervous system. Brain Res 448:106 – 114
Prote D Jr, Woods SC (1981) Regulation of food intake and body weight by insulin. Diabetologia 20:274 – 280
Puthuraya KP, Oomura Y, Shimizu N (1985) Effects of endogenous sugar acids on the ventromedial hypothalamic nucleus of the rat. Brain Res 332:165 – 168
Sakamoto Y, Oomura Y, Kita H, Shibata S, Suzuki S, Kuzuya T, Yoshida S (1980) Insulin content and insulin receptors in the rat brain. Effect of fasting and streptozotocin treatment. Biomed Res 1:334 – 340
Sakata T, Ookuma K, Fukagawa K, Fujimoto K, Yoshimatsu H, Shiraishi T, Wada H (1988) Blockade of the histamine H_1-receptor in the rat ventromedial hypothalamus and feeding elicitation. Brain Res 441:403 – 407
Saller CF, Stricker EM (1976) Hyperphagia and increased growth in rats after intraventricular injection of 5,7-dihydroxytryptamine. Science 192:385 – 387
Shibata R, Ono T, Nakamura K, Muramoto K (1988) Relationship between hypothalamic paraventricular neuronal activity and behaviors in reely moving rat. Neurosci Res [Suppl] S68
Shimizu N, Oomura Y (1986) Calcitonin-induced anorexia in rats: evidence for its inhibitory action on lateral hypothalamic chemosensitive neurons. Brain Res 367:128 – 140
Shimizu N, Oomura Y, Novin D, Grijalva C, Cooper P (1983) Functional correlations between lateral hypothalamic glucose-sensitive neurons and hepatic portal glucose-sensitive units in rat. Brain Res 265:49 – 54
Shimizu N, Oomura Y, Sataka T (1984) Modulation of feeding by endogenous sugar acids acting as hunger or satiety factors. Am J Physiol 246:R542 – R550
Shiraishi T (1987) Gastric related properties of rat paraventricular neurons. Brain Res Bull 18:315 – 323
Shiraishi T, Inoue A, Yanaihara N (1980) Neurotensin and bombesin effects on LHA-gastrosecretory relations. Brain Res Bull 5 [Suppl] 4:133 – 142
Shiraishi T, Kawashima M, Oomura Y (1985) Endogenous sugar acid control of hypothalamic neuron activity and gastric acid secretion in rats. Brain Res Bull 14:431 – 438
Sikdar SK, Oomura Y (1985) Selective inhibition of glucose-sensitive neurons in rat lateral hypothalamus by noxious stimuli and morphine. J Neurophysiol 53:17 – 31
Smith GP, Epstein AN (1969) Increased feeding in response to decreased glucose utilization in the rat and monkey. Am J Physiol 217:1083 – 1087
Smith GP, Jerome C, Cushin BJ, Eterno R, Simansky KJ (1981) Abdominal vagotomy blocks the satiety effect of cholecystokinin in the rat. Science 213:1036 – 1037
Smith GP, Jerome C, Norgren R (1985) Afferent axons in abdominal vagus mediate satiety effect of cholecystokinin in rats. Am J Physiol 249:R638 – R641
Stanley BG, Eppel NE, Hoebel BG (1982) Neurotensin injected into the paraventricular hypothalamus suppresses feeding in rats. In: Neurotensin, a brain and gastrointestinal Peptide. Ann NY Acad Sci 400:425 – 527
Stanley BG, Hoebel BG, Leibowitz SF (1983) Neurotensin: effects of hypothalamic and intravenous injections on eating and drinking in rats. Peptides 4:493 – 500
Stanley BG, Chin AS, Leibowitz SF (1985a) Feeding and drinking elicited by central injection of neuropeptide Y: evidence for a hypothalamic site(s) of action. Brain Res Bull 14:521 – 524
Stanley BG, Leibowitz SF, Eppel N, St-Pierre S, Hoebel BG (1985b) Suppression of norepinephrine-elicited feeding by neurotensin: evidence for behavioral, anatomical and pharmacological specificity. Brain Res 343:297 – 304

Stein LJ, Hjeresen DL, Porte D Jr, Woods SC (1982) Genetically obese Zucker rats have inappropriately low immunoreactive insulin levels in cerebrospinal fluid. Soc Neurosci Abstr 8:273

Strubbe JG, Mein CG (1977) Increased feeding in response to injection of insulin antibodies in the VMH. Physiol Behav 19:309 – 313

Szabo O, Szabo AJ (1975) Studies on the nature and mode of action of the insulin-sensitive glucoregulator receptor in the central nervous system. Diabetes 24:328 – 336

Terada K, Sakata T, Oomura Y, Fujimoto K, Arase K, Osanai T, Nagai Y (1986) Hypophagia induced by endogenous or liposome-encapsulated 3,4-dihydroxybutanoic acid. Physiol Behav 38:861 – 869

Terasawa E, Kawakami M (1974) Positive feedback sites of estrogen in the brain on ovulation: possible role of the bed nucleus of stria terminals and the amygdala. Endocrinol Jpn 21:51 – 60

VanderWeele DA, Pi-Sunyer FX, Novin D, Bush MJ (1980) Chronic insulin infusion suppresses food ingestion and body weight gain in rats. Brain Res Bull [Suppl] 1:7 – 11

Verbalis JG, McCann MJ, McHale CM, Stricker EH (1986) Oxytocin secretion in response to cholecystokinin and food: differentiation of nausea from satiety. Science 232:1417 – 1419

Wade GN, Zucker I (1970) Modulation of food intake and locomotor activity in female rats by diencephalic hormone implants. J Comp Physiol Psychol 72:328 – 336

Woods SC, Porte D Jr (1977) Relationship between plasma and cerebrospinal fluid insulin levels of dogs. Am J Physiol 233:E331 – E334

Woods SC, Lotter EC, McKay LD, Porte D Jr (1979) Chronic intracerebroventricular infusion of insulin reduced food intake and body weight of baboons. Nature (Lond) 282:503 – 505

Yoshimatsu H, Niijima A, Oomura Y, Yamabe K, Katafuchi T (1984) Effects of hypothalamic lesion of pancreatic autonomic nerve activity in the rat. Brain Res 303:147 – 152

Yoshimatsu H, Oomura Y, Katafuchi T, Niijima A (1987) Effects of hypothalamic stimulation and lesion on adrenal nerve activity. Am J Physiol 253:R418 – R424

Structure and Function of the Vomeronasal System – The Vomeronasal Organ as a Priming Pheromone Receptor in Mammals*

J.E. Sánchez-Criado [1], O.A. Mora [2], and A. Gallego [2]

[1] Department of Physiology, Faculty of Medicine, University of Córdoba, 14004 Córdoba, Spain
[2] Department of Physiology, Faculty of Medicine, Complutense University, 28040 Madrid, Spain

1 Introduction and Historical Background

The vomeronasal organ (VNO), the peripheral receptor of the vomeronasal system (VNS) or accessory olfactory system (AOS), was first described as a nonsensory organ with secretory functions by Jacobson (1811). McCotter (1912) was the first to mention the VNO as an organ with an olfactory character. Almost 100 years after the description of the VNO by Jacobson, Cajal (1902, 1904) described the accessory olfactory bulb (AOB; the target organ of the axons coming from the VNO neurons) in mice. On the basis of its structure, Cajal postulated that the AOS could have a special sensitivity to some olfactory cues. From its structure, Broman (1920) considered the VNO to be an organ that received odors of substances dissolved in liquids, while Milstein (1929), by

* This study was supported in part by grants: Fondo de Investigaçion de la Seguridad Social (FISS) 87/1465 and Comision Asesora de Investigaçion Cientifica (CAICYT) 0371.

closing the vomeronasal ducts in rabbits, was led to conclude that the VNO is the most peripheral receptor of the olfactory sense with no special function. Although Milstein's studies as well as those by others (Milhalkovics 1899) determined no special function for the VNO, they were the first to investigate the function of the VNO using an experimental procedure.

Adrian (1955) studied, in the rabbit, the electrical responses of the vomeronasal nerves (VNN) to different odors. He explained that the VNO did not react to air-borne smells in the nasal cavity, a fact also verified by Gallego (1957). In Adrian's experiments, only a localized pressure on the capsule of the VNO was effective in evoking a response. This fact allowed Adrian to conclude that the VNO is nonolfactory in character. Tucker (1963, 1971) did not find signs of response to odorants in the VNN of the rabbit except for some obtained from odorants in aqueous solution; however, he reported that the VNN of the tortoise were highly sensitive to air-borne odorants.

Some 30 years ago a substantial change in the experimental approach to the physiology of the VNS took place. It consisted in the association of the VNS to limbic functions. By sectioning the VNN of male guinea pigs, Planel (1953) showed that vomeronasal impairment did not influence sexual behavior consisting of orientation to an estrous female. Negus (1956), from anatomical studies, concluded that the VNO was implicated in the detection of food odors. By sectioning the VNN of female mice, Whitten (1963) showed that the shortening of the estrous cycle due to presence of the male was not present in animals bearing vomeronasal deficits. Because sham-operated females also failed in the estrous cycle shortening, no definitive conclusion could be drawn from this experiment; it was, however, the first report regarding the function of the VNO as a sex chemoreceptor. The concept of sexual pheromones playing a very important role in mammalian reproduction (Estes 1972) aroused increasing interest in the VNO as a sensor for both signalling and priming pheromones.

Most studies on the physiology of the VNS have used rodents, although other animal species have been also studied. In the male rhesus monkey, Michael and Keverne (1968) used a reversible method for producing anosmia in combination with surgical cutting of the VNN and found that the sexual behavior of the male monkey is not mediated by the VNS. In the cat, it has been shown that sexual behavior induced by the urine of the male is not prevented by closing the vomeronasal ducts (Verberne 1976). Based on anatomy and on the "flehmen" response exhibited by different mammals, Estes (1972) concluded that the VNO plays an important role in the perception of sexual pheromones.

Most of the evidence in nonrodent species is based on anatomical (Stephan et al. 1982), behavioral (Dagg and Taub 1970), and ethological (Estes 1972) studies. They represent 19% of the studies of the VNS in sharp contrast with 81% performed on rodents (Wysocki 1979). The majority of the experimental studies on the physiology of the VNS support the old hypothesis of Gratiolet (1845) relating to the function of the VNS as a sex chemoreceptor.

In recent years several excellent reviews have appeared discussing the structure and function of the VNS in different animal species (Wysocki 1979; Meredith 1983; Wysocki et al. 1986; Wysocki and Meredith 1987; Halpern 1987). Consequently, certain portions of the present treatise on the morphological and functional aspects of the VNS may appear redundant. Because of the literature on the functions of the VNS, some selection of the research covered has been necessary. This review will, therefore, concentrate on the function of the VNS in mammals, with particular emphasis on the role of the VNS in the pheromonal reception in rodents.

2 Structure of the Vomeronasal System

The study of the structure of the VNS comprises at least three sections: first, the VNO and its projections through the VNN; second, the accessory olfactory bulb (AOB); and third, the central projections of the AOB. The first and second have been described and morphologically studied in several species since the discovery of the organ by Ruysch (1724; cited by Vaccarezza et al. 1981) and its further description by Jacobson (1811) in other mammals (Gratiolet 1845; Balogh 1861 (cited by Stephan et al. 1982); Cajal 1902, 1904; Zuckerkandl 1910; Parker 1922; Mann 1961; Kratzing 1971; Stephan et al. 1982; Wöhrman-Repenning 1984). On the contrary, however the central projections of the AOB have been studied mainly in rodents through the use of histological techniques (Winans and Scalia 1970; Raisman 1972; Scalia and Winans 1975; De Olmos et al. 1978).

2.1 The Vomeronasal Organ

The VNO is a paired chemosensory structure localized inside the nose of most mammals (Negus 1958; Parson 1970) with a similar structure in different species (Estes 1972). The organ originates from the olfactory placoda and it is equally developed in both sexes (Estes 1972; Stephan et al. 1982; Garrosa et al. 1986). The two organs are situated on the two sides of the inferior part of the septum nasi and surrounded by a bony or cartilaginous capsule (Negus 1958; Fig. 1). The VNO opens to the nose through the nasal duct and/or the mouth through the nasopalatine duct (Bertmar 1981). In monotremate and marsupials, the VNO opens to the nose by the nasal duct and in eutherians (insectivora, carnivora, and ungulates) the VNO opens also to the mouth through the nasopalatine duct (Negus 1958). In some rodents the nasal cavity communicates with the mouth through the nasopalatine duct (e.g., guinea pig, hamster; Wysocki 1979; Mackay-Simand Rose 1986).
The structure and ultrastructure of the VNO is widely known at the present time (Schilling 1970; Kratzing 1971; Seiffert 1971; Mendoza and Breipohl 1983). In the golden hamster, rat, and mouse, it is a tubular structure 5 – 7 mm in length. The posterior part of the organ is blind and, in related animals, the organ opens into the nasal cavity. In the hamster, the nasal fossa communicates with the mouth through the nasopalatine duct (Vaccarezza et al. 1981, Taniguchi and Mochizuki 1982; Taniguchi et al. 1982; Garrosa et al. 1986). In transverse section, the VNO has a semicircular shape (Fig. 2). The lateral portion of the lumen is covered by pseudostratified ciliated respiratory epithelium and the medial wall by sensory epithelium. The vomeronasal epithelium consists of four distinct layers (Vaccarezza et al. 1981; Fig. 3): (a) superficial, formed by the prolongation of underlying cells; (b) a layer of supporting cells; (c) a layer of

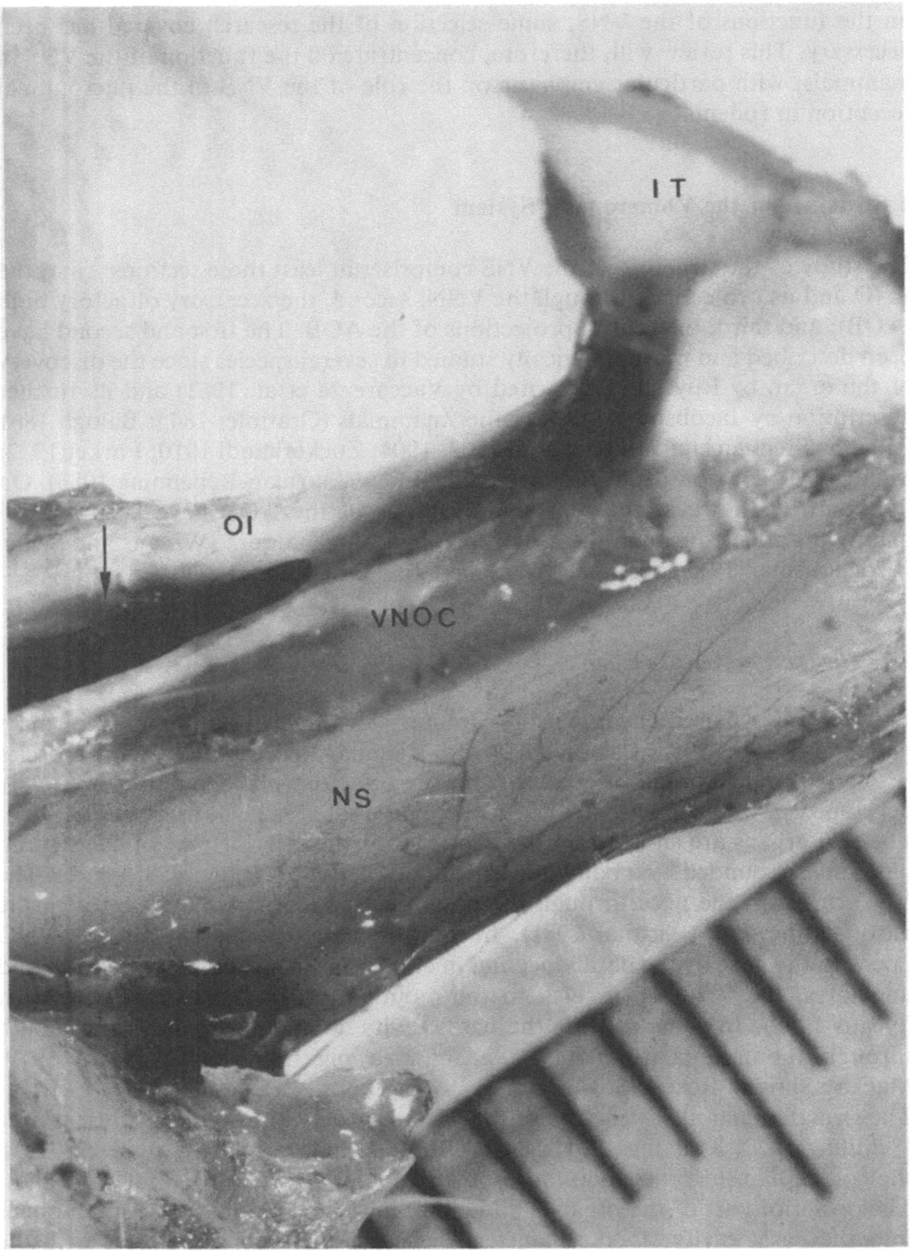

Fig. 1. Fresh dissection of the right part of the nasal septum of an adult rat. *NS*, nasal septum; *OI*, os incisivi (hard palate); *IT*, incisive tooth; *VNOC*, vomeronasal organ capsule. Foramen incisivi (*arrow*). 1 division = 1 mm. The *right part* of the picture is anterior

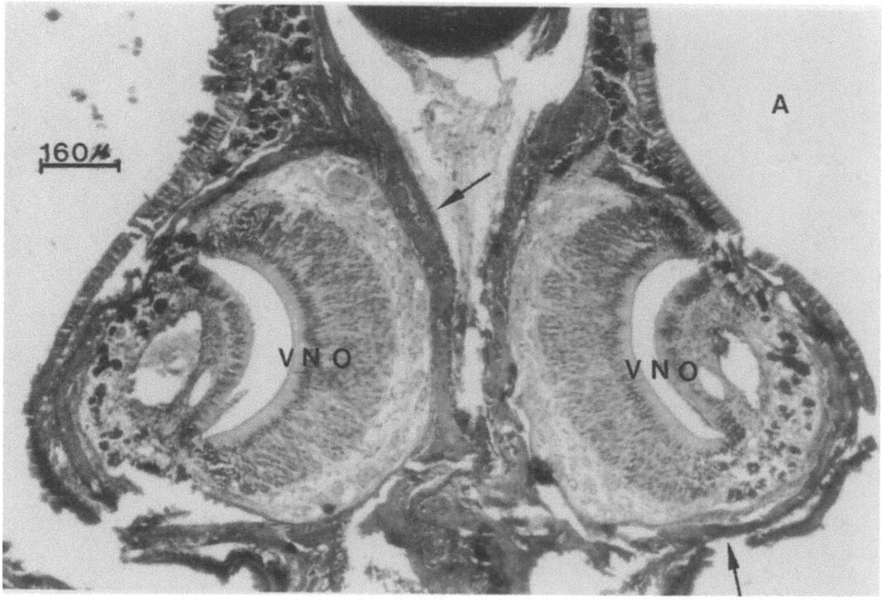

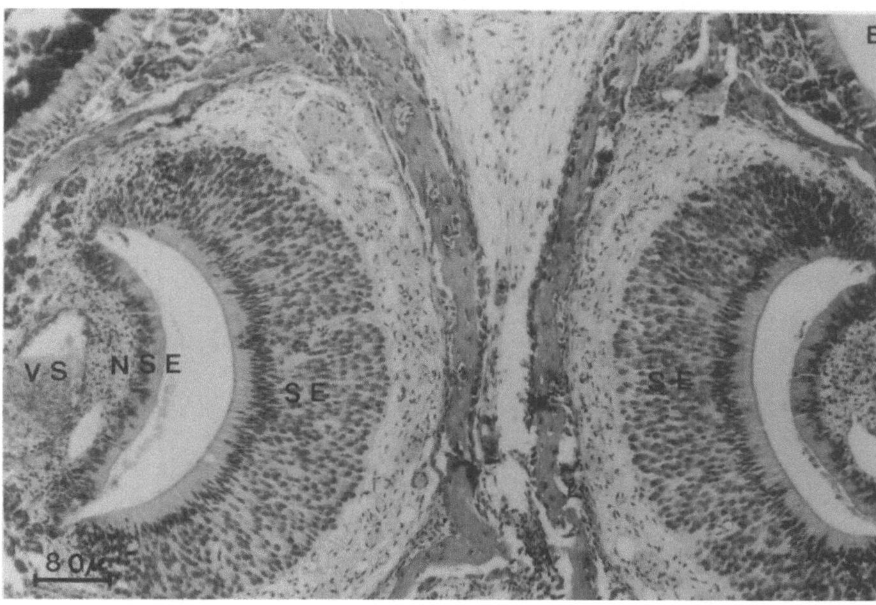

Fig. 2A, B. Frontal section of the nasal fossa of a 32-day-old rat. **A** Inferior part of the septum nasi where the paired vomeronasal organ (VNO) can be seen surrounded by its bony capsule (*arrows*). **B** Detail of VNO of an adult rat. *VS*, venous sinus; *SE*, sensory epithelium; *NSE*, nonsensory epithelium

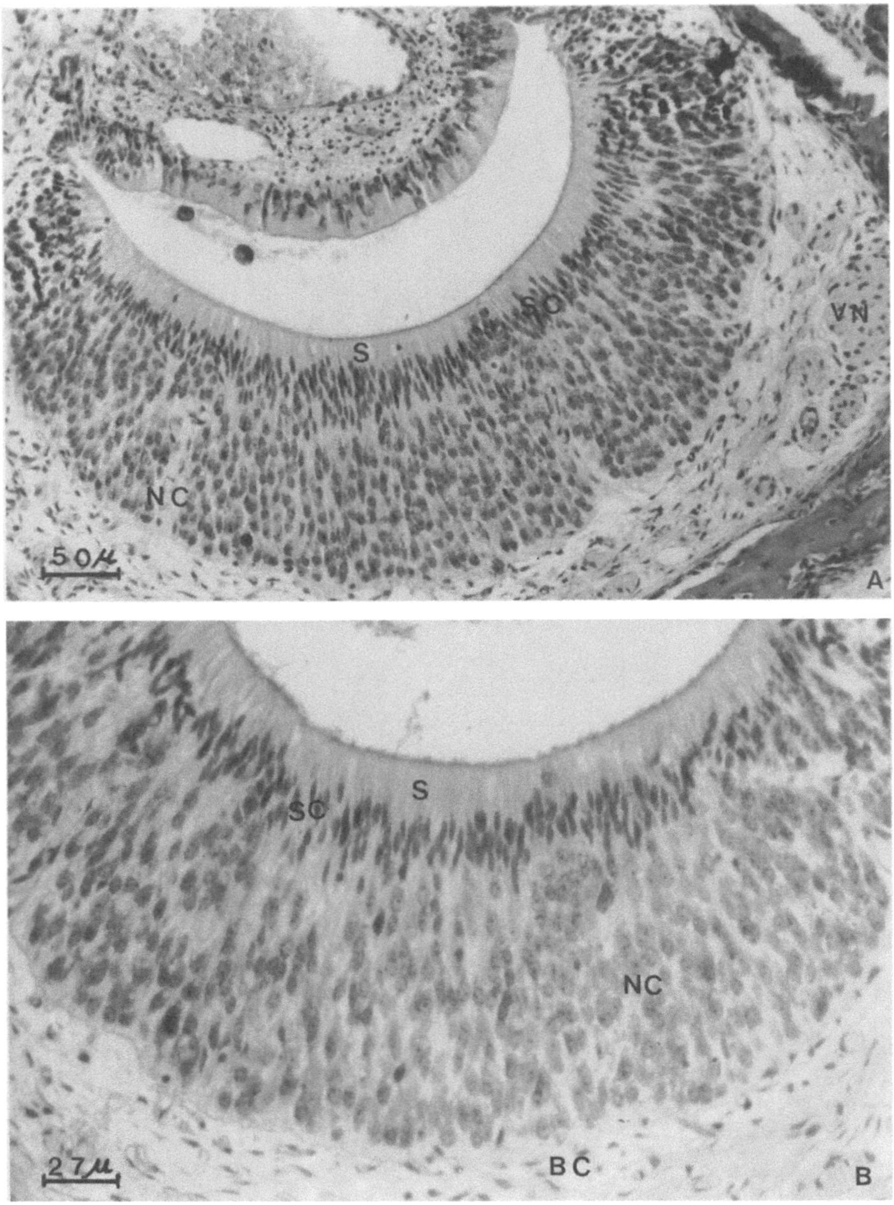

Fig. 3 A, B. Frontal section of the VNO of a 32-day-old rat, showing layers of the sensory epithelium and associated structures. *S*, superficial layer; *SC*, supporting cells; *NC*, bipolar neuronal cells; *BC*, basal cells; *VN*, transversal section of vomeronasal nerves. **B** Supporting cells and sensory neuronal layers can be clearly differentiated

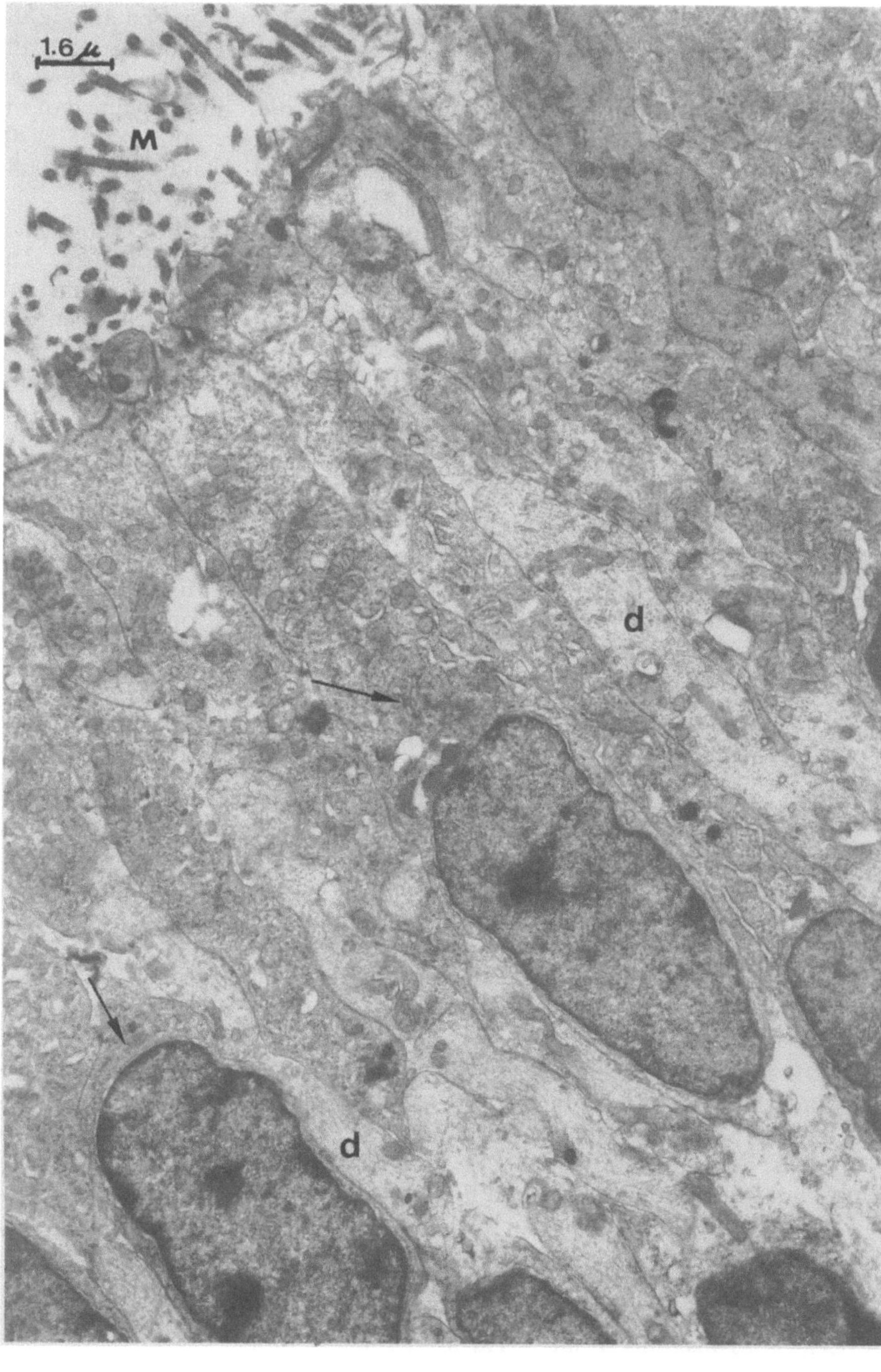

Fig. 4. Ultrastructure of the superficial layer of the sensory epithelium of the VNO. Supporting cells (*arrows*) and dendrites of underlying bipolar sensory neurones (*d*). *M*, superficial microvilli

bipolar cells (sensory neurones); and (d) scarce basal cells dispersed among the deepest bipolar cells. The bipolar cells have a dendrite extending from the supranuclear portion to the lumen of the organ. These dendrites show an expansion where numerous microvilli originate and project into the lumen of the VNO (Taniguchi and Mochizuki 1982, 1983; Vaccarezza et al. 1981; Fig. 4). There are mucous and serous glands situated dorsolaterally and ventromedially inside the organ (Taniguchi and Mochizuki 1983; Garrosa et al. 1986).

Many capillaries, several arterioles and venules, and a large central venous sinus make the VNO a highly vascularized structure (Garrosa et al. 1986). The intraepithelial vessels play an important role in olfactoendocrine functions (Breipohl et al. 1981) and in the mechanism providing access for chemicals to the organ (Meredith and O'Connell 1979; Meredith 1980).

The VNO is innervated by efferents from the autonomic nervous system running through the nasopalatine nerve. Both sympathetic (vasoconstrictor) fibers from the superior sympathetic ganglion, and, probably, parasympathetic (vasodilator) fibers (Meredith and O'Connell 1979; Eccles et al. 1981) reach the organ.

The VNO in the human species is present in the fetus, the newborn, and in a variable percentage of adults (Gabriele 1967; Wysocki 1979; Kreutzer and Jafek 1980; Johnson et al. 1985). In the human olfactory mucosa, some cells have been described as being similar to the vomeronasal sensory cells (Moran et al. 1982). The authors wondered if those cells had the same function as vomeronasal receptors in other species. No sensory cells have been reported in adult human VNO.

2.2 The Vomeronasal Nerves

The axons of the VNO bipolar cells enter three or four bundles as they exit from the organ (the vomeronasal nerves; VNN) situated deep in the septal mucosa. They course obliquely through the septum, pass through the cribiform plate and reach the AOB (vomeronasal bulb) between the main olfactory bulb and the frontal pole (Böjsen-Moller 1975; Naguro and Breipohl 1982).

2.3 The Accessory Olfactory Bulb

The AOB appears dorsocaudally as a small structure between the main olfactory bulb and the olfactory peduncle. The AOB was described anatomically by Balogh (1861; cited by Stephan et al. 1982) who named it "Jocobson's Hill" (*Jacobsonscher Hügel*). Later, the AOB was described by Cajal (1902, 1904) as being similar in structure to the main olfactory bulb. The axons of the sensory cells make contact with the primary dendrites of the mitral cells in characteristic glomeruli (Raisman 1972; Figs. 5 and 6).

2.4 Central Projections of the Accessory Olfactory Bulb

In mammals, the afferents from the AOB were first shown by Winans and Scalia (1970) in rabbits, and later demonstrated in opposum and rats (Scalia and

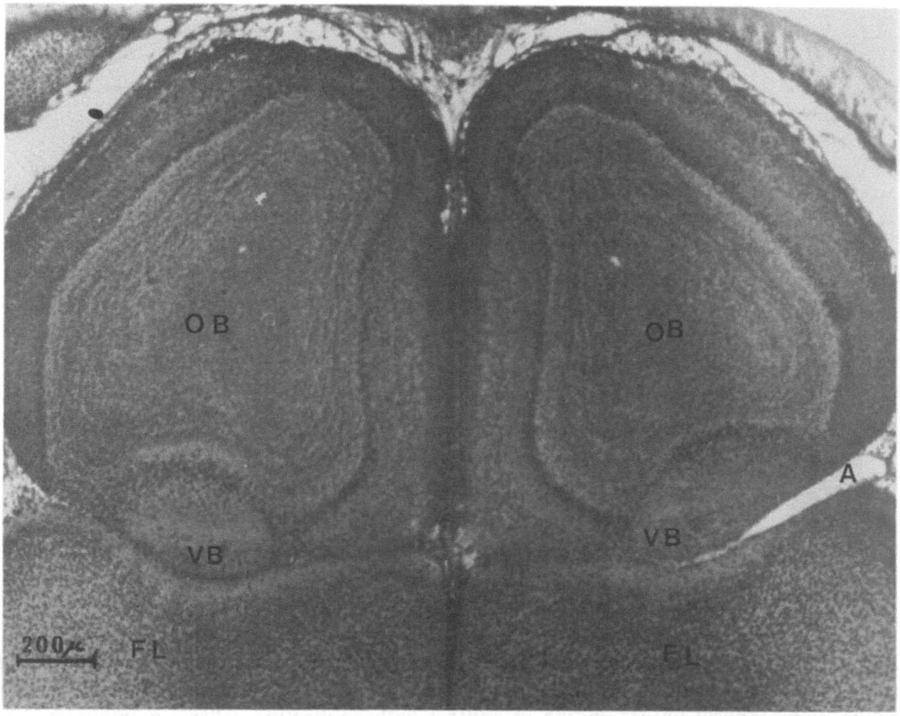

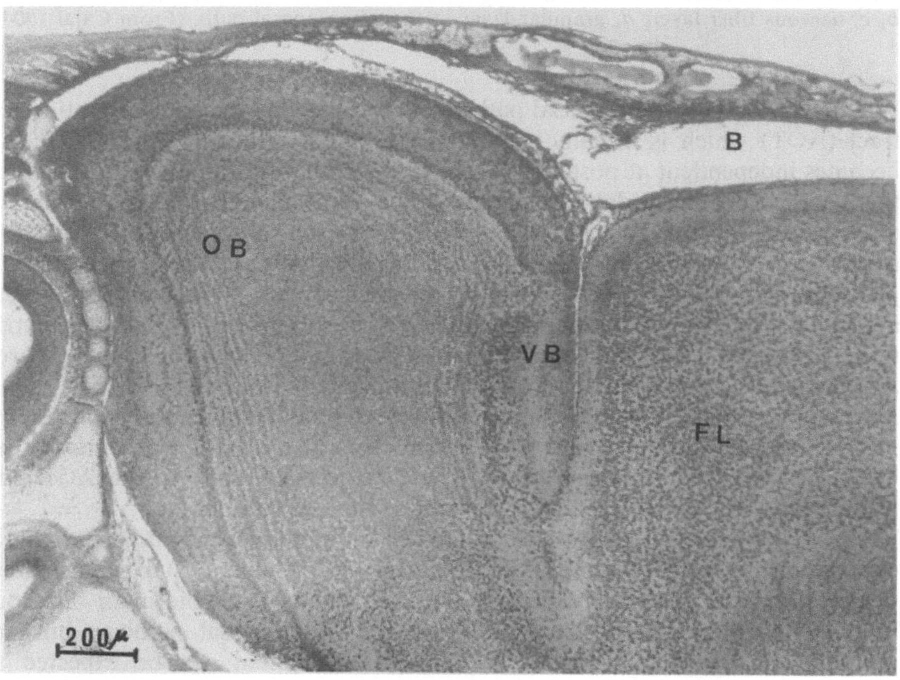

Fig. 5A, B. Horizontal (**A**) and sagittal (**B**) sections through olfactory bulbs (*OB*) of a 5-day-old mouse. *VB*, vomeronasal bulb (accessory olfactory bulb); *FL*, frontal lobe of the brain

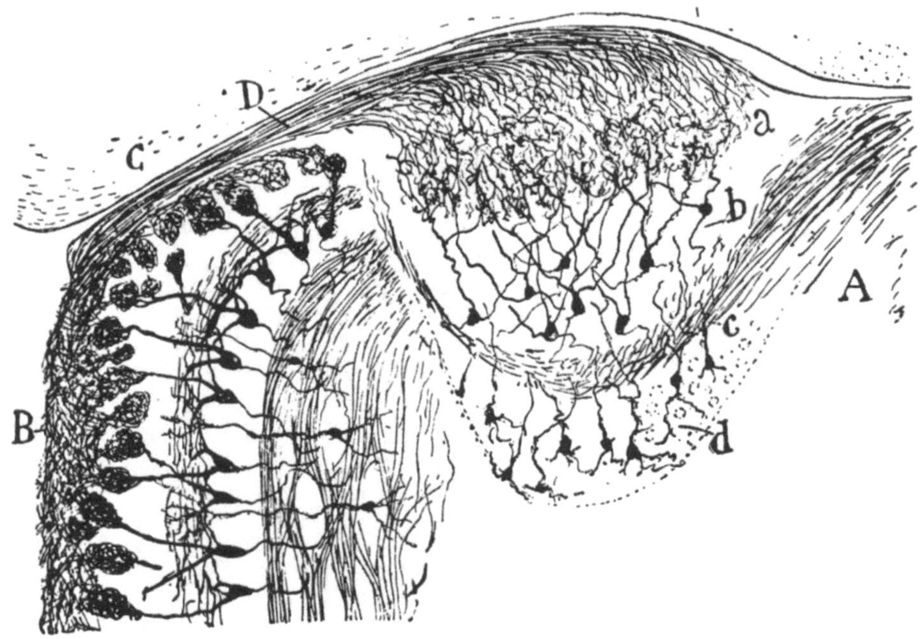

Fig. 6. Drawing of a sagital section of the olfactory bulb (*B*) showing the vomeronasal bulb (accessory olfactory bulb; *A*) in a 20-day-old mouse. *C*, cerebral frontal lobe; *D*, nervous fibers (vomeronasal nerves ending in contact with connecting cells; *a* glomerular layer; *b*, *c*, nervous fiber layer; *d*, granular layer of the vomeronasal bulb. (From Cajal 1904)

Winans 1975). The vomeronasal pathway is conveyed by the accessory olfactory tract (AOT) which is initially a part of the lateral olfactory tract (LOT) but becomes independent at posterior levels. In the species so far studied, it leads to the mediocortical amygdaloid region and one portion enters the stria terminalis to terminate in the bed nucleus. The distribution of projections was also confirmed in the hamster (Kevetter and Winans 1981 a). The principal AOB projections are to the AOT nucleus, the medial and posteromedial cortical amygdaloid nuclei ("vomeronasal amygdala"), and to the bed nucleus of the stria terminalis. As a result, the vomeronasal pathway enters the medial preoptic area and the medial hypothalamus through the amygdaloid nucleus.

At the same time, the AOB is innervated by efferents from medial and posteromedial amygdaloid nuclei and from the nucleus of the AOT (De Olmos et al. 1978; Kevetter and Winans 1981 a). These projections back to the AOB and presumably provide for feed-back control of the bulb output.

The other parts of the cortical amygdala, the anterior cortical and posterolateral cortical nuclei ("olfactory amygdala"; Kevetter and Winans 1981 b) receive direct projections from the main olfactory bulb. Afferents from the main olfactory bulb make contact with structures from the vomeronasal pathway but not mutually. From these data the main olfactory system can be expected to influence VNS functions but not vice-versa.

3 Function of the Vomeronasal System

The present knowledge of the physiology of the VNS in mammals has been arrived at through the study of hormonal, behavioral, and reproductive parameters in animals deprived of either VNO or some structure in the vomeronasal pathway. The parameters studied derive from pheromonal effects, or pheromonal-like effects described in the literature in the past 30 years. In general, the mammalian VNS appears to be involved in physiological and behavioral functions related mainly to reproductive fitness (Wysocki 1979) and it has been pointed out that the AOS mediates the majority of the priming or neuroendocrine effects of chemical signals (Beauchamp et al. 1985). Because of this, it is considered necessary to define briefly what the "chemical signals" and their biological actions are, before discussing the role of the VNS as a receptor of such substances.

In the ample field of biological chemical communication, pheromonal effects in mammals have received widespread consideration in the past few years (for reviews see: Birch 1974; Aron 1979; Müller-Schwarze and Silverstein 1980; Brei pohl 1982; Steiner and Ganchrow 1982; Vandenbergh 1983). The term pheromone was proposed by Karlson and Lüscher (1959) and discussed later by Kalmus (1964); it is now universally accepted. It is used to denominate "semiochemicals" (chemical communication among individuals; Law and Regnier 1971). The pheromones are substances or a blend of substances produced and released to the exterior by one individual and received by another of the same species producing one or several specific actions (Kalmus 1964). Wilson (1963) classified these substances according to the effects that they produce. According to this classification there are two kinds of pheromones: (a) *signalling pheromones* (or signalling effects) and (b) *priming pheromones* (or priming effects). The term signalling (releaser) is used when the pheromone elicits immediate changes in behavior (social or sexual) in the receptor animal, while "priming" is used if the pheromone produces neuroendocrine or endocrine reactions (Whitten 1956a; Bronson 1971, cited by Aron 1979) after a latency period from minutes to days (Wilson 1963).

The production of pheromones seems to be influenced by gonadal and/or hypophyseal hormones. This is true for several male odors (Marchlewska-Koy 1972) and for pheromonal capacity of urine and preputial extracts (Davies and Bellamy 1972; Jones and Nowell 1973; Lombardi et al. 1976). The utilization of androgenized females to induce male pheromonal effects is also effective (Dominic 1965; Bronson and Whitten 1968). Likewise, female pheromones are found to be influenced by estrogens (Stern 1970), progesterone (Bloch 1976), and prolactin (Moltz and Leon 1973). Michael et al. (1971, 1974) reported the hormonal dependence of vaginal secretions in female monkeys and women. Johnston and Bronson (1982) have shown that hypophyseal hormones control the urinary pheromone that induces LH secretion (priming effect), while the ovarian hormones control the female attractant pheromone (signalling effect) in mice.

3.1 Mechanism of Pheromone Access to the Vomeronasal Organ

The VNO is a tubular structure blind at one of its ends and communicating to the exterior by a narrow duct that opens through a fine pore. Thus, the access of chemical stimuli to the organ is slow if the substances are brought in by diffusion. On the other hand, stimulating substances must be brought out by the same duct utilized to bring them in. Therefore, the necessity of a pumping mechanism becomes evident.

It can be said that there are two types of access mechanisms: (a) a passive mechanism (diffusion) and (b) an active mechanism. The active mechanism is accompanied in some species by external motor patterns of structures surrounding the mouth and nose (lips, tongue, etc.). There is no doubt that the passive mechanism, although possible, must be of minor importance when compared with the active one because there is unlikely to be a high concentration of chemical stimuli at the opening of the VNO and also because the chemicals proceed through a long and narrow nasal duct with relative difficulty. On the contrary, the active mechanism demonstrates characteristics necessary to ideally introduce chemicals into the VNO.

As early as 1899, Mihalkovics observed the muscular tissue and wide veins inside the organ. He inferred that the muscle tissue served to dilate the lumen of the organ, and the veins acted to compress it and eject its content into the nasal cavity. Broman (1920) thought that the VNO was adapted to smell odors dissolved in aqueous solution from nasal glands. He suggested the "pumping" ability of the mammalian organ. This mechanism was demonstrated later by Hamlim (1930) who observed a bubble of gas or a drop of liquid that appeared at the orifice of the VNO when previously injected adrenalin elevated arterial pressure. When the pressure fell the drop of liquid was drawn back into the VNO. Schneider (1930, 1931, 1932) described the "flehmen" as a behavioral pattern in ungulates and other species which is observed when animals detect the odor of some biological products (urine, feces, female vulva, etc.) or some chemical substances.

From the time of the above-mentioned pioneer studies the active mechanism to bring substances into the VNO has thus far been studied in three models: (a) tongue movements; (b) the "flehmen"; and (c) the "pumping mechanism" controlled by the autonomic nervous system.

The utilization of tongue movements seems to be clear in snakes and other reptiles with bifid tongues (Burghardt and Pruitt 1975; Halpern 1987) and in some vertebrates in which the organ opens only in the mouth (Tucker 1971; Bertmar 1981; Halpern 1987).

Flehmen behavior has been considered as a mechanism that introduced "interesting" odors into the VNO, where they can be assessed (Knappe 1964). On the other hand, due to their observations in giraffes, Dagg and Taub (1970) thought that the flehmen might be related to the taste of urine and to the movements of the tongue. Both the flehmen and the tongue movements can also be provoked by olfactory stimuli in ungulates (Dagg and Taub 1970). Estes (1972) suggested that the flehmen could be a response that introduced sexual hormones or their derivatives contained in urine into the VNO. In the boar it has been shown that flehmen increase transport of chemicals from the mouth to the vomeronasal organ (Ladewig and Hart 1980, 1982; Melese-d'Hospital and Hart 1985). Flehmen behavior has also been observed in some felids (Verberne 1976; Brahmachary 1986).

Meredith (1980), Meredith and O'Connell (1979), and Meredith et al. (1980) demonstrated the existence of a vascular pump mechanism for the introduction of chemicals into the VNO in hamsters. The pump is under the control of the sympathetic innervation conveyed by the nasopalatine nerve. Electrical stimulus to the nasopalatine nerve produced a vasoconstriction that introduced liquids and their solutes into the VNO lumen. A parasympathetic vasodilator mechanism probably also exists. A similar pumping mechanism has been demonstrated in cats by Eccles et al. (1981) who also observed a clear secretion from the organ elicited by parasympathetic stimulation. Likewise, electrical stimulation of the cut peripheral end of the cervical sympathetic nerve caused a rapid suction of secretion back into the VNO.

3.2 Methods for Selective Demonstration of VNS Involvement in Pheromonal Chemoreception

The nasal fossa is a structure with conspicuous innervation. Inside it, the olfactory epithelium and nerves, the nasal trigeminal branches, the terminal nerve, the Masera organ, and the vomeronasal epithelium and nerves come together (Tucker 1963, 1971; Böjsen-Moller 1975). All these afferent or efferent fibers, except trigeminal ones, pass through the ethmoid cribriform plate and they run between the main and accessory olfactory bulbs.

Interactions between olfaction and reproduction were first described for human beings in 1856 by Maestre de San Juan. The participation of olfaction in reproductive events in rats has been clearly demonstrated for about 50 years (Rosen and Shelesnyak 1937; Shelesnyak and Rosen 1938); studies have been carried out by lesioning nasal mucosa or the olfactory bulbs. Such lesions also undoubtedly damage the AOS (Alberts 1974; Wysocki 1979). Therefore, and in the absence of complete knowledge of the chemical structure of specific stimuli (pheromones) for the receptor, the study of the physiology and the participation of the VNS on the pheromonal reception has been carried out using two basic methods: (a) electrical stimulation of some structures in the vomeronasal pathway and (b) deficits provoked by manipulation of specific parts of the VNS. In the first case electrical or hormonal responses were studied and in the second the participation of the VNS was tested in the perception of signalling and priming stimuli (Table 1).

3.3 Role of the Vomeronasal System as a Signalling Pheromonal Receptor

Signalling effects refer to the immediate changes in behavior elicited in most mammals by olfactory stimuli. These effects extend from behavior related to reproduction, aggression, social dominance, etc. (Stoddart 1974), to recognition of species and sexes, physiological status, marking space, etc. (Müller-Schwartze 1974; Mykytowycz 1974). A very important signalling effect is the elicitation of courtship behaviors (mating or copulatory behavior) by odors as can be seen in

Table 1. Specific methods and techniques used to study both the general physiology and the participation of the vomeronasal system in pheromonal perception

Experimental method	Parameters studied	References[b]
Stimulation of VNO[a], AOB, and the medial part of the nucleus of the stria terminalis	Evoked electrical responses and priming effects	1, 8, 9, 17, 33, 56, 57
Closure of the nasal or nasopalatine duct	Signalling and priming effects	21, 31, 36, 59
Surgical, electrical, or thermal ablation of VNO	Signalling and priming effects	4–7, 10–14, 19, 20, 22, 27–29, 31, 32, 33, 37, 39, 48, 62, 63
Intranasal or intra-craneal cut of the VNN	Signalling and priming effects	2, 3, 16, 18, 19, 30, 40, 42–46, 51–54, 60, 61
Electrical or surgical ablation of AOB	Signalling and priming effects	10, 19, 24, 40, 50, 52, 55
Sectioning lateral accessory olfactory tract	Priming effects	46
Electrical lesion of the corticomedial and basolateral nucleus of the amygdala	Signalling and priming effects	25, 26, 58

[a] For abbreviations see text.

[b] References: 1, Adrian 1955; 2, Barber and Raisman 1977; 3, Bean 1982a; 4, 5, 6, Beauchamp et al. 1982a, b, 1985; 7, Bellringer et al. 1980; 8, 9, 10, Beltramino and Taleisnik 1979, 1980, 1983; 11, 12, Clancy et al. 1984a, b; 13, Coquelin and Bronson 1979; 14, Coquelin et al. 1984; 15, Eisthen et al. 1987; 16, Fleming et al. 1979; 17, Gallego 1957; 18, Gallego and Sánchez-Criado 1979; 19, Gallego et al. 1982; 20, Hudson and Distel 1986; 21, Johns et al. 1978; 22, Johnston and Rasmusen 1984; 23, Kaneko et al. 1980; 24, Kelche and Aron 1984; 25, Lehman et al. 1980; 26, Lehman and Winans 1982; 27, Lepri and Wysocki 1987; 28, Lloyd-Thomas and Keverne 1982; 29, Lomas and Keverne 1982; 30, Marques 1979; 31, Mackay-Sim and Rose 1986; 32, Meredith 1986; 33, Meredith and O'Connell 1979; 35, Mihalkowics 1899; 36, Milstein 1929; 37, Mora and Gallego 1977; 38, 39, Mora et al. 1984, 1985; 40, Murphy 1980; 42, O'Connell and Meredith 1984; 43, Planel 1953; 44, Powers and Winans 1975; 45, Powers et al. 1979; 46, Rajendren and Dominic 1985; 48, Reynolds and Keverne 1979; 49, Saito and Moltz 1986; 50, 51, 52, Sánchez-Criado 1979a, b, 1982; 53, Sánchez-Criado and Gallego 1979; 54, Sánchez-Criado et al. 1984; 55, Schaeffer et al. 1986; 56, 57, Tucker 1963, 1971; 58, Velasco and Taleisnik 1969; 59, Verberne 1976; 60, Whitten 1963; 61, Winans and Powers 1977; 62, 63, Wysocki et al. 1982, 1983.

Table 2. Animals used and sex pheromonal effects studied to elucidate the role of the vomeronasal system in mammals

Animal	Effects	Results	References[a]
Monkey	Signalling	VNNx did not affect the sexual behavior	34
Cat	Signalling	"Flehmen" behavior decreased due to cVND	59
Rabbit	Signalling	VNOx did not affect the sexual behavior in adult or suckling behavior in newborn rabbits	20, 35, 37
Guinea pig	Signalling	Lesions did not affect the sexual behavior and had no effect on general behavior; also depressed the response to urine odors	4–6, 15, 43
Hamster	Signalling	The experimental lesions affected the sexual behavior in males and females to different degrees, but did not affect nonreproductive behavior	11, 22, 25, 26, 30–32, 40, 42, 44, 45, 61
Mouse	Priming and Signalling	Lesions prevented all the priming effects studied, and reduced the signalling ones	2, 3, 7, 12, 14, 23, 27–29, 47, 48, 60, 62, 63
Rat	Priming and Signalling	The experimental maneuvers blocked or reproduced all the priming effects studied. Lesions did not affect the signalling responses in female rats and reduced the sexual arousal in males	8–10, 16, 18, 19, 21, 24, 37–39, 41, 49–55

VNNx, section of the vomeronasal nerves; cVND, closure of vomeronasal duct; VNOx, removal of vomeronasal organ (surgically or electrically).
[a] The numbered references correspond to those in Table 1, except: 34, Michael and Keverne 1968; 41, Nelson et al. 1985; 47, Rajendren and Dominic 1986.

several mammals (Johnstone 1983). The animals used to study the role of the VNS in signalling responses are, mainly, the hamster (golden or sirian hamster) and mouse, and to a lesser extent, the guinea pig. In these animals, in most cases, either some aspect of sexual behavior such as male orientation to an estrous female, female lordosis behavior elicited by a male, or mating behavior (Table 2) has been tested. The participation of the VNS on nonsexual behaviors aggression, aggregation, feeding) has been largely studied (Bean 1982b; Wysocki et al. 1982; Clancy et al. 1984a; for reviews see Wysocki et al. 1986; Halpern 1987).
The idea of vomeronasal participation in signalling sex pheromone reception appears with the concept of VNO as a receptor for special chemicals (Mihalkowics 1899). This idea emerged even before the development of the "signalling" concept, which is a modern term (Kalmus 1964; Whitten 1966; Bronson 1971).

Despite early experimental approaches to the problem (see Introduction) and the considerable effort shown in the past 20 years, knowledge of the participation of the VNS in signalling perception is far from conclusive. In the studies carried out to test for such participation, the vomeronasal lesions alone never produce absolute behavioral deficits. In some cases (i.e., the attraction bioassay to hamster vaginal discharge) the behavior is mediated by the main olfactory deficits (Meredith 1980) and in other cases concomitant lesions of other structures such as nasal olfactory mucosa, are necessary to induce significant changes in behavior (Powers and Winans 1975; Winans and Powers 1977; Powers et al. 1979; Meredith 1980; Wysocki et al. 1982; Clancy et al. 1984b; Eisthen et al. 1987).

Experiments in Rabbits

Scalia and Winans (1975) showed connections of the VNO with amygdaloid nucleus in rabbits using histological techniques. The results led them to conclude that the VNO could be implicated in sexual and feeding behavior. We do not know of any other studies on rabbits to prove such a hypothesis, except for the attempt of Hudson and Distel (1986) who showed that the ablation of VNO did not affect suckling behavior in newborn rabbits.

Experiments in Guinea Pigs

There have been experiments trying to determine the role of the VNO in the signalling perception of guinea pigs. Planel (1953) found that sectioning the VNN did not influence the behavior orientation of a male to an estrous female. On the contrary, Beauchamp et al. (1982 a, b, 1985; Eisthen et al. 1987) reported that the VNO could be critical in some signalling responses, since it was shown that animals lacking VNO exhibited a lack of interest in the investigation of urine odors, attenuation in responsiveness to low-volatile urine compounds, and a relatively poor response to female urine odors. In spite of that, the animals were able to discriminate classes of urine and their mating behavior was normal.

Experiments in Rats and Mice

Clancy et al. (1984a, b) described a singificantly reduced, though not abolished, copulatory behavior in mice lacking the VNO, and Bean (1982a) and Wysocki et al. (1982) showed that the ablation of the VNO led to a consistent reduction in ultrasonic vocalizations in male mice in response to female stimulus. These are the only reports available concerning the vomeronasal system in sex-signalling responses in mice and the results need further experimentation for confirmation. In rats, attempts to demonstrate the role of VNS in signalling responses have been unsuccessful (Kelche and Aron 1984; Schaeffer et al. 1986), although the VNO seems to contribute, but is not indispensable, to sexual arousal in males (Saito and Moltz 1986).

Experiments in Hamsters

The participation of the VNS in signalling pheromonal perception in hamsters has been studied by Powers and Winans (1975). They observed severe deficits in mating tests after the sectioning of the VNN. In these experiments, only 10 out of 26 (about 38%) male animals showed deficits; the other 16 only showed deficits after cauterization of olfactory mucosa with zinc sulfate. These results were further analyzed by Winans and Powers (1977) and extended by Powers et al. (1979), in which the lesion of the main olfactory bulb resulting from vomeronasal nerve cutting did not account for behavioral differences; however, the treatment of olfactory mucosa with zinc sulfate in animals with bilaterally sectioned VNN eliminated mating behavior. The data of these authors emphasize that both the olfactory and the vomeronasal systems participate in some animals in signalling male responses to female stimuli. They also describe how the impairment of behavior is clear when analyzed 2 days after the operation but not so clear 9 days after, probably due to recovery of the peripheral receptors of the olfactory system from zinc sulfate lesions (Alberts 1974). Animals with deficits after VNN cut did not recover. The results obtained by Murphy (1980) on sectioning the VNN and by Johnston and Rasmusen (1984) lesioning the VNO are in line with those described above.

Electrolytic lesions of the "vomeronasal amygdala" disrupt copulatory behavior in some animals (Lehman et al. 1980) but mating deficits seen after rostral corticomedial amygdala lesions seem to be due primarily to the destruction of medial nucleus and not to the interruption of vomeronasal or olfactory inputs (Lehman and Winans 1982). The authors think that the vomeronasal and olfactory inputs to the amygdala may mediate both signalling and priming events associated with male hamster reproduction.

O'Connell and Meredith (1984) showed that lesions restricted to the VNS did not impair a male's interest or relative preferences for female vaginal discharge, but interfered significantly in the sexual behavior of the male in certain circumstances. The lack of effects of VNS lesions on male hamster sexual behavior in most circumstances (Murphy 1980; Johnston and Rasmusen 1984) may be due to the result of sexual experience. VNS lesions led to complete failure to mate in some inexperienced male hamsters, but if the lesion was performed after sexual experience, removal of the VN receptors had no effect (Meredith 1986).

As a whole, the above-reported data, together with other studies (Marques 1979; Clancy et al. 1984b; Mackay-Sim and Rose 1986) seem to implicate the VNS in signalling pheromone reception in hamsters; but the participation of the olfactory system seems necessary to elicit the entire signalling pheromonal response (Eisthen et al. 1987).

3.4 Role of the Vomeronasal System as a Priming Pheromonal Receptor

Planel (1951; cited by Schilling 1970) showed in the guinea pig that the destruction of the VNO produces a delay in testicular descent in males and a prolongation of estrous cycle duration in females. With the exception of this study and those of Lepri and Wysocki (1987) in female prairie voles, experiments on the

involvement of the VNS in perception, transmission and integration of the priming pheromonal stimuli have been carried out on those laboratory rodents (mouse and rat) in which the priming pheromonal effects and their associate endocrinological mechanism have been most extensively studied (Table 2).
The first and most immediate consequence of a priming pheromonal stimulus in the rat is the release of the luteinizing hormone (LH) from the anterior pituitary gland. This effect takes place by previous modification of the hypothalamic mechanism regulating gonadotropin secretion. An increase in LH secretion induced by male odors was described in female rats by Beltramino and Taleisnik (1983). Such an increase did not take place if the female rats had the VNO or the AOB ablated. Likewise, the stimulation of the AOB in overiectomized estrogen-primed rats, as the stimulation of the medial part of the bed nucleus of the stria terminalis (BNST), induces a release of LH (BNST received afferents from the vomeronasal amygdala; Beltramino and Taleisnik 1979, 1980).
Velasco and Taleisnik (1969) have already described how the stimulation (electrical or chemical) of the medial amygdaloid complex induces ovulation in rats made anovulatory by either constant light or by pharmacological methods (atropine, reserpine, and urethrane administration).
The priming pheromonal effects in rodents triggered by vomeronasal stimulation of the release of gonadotropins depend on the species, age, and physiological status. Thus, if the receptor animal is a prepubertal female, a precocious puberty takes place in both mice and rats (Vandenbergh 1967, 1976; Sanchez-Criado and Gallego 1979). If they are of adult age, the effect consists of a synchronization and shortening of the estrous phase in mice (Whitten 1956a) and a shortening of the estrous cycle in the rat (Gallego and Sánchez-Criado 1979). When the female mice are pregnant, the neuroendocrinological changes associated with vomeronasal system activation in some circumstances produce a block of pregnancies (Keverne 1982). Such an effect is not seen in the rat (Davies and de Groot 1964). If female mice are grouped without the influence of males, they become pseudopregnant due to the female odors which stimulate the vomeronasal system (Keverne 1982). The release of LH produces ovulations in female rats made anovulatory by constat light (Johns et al. 1978). And finally if the receptor animal is a male (mouse or rat), the activation of the vomeronasal system produces an enhancement of the reproductive functions through an increased testosterone secretion (Wysocki et al. 1983; Sanchez-Criado et al. 1984).

Experiments in Mice

The exteroceptive olfactory stimuli associated with the presence of males or their odors accelerate the sexual development of female mice (precocious vaginal opening and the age of first estrus; Vanderbergh 1967, 1969, 1971). On the other hand, the stimuli from castrated males appear unable to produce the Vanderberg pheromonal effect, which supports hormonal dependence on such stimuli.
From an endocrinological point of view, precocious puberty induced by pheromones seems to be produced mainly by an increase in the circulating levels of luteinizing hormone (LH). The increase occurs within the first 3 days after the beginning of the stimuli, and is accompanied by modifications in FSH, estradiol,

progesterone, and by an increased uterine weight (Zarrow et al. 1970, 1971; Bronson and Desjardins 1974; Bronson and Stetson 1973; Bronson 1975, 1981; Bronson and Maruniak 1976; Bronson and Saal 1979) and also in prolactin (Keverne 1982).

The sectioning of the vomeronasal nerves blocks the pheromonal effect on female puberty (Kaneko et al. 1980) and the ablation of the vomeronasal organ also prevents the acceleration of puberty in female mice (Lomas and Keverne 1982). These authors' data support the idea that the VNO is the receptor of the male primer pheromone which accelerates puberty in female mice.

The presence of male mice or their odors induces synchronization of estrus in females (Whitten 1956a, b; Whitten pheromonal effect). This effect is associated with changes in the pattern of LH secretion (Bronson 1981).

To the authors' knowledge, only two experiments have been carried out in mice to test the participation of the VNO in the perception of pheromones responsible for the Whitten effect. One of them (Barber and Raisman 1977) offers no clear results, since the sectioning of the VNN failed to prevent the Whitten effect. The other experiment (Whitten 1963) showed that the section of the VNN blocks estrus synchronization induced by males. Whitten, then, suggested that the VNO must be preferentially concerned with the chemoreception of priming pheromones.

Exteroceptive stimuli from strange males produce block pregnancy in recently mated female mice (Bruce 1959). This phenomenon (Bruce pheromonal effect) is probably due to increase in hypophyseal and circulating LH (Ryan and Schwartz 1980) followed by a decrease in prolactin (Ryan and Schwartz 1980; Bellringer et al. 1980).

Bilateral destruction of the VNO in female mice significantly reduces the pregnancy block caused by the exposure of mated females to strange males (Bellringer et al. 1980; Lloyd-Thomas and Keverne 1982). Likewise, bilateral transection of the VNN blocks the Bruce effect, as does sectioning of the lateral accessory tract, which carries afferents from the AOB (Rajendren and Dominic 1985, 1986).

Grouped female mice exhibit a significant increase of pseudopregnancies as demonstrated by prolongation of diestrous and by the presence of decidual uterine reactions (Van der Lee and Boot 1955, 1956). This fact (Lee-Boot pheromonal effect) is probably due to pheromonal influences from other females (Kimura 1971), since the effect is blocked in females caged individually or in the presence of males (Lamond 1958, Whitten 1959).

The most probable cause of this pheromonal effect involves the release of prolactin (Ryan and Schwartz 1980). However, it is also probable that the first endocrinological change after vomeronasal stimulation by other females is the release of LH (Ryan and Schwartz 1980) followed by an increase in estradiol secretion which may produce a prolactin release from the pituitary (Krey and Everett 1973).

Reynolds and Keverne (1979) demonstrated that either the removal of the VNO, which suppresses vomeronasal inputs, or the pharmacological suppression of prolactin secretion blocks the Lee-Boot effect.

In spite of the fact that the majority of the studies on priming pheromonal effects have been done in females, priming effects have been also described in males

(Fox 1968; Vandenbergh 1971). One priming effect in male mice is the acceleration of prepubertal male sexual development in the presence of pheromonal odors originating in females. Female pheromones produce an increase in the rate of LH release (Maruniak and Bronson 1976; Coquelin and Bronson 1979; Coquelin et al. 1984) followed by a surge of plasma testosterone (Macrides et al. 1975). Such a surge of testosterone and LH is abolished by the surgical destruction of the VNO (Wysocki et al. 1983; Coquelin et al. 1984).

Experiments in Rats

The precocious puberty induced by priming pheromones has also been tested in rats. Vandenbergh (1976) showed that rearing female rats in the presence of an adult male from the age of weaning induced an advance in the onset of puberty of 5 to 9 days together with an acceleration of sexual development. This fact has also been shown by Sánchez-Criado and Gallego (1979) and Sánchez-Criado (1982) who described an advancement in the onset of puberty of about 6 days with reference to vaginal opening and of 7 days with respect to first estrous, without differentiation in body weight.

Raisman (1972) studied the connections between the AOB and the amygdala in the rat. He concluded that by virtue of its neural connections the AOS in them rat, as in the mouse, is ideally placed to mediate the effects of olfactory stimuli on reproduction. Confirming that deduction, it has been demonstrated that surgical destruction of the VNO, the sectioning of the VNN, and the bilateral destruction of the vomeronasal bulb block precocious puberty in rats (Sánchez-Criado and Gallego 1979; Sánchez-Criado 1982; Gallego et al. 1982).

Female rats normally display estrous cycles of 4 to 5 days in length. It has been described that 5-day estrous cycle rats display 1-day shortening of the cycle under the olfactory influence of male or female rat odors (Aron 1973, 1979; Aron pheromonal effect). This effect has also been demonstrated in rats by Mora and Gallego (1977), Gallego et al. (1982), and Sánchez-Criado (1982). As in precocious puberty, the lesions of the three first steps of the VNS (VNO, VNN and AOB) block the shortening of the estrous cycle induced by male odors (Mora and Gallego 1977; Sánchez-Criado 1979a; Sánchez-Criado 1982; Mora et al. 1985).

The shortening of the estrous cycle is probably produced by a reduction of corpus luteum progesterone secretion (Aron 1979; Chateau et al. 1976). The pheromonal shortening of the cycle in these rats may be the result of inhibition of estrous afternoon prolactin secretion and/or of an accentuation of the luteolytic effect of LH during diestrous. In this way, vomeronasal stimuli modulate the pattern of tonic secretion during the diestrous period, responsible for the 24-h shortening of the estrous cycle (Sánchez-Criado et al. 1986).

When normal cyclic female rats are submitted to constant illumination, they became anovulatory with a permanent period of estrous. In such circumstances, ovulation can be induced by some social or environmental stimuli, among them pheromones from male urine (Johns et al. 1978). This ovulatory reflex is prevented if the vomeronasal ducts are closed (Johns et al. 1978) or the VNN cut (Sánchez-Criado 1980).

In male rats some priming effects induced by pheromones from females have also been described. The endocrinology of such effects has not been studied extensively, but it has been shown that pheromonal stimulation modifies the pattern of LH secretion with the consequent effects on testosterone circulating levels (Sánchez-Criado et al. 1984). The sexual development of male rats is stimulated by female odors. Sectioning the VNN delays sexual maturation as indicated by lower accessory sexual organ weights in lesioned rats compared with intact ones (Sánchez-Criado 1979b; Sánchez-Criado et al. 1984). Furthermore, the destruction of the VNO in newborn male rats also produces lower testis and accessory sexual organ weights than in intact rats (Mora et al. 1984). The above data strongly support the important role of the vomeronasal system in priming pheromonal reception in both female and male rats.

4 Concluding Remarks

From a structural point of view, the vomeronasal system (accessory olfactory system) is made up of a peripheral and central part. The peripheral part consists of the receptor (the vomeronasal organ or Jacobson organ) and its projections to the central nervous system (vomeronasal nerves). The central part consists of the accessory olfactory bulb (vomeronasal bulb), where the first sensorial relay takes place, and the vomeronasal projection to the amygdale (vomeronasal amygdala), where the second sensorial relay takes place. Finally, the vomeronasal system ends in the medial preoptic area and in the medial hypothalamus. As compared with the main olfactory system, the vomeronasal constitutes: (a) a separate pathway for inputs from the environment to the central nervous system (CNS); (b) an extremely simple system with only two synapses, and (c) a system aimed directly at the hypophysiotropic area of the hypothalamus (Scalia and Winans 1975; Kevetter and Winans 1981a, b).

The vomeronasal system not only constitutes a separate structural unit, but also a functional one. Of the two main pheromonal effects (signalling and priming), the vomeronasal system is clearly involved in priming pheromonal effects, since the interruption of the vomeronasal inputs blocks all the priming effects so far described in rats and mice and female voles. The role of the VNS in sex signalling pheromone reception also seems clear, overall, in hamsters and mice. Nevertheless, the olfactory system seems neccesary, together with the VNS, to elicit the entire signalling effect (Table 2; Eisthen et al. 1987).

Because of the special characteristics of the VNO pumping mechanism, it is thought that the chemical stimulating substances are of low volatility and relatively high molecular weight (Beauchamp et al. 1980, 1982a, b; Gangrade and Dominic 1984; Drickamer 1986). These facts are in agreement with the opinion that olfactory and vomeronasal responses may be preferentially responsive to volatile and nonvolatile odorants respectively (Powers et al. 1979; Meredith 1982; Clancy et al. 1984b; Eisthen et al. 1987).

Since the urinary pheromones from females that elicit LH secretion in males are pituitary dependent, while the female attractant pheromones are gonadal dependent, a different origin and control for signalling and priming pheromones has been suggested (Johnston and Bronson 1982).

The luteinizing hormone (LH) seems to be the "inevitable ending point" after the activation of the vomeronasal system. The vomeronasal release of LH (Bronson and Macmillan 1983) together with some associated changes in the regulation of prolactin secretion (Keverne 1982) are the main neuroendocrinological events responsible for priming pheromonal effects.

Pituitary and gonadal hormones act directly on the central nervous system (CNS) structures inducing behavioral and neuroendocrinological regulatory functions. The above-cited evidence that pheromones can also affect CNS functions (behavioral and neuroendocrinological) through the stimulation of peripheral projections of the same CNS structures that are affected by hormones (Segovia et al. 1984) raised the idea of a dual hormonal control of the CNS in mammals. One would be mediated by hormones acting through the blood stream, and the other by hormone-dependent substances (pheromones; see Function of the VNS) or hormone derivatives (Gower 1972; Booth 1975) that act mainly through the peripheral olfactory and vomeronasal receptors.

From an ethological point of view, the signalling pheromones (high volatility, low molecular weight, gonadal-dependent) would stimulate mainly the olfactory system (Eisthen et al. 1987) to trigger behavioral patterns (signalling effects) facilitating both sexual encounters and (Wysocki and Meredith 1987) the mechanism of access of the priming pheromones (low volatility, high molecular weight, pituitary-dependent) to the vomeronasal epithelium, and thus producing neuroendocrinological events (priming effects; Estes 1972).

5 References

Adrian ED (1955) Synchronised activity in the vomero-nasal nerves with a note on the function of the organ of Jacobson. Pflügers Arch Ges Physiol 260:188–192

Alberts JR (1974) Producing and interpreting experimental olfactory deficits. Physiol Behav 12:657–670

Aron CL (1973) Phéromone et régulation de la durée du cycle oestral chez la ratte. Arch Anat Histol Embryol (Strasb) 56:209–216

Aron C (1979) Mechanisms of control of the reproductive function by olfactory stimuli in female mammals. Physiol Rev 59:229–284

Balogh C (1861) Das Jacobson'sche Organ des Schafes. Sitzungsber Kais Akad Wissensch Wien Mathem-Naturw Cl 42:449–476

Barber PC, Raisman G (1977) Experimental investigation of the possible involvement of the accessory olfactory system in response to primer pheromones in female mice. Proc Int Un Physiol Sci XIII:51

Bean NJ (1982a) Olfactory and vomeronasal mediation of ultrasonic vocalization in male mice. Physiol Behav 28:31–37

Bean NJ (1982b) Modulation of agonistic behavior by the dual olfactory system in male mice. Physiol Behav 29:433–437

Beauchamp GK, Wellington JL, Wysocki CJ, Brand JG, Kubie JL, Smith AB (1980) Chemical communication in the guinea pig: urinary components of low volatility and their access to the vomeronasal organ. In: Müller-Schwarze D, Silverstein RM (eds) Chemical signals. Vertebrate and aquatic invertebrates. Plenum New York

Beauchamp GK, Martin IG, Wysocki CJ, Wellington JL (1982a) Chemoinvestigatory and sexual behavior of male guinea pigs following vomeronasal organ removal. Physiol Behav 29:329–336

Beauchamp GK, Wysocki CJ, Wellington JL, Martin IG (1982b) Chemical communication and the accessory olfactory system in guinea pigs. In: Steiner JE, Ganchrow JR (eds) Determination of behavior by chemical stimuli. IRL, London

Beauchamp GK, Wysocki CJ, Wellington JL (1985) Extinction of response to urine odor as a consequence of vomeronasal organ removal in male guinea pigs. Behav Neurosci 99:950–955

Bellringer JF, Hester PM, Pratt PM, Keverne EB (1980) Involvement of the vomeronasal organ and prolactin in pheromonal induction of delayed implantation in mice. J Reprod Fertil 59:223–228

Beltramino C, Taleisnik S (1979) Effect of electrochemical stimulation in the olfactory bulbs on the release of gonadotropin hormones in the rat. Neuroendocrinology 28:320–328

Beltramino C, Taleisnik S (1980) Dual action of electrochemical stimulation of the bed nucleus of the stria terminalis on the release of LH. Neuroendocrinology 30:238–242

Beltramino C, Taleisnik S (1983) Release of LH in the female rat by olfactory stimuli. Neuroendocrinology 36:53–58

Bertmar G (1981) Evolution of vomeronasal organs in vertebrates. Evolution 35:359–366

Birch MC (ed) (1974) Pheromones. North-Holland, Amsterdam

Bloch S (1976) A progesterone-dependent pheromone of the female mice. Experientia 32:937–938

Böjsen-Moller F (1975) Demonstration of terminalis, olfactory, trigeminal and perivascular nerves in the rat nasal septum. J Comp Neurol 159:245–256

Booth WD (1975) Changes with age in the occurrence of C_{19} steroids in the testis and submaxillary gland of the boar. J Reprod Fertil 42:459–472

Brahmachary RL (1986) Ecology and chemistry of mammalian pheromones. Endeavour, new series 10:65–68

Breipohl W (ed) (1982) Olfaction and endocrine regulation. IRL, London

Breipohl W, Bhatnagar KP, Blank M, Mendoza AS (1981) Intraepithelial blood vessels in the vomeronasal neuroepithelium of the rat. Cell Tissue Res 215:465–473

Broman I (1920) Das Organon vomero-nasale Jacobsoni – ein Wassergeruchsorgan. Anat Hefte Abt I 58:143–191

Bronson FH (1971) Rodent pheromones. Biol Reprod 4:344–357

Bronson FH (1975) Male-induced precocial puberty in female mice: confirmation of the role of estrogen. Endocrinology 96:511–514

Bronson FH (1981) The regulation of luteinizing hormone secretion by estrogen: relationships among negative feedback, surge potential and male stimulation in juvenile, peripubertal and adult female mice. Endocrinology 108:506–516

Bronson FH, Desjardins C (1974) Circulating concentration of FSH, LH, estradiol and progesterone associated with acute male-induced puberty in female mice. Endocrinology 94:1658–1668

Bronson FH, Macmillan B (1983) Hormonal responses to primer pheromones. In: Vandenberg JG (ed) Pheromones and reproduction in mammals. Academic, New York

Bronson FH, Maruniak JA (1976) Differential effects of male stimuli on follicle stimulating hormone, luteinizing hormone, and prolactin secretion in prepubertal female mice. Endocrinology 98:1101–1108

Bronson FH, Saal FS (1979) Control of the preovulatory release of luteinizing hormone by steroids in the mouse. Endocrinology 104:1247–1255

Bronson FH, Stetson MH (1973) Gonadotropin release in prepubertal female mice following male exposure. A comparison with the adult cycle. Biol Reprod 9:449–459

Bronson FH, Whitten WK (1968) Oestrus-accelerating pheromone of mice: assay, androgen-dependency and presence in bladder urine. J Reprod Fertil 15:131–134

Bruce HM (1959) An exteroceptive block to pregnancy in the mouse. Nature 1984:105

Burghardt GM, Pruitt CH (1975) Role of the tongue and senses in feeding of naive and experienced garter snakes. Physiol Behav 14:185–192
Cajal SR (1902) Textura del lóbulo olfativo accesorio. Trab Lab Invest Biol Univ Madrid 1:141–150
Cajal SR (1904) Textura del sistema nervioso del hombre y de los vertebrados, vol 2, N. Moya, Madrid
Chateau D, Roos J, Plas-Roser S, Roos M, Aron CL (1976) Hormonal mechanism involved in the control of oestrus cycle duration by the odour of urine in the rat. Acta Endocrinol (Copenh) 82:426
Clancy AN, Coquelin A, Macrides F, Gorski RA, Noble EP (1984a) Sexual behavior and aggression in male mice: involvement of the vomeronasal system. J Neurosci 4:2222–2229
Clancy AN, Macrides F, Singer AG, Agosta WC (1984b) Male hamster copulatory responses to a high molecular weight fraction of vaginal discharge: effects of vomeronasal organ removal. Physiol Behav 33:653–660
Coquelin A, Bronson FH (1979) Release of luteinizing hormone in male mice during exposure to females: habituation of the response. Science 206:1099–1101
Coquelin A, Clancy AN, Macrides F, Noble EP, Gorski RA (1984) Pheromonally induced release of LH in male mice: involvement of the vomeronasal system. J Neurosci 4:2230–2236
Dagg AI, Taub A (1970) Flehmen. Mammalia 34:686–695
Davies DL, de Groot J (1964) Failure to demonstrate olfactory inhibition of pregnancy (Bruce effect) in the rat. Anat Rec 148:366
Davies VJ, Bellamy D (1972) The olfactory response of mice to urine, effects of gonadectomy. J Endocrinol 55:11–20
De Olmos J, Hardy H, Heimer L (1978) The afferent connections of the accesory olfactory bulb formations in the rat: an experimental HRP study. J Comp Neurol 181:213–243
Dominic CJ (1965) The origin of the pheromones causing pregnancy block in mice. J Reprod Fertil 10:469–472
Drickamer LC (1986) Effects of urine from females in oestrus on puberty in female mice. J Reprod Fertil 77:613–622
Eccles R, Malm L, Tributsch C (1981) Autonomic control of stimulus access to the cat vomeronasal organ. J Physiol (Lond) 320:59P
Eisthen HL, Wysocki CJ, Beauchamp GK (1987) Behavioral responses of male guinea pigs to conspecific chemical signals following neonatal vomeronasal organ removal. Physiol Behav 41:445–449
Estes RD (1972) The role of the vomeronasal organ in mammalian reproduction. Mammalia 36:315–341
Fleming A, Vaccarino F, Tambosso L, Ches P (1979) Vomeronasal and olfactory system modulation of maternal behavior in the rat. Science 203:372–374
Fox KA (1968) Effects of prepubertal habitation conditions on the reproductive physiology of the male house mouse. J Reprod Fertil 17:75–85
Gabriele OF (1967) Persistent vomeronasal organ. Am J Roentgenol Radium Ther Nucl Med 99:697–699
Gallego A (1957) Electrofisiologia de la olfación. An Inst Farmacol Esp VI:11–43
Gallego A, Sánchez-Criado JE (1979) Pheromonal regulation of the oestrus cycle in the female rat by the accessory olfactory system. Acta Endocrinol (Copenh) [Suppl] 225:256
Gallego A, Sánchez-Criado JE, Mora O (1982) Role of the vomeronasal system in pheromonal reception in the rat. In: Steiner JE, Ganchrow JR (eds) Determination of behavior by chemical stimuli. IRL, London
Garrosa M, Coca S, Mora OA (1986) Histological development of the vomeronasal complex in the pre- and postnatal rat. Acta Otolaryngol (Stockh) 102:291–301

Gower DB (1972) 16-unsaturated C_{19} steroids. A review of their chemistry, biochemistry and possible physiological role. J Steroid Biochem 3:45 – 103
Grangrade BK, Dominic CJ (1984) Studies of the male originating pheromones involved in the Whitten effect and Bruce effect in mice. Biol Reprod 31:89 – 96
Gratiolet LP (1845) Recherches sur l'organe de Jacobson. Thesis Faculté de Medecine de Paris, no 164
Halpern M (1987) The organization and function of the vomeronasal system. Ann Rev Neurosci 10:325 – 362
Hamlim HE (1930) Working mechanisms for the liquid and gaseous intake and output of the Jacobson's organ. Am J Physiol 91:201 – 205
Hudson R, Distel H (1986) Pheromonal release of suckling in rabbits does not depend on the vomeronasal organ. Physiol Behav 37:123 – 128
Jacobson L (1811) Description anatomique d'un organe observé dans les mammifères. Ann Mus d'Hist Nat Paris 18:412 – 424
Johns MA, Feder HH, Komisaruk BR, Mayer AD (1978) Urine-induced reflex ovulation in anovulatory rats may be a vomeronasal effect. Nature 272:446 – 448
Johnson A, Josephson R, Hawke M (1985) Clinical and histological evidence for the presence of the vomeronasal (Jacobson's) organ in adult humans. J Otolaryngol 14:71 – 79
Johnston RE (1983) Chemical signals and reproductive behavior. In: Vandenbergh JG (ed) Pheromones and reproduction in mammals. Academic, New York
Johnston RE, Bronson F (1982) Endocrine control of female mouse odors that elicit luteinizing hormone surges and attraction in males. Biol Reprod 27:1174 – 1180
Johnston RE, Rasmusen K (1984) Individual recognition of female hamsters by males: role of chemical cues and of the olfactory and vomeronasal systems. Physiol Behav 33:95 – 104
Jones RB, Nowell NW (1973) The effect of urine on the investigatory behavior of male mice. Physiol Behav 11:35 – 38
Kaneko N, Debski EA, Wilson MC, Whitten WK (1980) Puberty acceleration in mice. II. Evidence that the vomeronasal organ is a receptor for the primer pheromone in male mouse urine. Biol Reprod 22:873 – 878
Kalmus H (1964) Possibilities and constraints of chemical telecommunication. Proc II Int Cong Endocrinol, pt I. Academic, London
Karlson P, Lüscher M (1959) Pheromones: a new term for a class of biologically active substances. Nature (London) 183:55
Kelche C, Aron C (1984) Olfactory cues and accessory olfactory bulb lesion: effects on sexual behavior in the cyclic female rat. Physiol Behav 33:45 – 48
Keverne EB (1982) The accessory olfactory system and its role in pheromonally mediated changes in prolactin. In: Breipohl W (ed) Olfaction and endocrine regulation. IRL, London
Kevetter GA, Winans SS (1981 a) Connections of the corticomedial amygdala in the golden hamster. I. Efferents of the vomeronasal amygdala. J Comp Neurol 197:81 – 98
Kevetter GA, Winans SS (1981 b) II. Efferents of the olfactory amygdala. J Comp Neurol 197:99 – 111
Kimura T (1971) Modifications of the oestrus cycle in mice housed together with normal or overiectomized females. Sci Pap Coll Gen Educ Univ Tokyo (Biol Part) 21:161 – 166
Knappe H (1964) Zur Funktion des Jacobsonschen Organs (Organon vomeronasale Jacobsoni). Zool Gart 28:188 – 194
Kratzing J (1971) The structure of the vomeronasal organ in the sheep. J Anat 108: 247 – 260
Kreutzer EW, Jafek BW (1980) The vomeronasal organ of Jacobson in the human embryo and fetus. Otolaryngol Head Neck Surg 88:119 – 123

Krey LC, Everett JW (1973) Multiple ovarian responses to single estrogen injection early in rat estrous cycle: impaired growth luteotropic stimulation and advanced ovulation. Endocrinology 93:377 – 384
Ladewig J, Hart BL (1980) Flehmen and vomeronasal organ function in male goats. Physiol Behav 24:1067 – 1071
Ladewig J, Hart BL (1982) Flehmen and vomeronasal function. In: Breipohl W (ed) Olfaction and endocrine regulation, IRL, London, pp 237 – 247
Lamond DR (1958) Spontaneous anoestrus in mice. Proc Aust Soc Anim Prod 2:97 – 101
Law JH, Regnier FE (1971) Pheromones. Annu Rev Biochem 40:533 – 548
Lehman MN, Winans SS (1982) Vomeronasal and olfactory pathways to the amygdala controling male hamster sexual behavior: autoradiographic and behavioral analyses. Brain Res 40:27 – 41
Lehman MN, Winans SS, Powers JB (1980) Medial nucleus of the amygdala mediates chemosensory control of male hamster sexual behavior. Science 210:557 – 560
Lepri JJ, Wysocki CJ (1987) Removal of the vomeronasal organ disrupts the activation of reproduction in female voles. Physiol Behav 40:349 – 355
Lloyd-Thomas A, Keverne EB (1982) Role of the brain and accessory olfactory system in the block to pregnancy in mice Neuroscience 7:907 – 913
Lomas DE, Keverne EB (1982) Role of the vomeronasal organ and prolactin in the acceleration of puberty in female mice. J Reprod Fertil 66:101 – 107
Lombardi JR, Vandenbergh MJG, Whitsett JM (1976) Androgen control of sexual maturation pheromone in house mouse urine. Biol Reprod 15:179 – 186
Mackay-Srin A, Rose JD (1986) Removal of the vomeronasal organ impairs lordosid in female hamsters: efffect is reversed by luteinizing hormone releasing hormone. Neuroendocrinology 42:489 – 493
Macrides F, Bartke A, Dalterio S (1975) Strange females increase plasma testosterone levels in male mice. Science 189:1104 – 1106
Maestre de San Juan A (1856) Falta total de los nervios olfactorios con anosmia en un individuo en quien existía una atrofia congénita de los testículos y miembro viril. El Siglo Médico 131:211 – 214
Mann G (1961) Bulbus olfactorius accessorius in chiroptera. J Comp Neurol 116: 135 – 144
Marchlewska-Koj A (1972) Pheromones in rodents. Kosmos Ser A Biol (Warsaw) 21:579 – 588
Marques DM (1979) Roles of the main olfactory and vomeronasal systems in the response of the female hamster to young. Behav Neural Biol 26:311 – 329
Maruniak JA, Bronson FH (1976) Gonadotropic responses of male mice to female urine. Endocrinology 99:963 – 969
McCotter RE (1912) The conection of the vomeronasal nerves with the accessory olfàctory bulb in the opposum and other mammals. Anat Rec 6:299 – 318
Melese-d'Hospital PY, Hart BL (1985) Vomeronasal organ cannulation in male goats: evidence for transport of fluid from oral cavity to vomeronasal organ during flehmen. Physiol Behav 35:941 – 944
Mendoza AS, Breipohl W (1983) The cell coat of the olfactory epithelium proper and vomeronasal neuroepithelium of the rat revealed by means of the ruthenium red reaction. Cell Tissue Res 230:139 – 146
Meredith M (1980) The vomeronasal organ and accessory olfactory system in the hamster. In: Müller-Schwarze D, Silverstein RM (eds) Chemical signals in vertebrates and aquatic animals. Plenum, New York
Meredith M (1982) Stimulus access and other processes involved in nasal chemosensory function: potential substrates for neural and hormonal influence. In: Breipohl W (ed) Olfaction and endocrine regulation. IRL, London

Meredith M (1983) Sensory physiology of pheromone communication. In: Vandenbergh JG (ed) Pheromones and reproduction in mammals. Academic, New York
Meredith M (1986) Vomeronasal organ removal befor sexual experience impairs male hamster mating behavior. Physiol Behav 36:737 – 743
Meredith M, O'Connell RS (1979) Efferent control of stimulus access to the hamster vomeronasal organ. J Physiol (Lond) 286:301 – 316
Meredith M, Marques D, O'Connell RJ, Stern FL (1980) Vomeronasal pump: significance for male hamster sexual behavior. Science 207:1224 – 1226
Michael RP, Keverne EB (1968) Pheromones in the communication of sexual status in primates. Nature 218:746 – 749
Michael RP, Keverne EB, Bonsall RW (1971) Pheromones: isolation of male sex attractants from a female primate. Science 172:964 – 966
Michael RP, Bonsall RW, Warner P (1974) Human vaginal secretion: volatile fatty acid content. Science 186:1217 – 1219
Mihalkovics von V (1899) Nasenhöhle und Jacobsonsches Organ – eine morphologische Studie. Anat Hefte 11:1 – 67
Milstein T (1929) Sur la physiologie de l'organe de Jacobson. Rev Laryngol Otol 50:705 – 712
Moltz H, Leon M (1973) Stimulus control of the maternal pheromone in the lactating rat. Physiol Behav 10:69 – 71
Mora OA, Gallego A (1977) The vomeronasal organ: a pheromonal chemoreceptor in the rat. Proc Int Union Physiol Sci vol XIII:525
Mora OA, Guisado S, Vaticon D, Sánchez-Criado JE (1984) Retraso en la maduración sexual de la rata macho inducido por vomeronasalectomía bilateral. Endocrinologia 31:150 – 153
Mora OA, Sánchez-Criado JE, Guisado S (1985) Role of the vomeronasal organ on the estral cycle reduction by pheromones in the rat. Rev Esp Fisiol 41:305 – 310
Moran DI, Rowley JC III, Jafek BW (1982) Electron microscopy of human olfactory eptihelium reveals a new cell type: the microvillar cell. Brain Res 253:39 – 46
Müller-Schwarze D (1974) Olfactory recognition of species, groups, individuals and physiological states among mammals. In: Birch MC (ed) Pheromones. Frontiers of biology, vol 32. North-Holland, New York
Müller-Schwarze D, Silverstein RM (eds) (1980) Chemical signals. Vertebrates and aquatic invertebrates. Plenum, New York
Murphy MR (1980) Sexual preference of male hamsters: importance of preweaning and adult experience, vaginal secretion and olfactory or vomeronasal sensation. Behav Neural Biol 30:323 – 340
Mykytowycz R (1974) Odor in the spacing behavior of mammals. In: Birch MC (ed) Pheromones. Frontiers of biology, vol 32. North-Holland, New York
Naguro T, Breipohl W (1982) The vomeronasal epithelia of the NMRI mouse. A scanning electron-microscopic study. Cell Tissue Res 227:519 – 534
Negus V (1956) The organ of Jacobson. J Anat 90:515 – 519
Negus V (1958) The comparative anatomy and physiology of the nose and paranasal sinuses. Livingstone, London
Nelson RJ, Fleming AS, Wysocki CJ, Snider TW, Zucker I (1985) Chemosensory and neural influences on photoperiodic responsiveness of laboratory rats. Neuroendocrinology 40:285 – 290
O'Connell RJ, Meredith M (1984) Effects of volatile and nonvolatile chemical signals on male sex behavior mediated by the main and accessory olfactory systems. Behav Neurosci 98:1083 – 1093
Parker GH (1922) Monographs on experimental biology: smell, taste and allied senses in the vertebrates. Lippincott, Philadelphia

Parson TS (1970) The nose and Jacobson's organ. In: Gans C (ed) Biology of the reptile: morphology. Academic, London
Planel H (1951) Etude anatomique et physiologique sur les fosses nasales des rongeurs. Medical thesis, Toulouse
Planal H (1953) Etudes sur la physiologie de l'organ de Jacobson. Arch Anat Histol Embryol (Strasb) 36:199–205
Powers JB, Winans S (1975) Vomeronasal organ: critical role in mediating sexual behavior of the male hamster. Science 187:961–963
Powers JB, Fields RB, Winans SS (1979) Olfactory and vomeronasal system participation in male hamsters' attraction to female vaginal secretion. Physiol Behav 22:77–84
Raisman G (1972) An experimental study of the projection of the amygdala to the accessory olfactory bulb and its relationship to the concept of a dual olfactory system. Exp Brain Res 14:395–408
Rajendren G, Dominic CJ (1985) Effect of transection of the vomeronasal nerve on the male-induced implantation failure (the Bruce effect) in mice. Indian J Exp Biol 23:635–637
Rajendren G, Dominic CJ (1986) Effect of lateral transection of the lateral olfactory tract on the male-induced implantation failure (the Bruce effect) in mice. Physiol Behav 36:587–590
Reynolds J, Keverne EB (1979) The accessory olfactory system and its role in the pheromonally mediated suppression of oestrus in grouped mice. J Reprod Fertil 57:31–35
Rosen S, Shelesnyak M (1937) Induction of pseudopregnancy in rat by silver nitrate on nasal mucosa. Proc Soc Exp Biol Med 36:832–834
Ruysch (1724) Thesaurus anatomicus, vol III. Jans Waesberg, Amsterdam, p 26
Ryan KD, Schwartz NB (1980) Changes in serum hormone levels associated with male-induced ovulation in group-housed adult female mice. Endocrinology 106:959–966
Saito TR, Moltz H (1986) Copulatory behavior of sexually naive and sexually experienced male rats following removal of the vomeronasal organ. Physiol Behav 37:507–510
Sánchez-Criado JE (1979a) Bloqueo de los efectos feromonales en la rata por deaferentación central del sistema olfatorio accesorio. Rev Exp Fisiol 35:137–142
Sánchez-Criado JE (1979b) Efecto de la deaferentación vomeronasal sobre la maduración sexual de la rata macho. Endocrinología 26:16–19
Sánchez-Criado JE (1980) Pheromonal male odors-induced oestrus cyclicity in anovulatory rats by continuous light through the vomeronasal system. Proc XXVIII Int Congr IUPS, vol XIV, p 679
Sánchez-Criado JE (1982) Involvement of the vomeronasal system in the reproductive physiology of the rat. In: Olfaction and endocrine regulation. Breipohl W (ed) IRL Press, London
Sánchez-Criado JE, Gallego A (1979) Male-induced precocious puberty in the female rat. Role of the vomeronasal system. Acta Endocrinol [Suppl] (Copenh) 225:255
Sánchez-Criado JE, Fernández-Galaz C, Vaticón D, Mora-Novaro OA (1984) Niveles plasmáticos de LH y testosterona en ratas macho sometidas a deaferentación periférica del sistema vomeronasal. Rev Esp Fisiol 40:359–364
Sánchez-Criado JE, López F, Aguilar E (1986) Pituitary regulation of corpus luteum progesterone secretion in cyclic rats. Endocrinology 119:1083–1088
Scalia F, Winans SS (1975) The differential projections of the olfactory bulb and accessory olfactory bulb in mammals. J Comp Neurol 161:31–56
Schaeffer CS, Roos J, Aron C (1986) Accessory olfactory bulb lessions and lordosis behavior in the male rat feminized with ovarian hormones. Horm Behav 20:118–127
Schilling A (1970) L'organe de Jacobson du Lemurien malgache microebus murinus (Miller, 1777). Mémoires du Muséum National d'Histoire Naturelle, serie A, tome LXI, fasc 4, pp 203–280

Schneider KM (1930) Das Flehmen, pt I. Zool Gart 3:183 – 198

Schneider KM (1931) Das Flehmen II. Zool Gart 4:349 – 364

Schneider KM (1932) Das Flehmen II, III. Zool Gart 5:200 – 226, 287 – 297

Segovia S, Orensanz LM, Valencia A, Guillamon A (1984) Effects of sex steroids on the development of the accessory olfactory bulb in the rat: a volumetric study. Dev Brain Res 16:312 – 314

Seiffert K (1971) Licht- und elektronenmikroskopische Untersuchungen am Jacobsonschen Organ (organon vomeronasale) der Katze. Arch Otorhinolaryngol (NY) 200:223 – 251

Shelesnyak MC, Rosen S (1938) Naso-genital relationship: induction of pseudopregnancy in the rat by nasal treatment. Endocrinology 23:58 – 63

Steiner JE, Ganchrow JR (eds) (1982) Determination of behavior by chemical stimuli. IRL Press, London

Stephan H, Baron G, Frahm HD (1982) Comparison of brain structure volumes in insectivora and primates. II. Accessory olfactory bulb (AOB). J Hirnforsch 23:575 – 591

Stern JJ (1970) Responses of male rats to sex odors. Physiol Behav 5:519 – 524

Stoddart DM (1974) The role of odor in the social biology of small mammals. In: Birch MC (ed) Pheromones, frontiers of biology. vol 32. North-Holland, New York

Taniguchi K, Mochizuki K (1982) Morphological studies on the vomeronasal organ in the golden hamster. Jpn J Vet Sci 44:419 – 426

Taniguchi K, Mochizuki K (1983) Comparative morphological studies on the vomeronasal organ in rats, mice and rabbits. Jpn J Vet Sci 45:67 – 76

Taniguchi Kazuyuki, Taniguchi Kazumi, Mochizuki K (1982) Developmental studies on the vomeronasal organ in the golden hamster. Jpn J Vet Sci 44:709 – 716

Tucker D (1963) Olfactory, vomeronasal, and trigeminal receptor responses to odorants. In: Zotterman Y (ed) Olfaction and taste. Pergamon, New York

Tucker D (1971) Nonolfactory responses from the nasal cavity: Jacobson's organ and the trigeminal system. In: Beidler LM (ed) Handbook of sensory physiology. Springer New York

Vaccarezza OL, Sepich LN, Tramezzani JH (1981) The vomeronasal organ of the rat. J Anat 132:167 – 185

Vandenbergh JG (1967) Effect of the presence of a male on the sexual maturation of the female mice. Endocrinology 81:345 – 349

Vandenbergh JG (1969) Male odor accelerates female sexual maturation in mice. Endocrinology 84:658 – 660

Vandenbergh JG (1971) The influence of the social environment on sexual maturation in male mice. J Reprod Fertil 24:383 – 390

Vandenbergh JG (1976) Acceleration of sexual maturation in female rats by male stimulation. J Reprod Fertil 46:451 – 453

Vandenbergh JG (1983) Pheromones and reproduction in mammals. Academic, New York

van der Lee S, Boot LM (1955) Spontaneous pseudopregnancy in mice. Acta Physiol Pharmacol Neerl 4:442 – 444

van der Lee S, Boot LM (1956) Spontaneous pseudopregnancy in mice II. Acta Physiol Pharmacol Neerl 5:213 – 214

Velasco ME, Taleisnik S (1969) Release of gonadotropins induced by amygdaloid stimulation in the rat. Endocrinology 84:132 – 139

Verberne G (1976) Chemocommunication among domestic cats, mediated by the olfactory and vomeronasal senses. II. The relation between the function of Jacobson's organ (vomeronasal organ) and flehmen behavior. Z Tierpsychol 42:113 – 128

Whitten WK (1956a) Modification of the oestrous cycle of the mouse by external stimuli associated with the male. J Endocrinol 13:399 – 404

Whitten WK (1956b) The effect of the removal of olfactory bulbs on the gonads of mice. J Endocrinol 14:160–163
Whitten WK (1959) Occurrence of anoestrus in mice caged in groups. J Endocrinol 18:102–107
Whitten WK (1963) Is the vomeronasal organ a sex chemoreceptor in mice? 2nd Proc Asia Oceania Congr Endocrinol, Sydney, session 11a, p 4
Whitten WK (1966) Pheromones and mammalian reproduction. Adv Reprod Physiol 1:155–177
Wilson EO (1963) Pheromones. Sci Am 208:100–114
Winans SS, Powers JB (1977) Olfactory and vomeronasal deafferentation of male hamsters: histological and behavioral analyses. Brain Res 126:325–344
Winans SS, Scalia F (1970) Amygdaloid nucleus: new afferent input from the vomeronasal organ. Science 170:330–332
Wöhrman-Repenning A (1984) Phylogenetische Aspekte zur Topographie der Jacobsonschen Organe und der Ductus nasopalatini bei Insectivora, Primates, Tupaia und Didelphis. Anat Anz 157:137–149
Wysocki CJ (1979) Neurobehavioral evidence for the involvement of the vomeronasal system in mammalian reproduction. Neurosci Biobehav Rev 3:301–341
Wysocki CJ, Meredith CJ (1987) The vomeronasal system. In: Finger TE, Silver WL (eds) Neurobiology of taste and smell. Wiley, New York
Wysocki CJ, Nyby J, Witney G, Beauchamp GK, Katz Y (1982) The vomeronasal organ: primary role in mouse in chemosensory gender recognition. Physiol Behav 29:315–327
Wysocki CJ, Katz Y, Bernhard R (1983) Male vomeronasal organ mediates female-induced testosterone surges in mice. Biol Reprod 28:917–922
Wysocki CJ, Bean NJ, Beauchamp GK (1986) The mammalian vomeronasal organ: its role in learning and social behaviors. In: Duvall D, Müller-Schwarze D, Silverstein RM (eds) Chemical signals in vertebrates, 4th edn. Plenum, New York
Zarrow MX, Estes SA, Denenberg VH, Clark JM (1970) Pheromonal facilitation of ovulation in the immature mice. J Reprod Fertil 23:357–360
Zarrow MX, Christenson CM, Eleftheriou BE (1971) Strain difference in the ovulatory response of immature mice to PMS and to the pheromonal facilitation of PMS induced ovulation. Biol Reprod 4:52–56
Zuckerkandl E (1910) Das Jacobsonsche Organ. Ergebnisse der Anatomie und Entwicklungsgeschichte 18:801–843

Subject Index

MIX
Papier aus verantwortungsvollen Quellen
Paper from responsible sources
FSC® C105338

FSC
www.fsc.org

If you have any concerns about our products,
you can contact us on
ProductSafety@springernature.com

In case Publisher is established outside the EU,
the EU authorized representative is:
Springer Nature Customer Service Center GmbH
Europaplatz 3, 69115 Heidelberg, Germany

Printed by Libri Plureos GmbH
in Hamburg, Germany